Example Code — Examples from This Book at Your Fingertips

You can access the example programs for this book by linking to its companion Web site at **support.sas.com/companionsites**. Select the book title to display its companion Web site, then select **Example Code** to display the SAS programs that are included in the book.

For an alphabetical listing of all books for which we offer example code, see **support.sas.com/bookcode**. Select a title to display the book's example code.

If you are unable to access the code through the Web site, send e-mail to **saspress@sas.com**.

Comments or Questions?

If you have comments or questions about this book, you may contact the author through SAS by

Mail: SAS Institute Inc.
SAS Press
Attn: <Author's name>
SAS Campus Drive
Cary, NC 27513

E-mail: saspress@sas.com

Fax: (919) 677-4444

Please include the title of the book in your correspondence.

See the last pages of this book for a complete list of books available through **SAS Press** or visit **support.sas.com/pubs**.

Praise for the Second Edition

"This book is an invaluable, proven asset for any student, researcher, or data analyst who wants to use SAS as a working tool. The step-by-step approach is extremely well structured. The end result is that even without any knowledge of SAS, you can complete sophisticated tasks without the pain that is often associated with computer-assisted data analysis. The approach is clear and ensures that the goals of the analysis are achieved. Each chapter is conveniently laid out, with clear examples covering every aspect of data entry and manipulation, analysis and interpretation. If you need a thorough reference for working with SAS, this book must be at the top of the list."

> Daniel Coulombe, Ph.D.
> School of Psychology
> University of Ottawa

"This book is detailed, systematic, and easy to follow. Each chapter is divided into subtopics that follow one another logically and give the reader a complete understanding of the concepts. Each topic is accompanied by vivid examples, figures, or diagrams that help the reader grasp the material. In this manner, a gradual and comprehensive knowledge is built. I will definitely use this text myself, and strongly recommend it to my students as well as my colleagues."

> Yaacov G. Bachner
> Department of Sociology of Health
> Ben-Gurion University of the Negev

A Step-by-Step Approach to

Using SAS®
for Univariate
& Multivariate
Statistics

Second Edition

Norm O'Rourke
Larry Hatcher
Edward J. Stepanski

The correct bibliographic citation for this manual is as follows: O'Rourke, Norm, Larry Hatcher, Edward J. Stepanski. 2005. *A Step-by-Step Approach to Using SAS® for Univariate and Multivariate Statistics, Second Edition*. Cary, NC: SAS Institute Inc.

A Step-by-Step Approach to Using SAS® for Univariate and Multivariate Statistics, Second Edition

Copyright © 2005, SAS Institute Inc., Cary, NC, USA
Jointly co-published by SAS Institute and Wiley 2005.

SAS Institute Inc. ISBN 1-59047-417-1
John Wiley & Sons, Inc. ISBN 0-471-46944-0

All rights reserved.

For a hard-copy book: No part of this publication may be reproduced, stored in a retrieval system, or transmitted, in any form or by any means, electronic, mechanical, photocopying, or otherwise, without the prior written permission of the publisher, SAS Institute Inc.

For a Web download or e-book: Your use of this publication shall be governed by the terms established by the vendor at the time you acquire this publication.

U.S. Government Restricted Rights Notice: Use, duplication, or disclosure of this software and related documentation by the U.S. government is subject to the Agreement with SAS Institute and the restrictions set forth in FAR 52.227-19, Commercial Computer Software-Restricted Rights (June 1987).

SAS Institute Inc., SAS Campus Drive, Cary, North Carolina 27513.

6th printing

SAS Publishing provides a complete selection of books and electronic products to help customers use SAS software to its fullest potential. For more information about our e-books, e-learning products, CDs, and hard-copy books, visit the SAS Publishing Web site at **support.sas.com/pubs** or call 1-800-727-3228.

SAS® and all other SAS Institute Inc. product or service names are registered trademarks or trademarks of SAS Institute Inc. in the USA and other countries. ® indicates USA registration.

Other brand and product names are registered trademarks or trademarks of their respective companies.

This book is dedicated to the memory of
Norman R. O'Rourke (1919–2004),
who taught his sons to love mathematics.

Contents

X

Acknowledgments

This book is an update of *A Step-by-Step Approach to Using the SAS System for Univariate and Multivariate Statistics* by Larry Hatcher and Edward J. Stepanski. First and foremost, I want to thank Drs. Hatcher and Stepanski for the opportunity to revise this comprehensive text. I would like also to acknowledge Dr. Gina Graci for facilitating our collaboration. It is amazing the opportunities in life that arise due to sheer serendipity!

I would like to extend special thanks to Lynne Bresler, Kathleen Kiernan, Paul Marovich, and Catherine Truxillo for their invaluable technical review and constructive input during the revision process.

Finally, I must acknowledge the consistent support provided by the SAS Press group at SAS. In particular, thanks to Patsy Poole for her cheerful and rapid responses to my regular deluge of major and not-so-major questions. Thanks also to Mike Boyd for careful final editing, to Joan Stout for index editing, and to Candy Farrell for final composition.

NO'R

Acknowledgments from the First Edition

This book began as a handout for a statistics course and never would have progressed beyond that stage were it not for the encouragement of my colleague, Heidar Modaresi and my department chair, Mel Goldstein. In particular, Mel has my thanks for making every possible accommodation in my teaching schedule to allow time for this book. Thanks also to my colleagues Bill Murdy and Reid Johnson who answered many statistical questions, and to my secretary Cathy Carter who performed countless helpful tasks.

I began work on this text during a sabbatical at Bowling Green State University in 1990. My thanks to Joe Cranny, who was chair of the Psychology Department at BGSU at the time, and who helped make the sabbatical possible.

All of the SAS Institute people with whom I worked were very encouraging, generous with their time, and helpful in their comments as they reviewed and edited this book. These include David Baggett, Jennifer Ginn, Jeff Lopes, Blanche Phillips, Jim Ashton, and Cathy Maah-Fladung. It was a pleasure to work with one and all.

Finally, my wife Ellen was unrelentingly supportive and understanding throughout the entire writing and editing process. I could not have done it without her.

LH

Using This Book

Purpose

This book provides you with virtually everything you need to know to enter data and to perform the statistical analyses that are most commonly used in research in the social sciences. *A Step-by-Step Approach to Using SAS for Univariate and Multivariate Statistics, Second Edition* shows you how to do the following:

- enter data;
- choose the correct statistic;
- perform the analysis;
- interpret the results;
- prepare tables, figures, and text that summarize the results according to the guidelines of the *Publication Manual of the American Psychological Association* (the most widely used format in the social science literature).

Audience

This text is designed for students and researchers whose background in statistics might be limited. An introductory chapter reviews basic concepts in statistics and research methods. The chapters on data entry and statistical analyses assume that the reader has little or no familiarity with SAS; all programming concepts are conveyed at an introductory level. The chapters that deal with specific statistics clearly describe the circumstances under which the use of each is appropriate. Finally, each chapter provides at least one detailed example of how the researcher should enter data, prepare the SAS program, and interpret the results for a representative research problem. Even users whose only exposure to data analysis was an elementary statistics course taken years previously should be able to use this guide to successfully perform statistical analyses.

Organization

Although no single text can discuss *every* statistical procedure, this book covers the statistics that are most commonly used in research in psychology, sociology, marketing, organizational behavior, political science, communication, and the other social sciences. Material covered in each chapter is summarized as follows:

Chapter 1, Basic Concepts in Research and Data Analysis, reviews fundamental issues in research methodology and statistics. This chapter defines and describes the differences between concepts such as *variables* and *values*, *quantitative variables* and *classification variables*, *experimental research* and *nonexperimental research*, and *descriptive analysis* and *inferential analysis*. Chapter 1 also describes the various scales of measurement (e.g., nominal, ordinal) and covers the basic issues in hypothesis testing. After completing this chapter, you should be familiar with the fundamental issues and terminology of data analysis, and be prepared to begin learning about SAS in subsequent chapters.

Chapter 2, Introduction to SAS Programs, SAS Logs, and SAS Output, introduces three elements that you will work with when using SAS: the SAS program; the SAS log; and the SAS output file. Chapter 2 presents a simple SAS program along with the log and output files produced by the program. By the conclusion of this chapter, you will understand the steps to follow to submit SAS programs for computation and to review the results of those programs.

Chapter 3, Data Input, shows you how to enter data and write the statements that create a SAS dataset. This chapter provides instructions on entering data with either the DATALINES or the INFILE statement, and shows you how information can be read either as raw data or as a correlation or covariance matrix. By the conclusion of the chapter, you should be prepared to enter any type of data that is commonly encountered in social science research.

Chapter 4, Working with Variables and Observations in SAS Datasets, shows you how to modify a dataset so that existing variables are transformed, or recoded, so that new variables are created. Chapter 4 shows you how to write statements that eliminate unwanted observations from a dataset so you can perform analyses on a specified subgroup or on participants who have no missing data. This chapter also demonstrates the correct use of arithmetic operators, IF-THEN control statements, and comparison operators. After completing this chapter, you will know how to transform variables and perform analyses on data from specific subgroups of participants.

Chapter 5, Exploring Data with PROC MEANS, PROC FREQ, PROC PRINT, and PROC UNIVARIATE, discusses the use of four procedures:

- PROC MEANS; it can be used to calculate means, standard deviations, and other descriptive statistics for quantitative variables;
- PROC FREQ; it can be used to construct frequency distributions;
- PROC PRINT; it allows you to create a printout of the dataset;
- PROC UNIVARIATE; it allows you to prepare stem-and-leaf plots to test for normality.

By the conclusion of the chapter, you will understand how these procedures can be used to screen data for errors as well as to obtain simple descriptive statistics.

Chapter 6, Measures of Bivariate Association, discusses procedures that can be used to test the significance of the relationship between two variables. Chapter 6 also gives guidelines for choosing the correct statistic based on the level of measurement used to assess the two variables. After completing this chapter, you will know how to use the following:

- PROC PLOT to prepare bivariate scattergrams;
- PROC CORR to compute Pearson and Spearman correlations;
- PROC FREQ to perform the chi-square test of independence.

Chapter 7, Assessing Scale Reliability with Coefficient Alpha, shows how PROC CORR can be used to compute the coefficient alpha reliability index for a multiple-item scale. This chapter reviews basic issues regarding the assessment of reliability and describes the circumstances under which a measure of internal consistency is likely to be high. The chapter also analyzes fictitious questionnaire data to demonstrate how the results of PROC CORR can be used to perform an item analysis, thereby improving the reliability of scale responses. After completing this chapter, you should understand how to construct multiple-item scales that are likely to have high reliability coefficients, and you should be able to use PROC CORR to assess and improve the reliability of responses to your scales.

Chapter 8, *t* Tests: Independent Samples and Paired Samples, shows you how to enter data and prepare SAS programs that perform an independent-samples *t* test using the TTEST procedure or a paired-samples *t* test using the MEANS procedure. This chapter introduces a fictitious study designed to test a hypothesis based on the investment model (Rusbult, 1980), a theory of interpersonal attraction. With respect to the independent-samples *t* test, the chapter shows how to use a folded form of the F statistic to determine whether the equal-variances or unequal-variances *t* test is appropriate for a given analysis. The chapter also provides a structured set of guidelines for interpreting the output from PROC TTEST and for summarizing the results of the analysis. With respect to the paired-samples *t* test, Chapter 8 shows how to create the necessary difference scores and how to test the mean difference score for statistical significance. By the chapter's conclusion, you will know when it is appropriate to perform either the independent-samples *t* test or the paired-samples *t* test, and you will understand what steps to follow when performing both analyses.

Chapter 9, One-Way ANOVA with One Between-Subjects Factor, shows you how to enter data and prepare SAS programs that perform a one-way analysis of variance (ANOVA) using the GLM procedure. Chapter 9 focuses on the *between-subjects* design in which each participant is exposed to only one condition under the independent variable. This chapter also describes the use of a multiple comparison procedure (Tukey's HSD test). By the chapter's conclusion, you should be able to do the following:

- identify the conditions under which this analysis is appropriate;
- prepare the necessary SAS program to perform the analysis;
- review the SAS log and the output for errors;
- prepare tables and text that summarize the results of your analysis.

Chapter 10, Factorial ANOVA with Two Between-Groups Factors, shows you how to enter data and prepare SAS programs to perform a two-way ANOVA using the GLM procedure. This chapter focuses on factorial designs with two *between-subjects* factors, meaning that each participant is exposed to only one condition under each independent variable. The chapter provides guidelines for interpreting results that do not display a significant interaction, and it provides separate guidelines for interpreting results that do display a significant interaction. After completing this chapter, you should be able to determine whether the interaction is significant and to summarize the results involving main effects in the case of a nonsignificant interaction. For significant interactions, you should be able to display the interaction in a figure and perform tests for simple effects.

Chapter 11, Multivariate Analysis of Variance (MANOVA) with One Between-Subjects Factor, shows you how to enter data and prepare SAS programs that perform a one-way multivariate analysis of variance (MANOVA) using the GLM procedure. MANOVA is an extension of ANOVA that allows for the inclusion of multiple criterion variables in a single analysis. This chapter focuses on the between-subjects design in which each participant is exposed to only one condition under the independent variable. By the completion of this chapter, you should be able to summarize both significant and nonsignificant MANOVA results.

Chapter 12, One-Way ANOVA with One Repeated-Measures Factor, shows you how to enter data and prepare SAS programs to perform a one-way repeated-measures ANOVA using the GLM procedure with the REPEATED statement. This chapter focuses on *repeated-measures* designs in which each participant is exposed to every condition under the independent variable. After completing this chapter, you should understand the following:

- necessary conditions for performing a valid repeated-measures ANOVA;
- alternative analyses to use when the validity conditions are not met;
- strategies for minimizing sequence effects.

Chapter 13, Factorial ANOVA with Repeated-Measures Factors and Between-Subjects Factors, shows you how to enter data and prepare SAS programs to perform a two-way mixed-design ANOVA using the GLM procedure and the REPEATED statement. This chapter provides guidelines for the hierarchical interpretation of the analysis along with an alternative method of performing the analysis that allows for use of a variety of post-hoc tests using SAS software. By the end of this chapter, you should be able to interpret the results of analyses both with and without significant interactions.

Chapter 14, Multiple Regression, shows you how to perform multiple regression analysis to examine the relationship between a continuous criterion variable and multiple continuous predictor variables. Chapter 14 describes the different components of the multiple-regression equation and discusses the meaning of R^2 and other results from a multiple-regression analysis. The chapter also shows how bivariate correlations, multiple-regression coefficients, and uniqueness indices can be reviewed to assess the relative importance of predictor variables. After completing the chapter, you should be able to use PROC CORR and PROC REG to conduct the analysis, and you should be able to summarize the results of a multiple regression analysis in tables and in text.

Chapter 15, Principal Component Analysis, introduces principal component analysis, a variable-reduction procedure similar to factor analysis. This chapter offers guidelines regarding the sample size requirements and recommended number of items per component. The chapter also analyzes fictitious data from two studies to illustrate these procedures. By the end of the chapter, you should be able to determine the correct number of components to retain, interpret the rotated solution, create component scores, and summarize the results.

This chapter deals only with the creation of orthogonal (uncorrelated) components; oblique (correlated) solutions are covered in the exploratory factor analysis chapter from *A Step-by-Step Approach to Using the SAS System for Factor Analysis and Structural Equation Modeling* (O'Rourke & Hatcher, in press), a companion volume.

Appendix A, Choosing the Correct Statistic, provides a structured approach to choosing the correct statistical procedure to use when analyzing data. This approach bases the choice of a specific statistic upon the number and scale of the criterion (dependent) variables in conjunction with the number and scale of the predictor (independent) variables. The chapter groups commonly used statistics into three tables based on the number of criterion and predictor variables in the analysis.

Appendix B, Datasets, provides datasets used in Chapters 7, 14, and 15.

Appendix C, Critical Values of the *F* Distribution, provides tables that show critical values of the *F* distribution with alpha levels of .05 and .01.

References

American Psychological Association (2001). *Publication manual of the American Psychological Association* (5th edition). Washington, DC: Author.

O'Rourke, N. & Hatcher, L. (in press). *A step-by-step approach to using SAS for factor analysis and structural equation modeling, 2nd ed.* Cary, NC: SAS Institute Inc.

Rusbult, C. E. (1980). Commitment and satisfaction in romantic associations: A test of the investment model. *Journal of Experimental Social Psychology, 16,* 172-186.

Basic Concepts in Research and DATA Analysis

> **Overview**. This chapter reviews basic concepts and terminology with respect to research design and statistics. Different types of variables that can be analyzed are described as well as the scales of measurement with which these variables are assessed. The chapter reviews the differences between nonexperimental and experimental research and the differences between descriptive and inferential analyses. Finally, basic concepts in hypothesis testing are presented. After completing this chapter, you should be familiar with the fundamental issues and terminology of data analysis. You will be prepared to begin learning about SAS in subsequent chapters.

Introduction: A Common Language for Researchers

Research in the social sciences is a diverse topic. In part, this is because the social sciences represent a wide variety of disciplines, including (but not limited to) psychology, sociology, political science, anthropology, communication, education, management, and economics. A further complicating matter is the fact that, within each discipline, researchers can use a number of very different *quantitative methods* to conduct research. These methods can include unobtrusive observation, participant observation, case studies, interviews, focus groups, surveys, *ex post facto* studies, laboratory experiments, and field experiments to name but a few.

Despite this diversity in methods and topics investigated, most social science research still shares a number of common characteristics. Regardless of field, most research involves an investigator gathering data and performing analyses to determine what the data mean. In addition, most social scientists use a common language when conducting and reporting their research findings. For instance, researchers in both psychology and management speak of "testing null hypotheses" and "obtaining statistically significant p values."

The purpose of this chapter is to review some of the fundamental concepts and terms that are shared across social science disciplines. You should familiarize (or refamiliarize) yourself with this material before proceeding to the subsequent chapters, as most of the terms introduced here are referred to repeatedly throughout the text. If you are currently taking your first course in statistics, this chapter provides an elementary introduction; if you have already completed a course in statistics, it provides a quick review.

Steps to Follow When Conducting Research

The specific steps to follow when conducting research depend, in part, on the topic of investigation, where the researchers are in their overall program of research, and other factors. Nonetheless, much research in the social sciences follows a systematic course of action that begins with the statement of a research question and ends with the researcher drawing conclusions about a null hypothesis. This section describes the research process as a planned sequence that consists of the following six steps:

1. developing a statement of the research question;
2. developing a statement of the research hypotheses (i.e., specific questions to be tested);
3. defining the instruments (e.g., questionnaires, unobtrusive observation measures);
4. gathering the data;
5. analyzing the data;
6. drawing conclusions regarding the null and research hypotheses.

The preceding steps are illustrated here with reference to a fictitious research problem. Imagine that you have been hired by a large insurance company to find ways of improving the productivity of its insurance agents. Specifically, the company would like you to find ways to increase the amount of insurance policies sold by the average agent. You will therefore begin a program of research to identify the determinants of agent productivity.

The Research Question

The process of research often begins with an attempt to arrive at a clear statement of the **research question** (or questions). The research question is a statement of what you hope to learn by the time you have completed the study. It is good practice to revise and refine the research question several times to ensure that you are very explicit and precise.

For example, in the present case, you might begin with the question, "What is the difference between agents who sell a lot of insurance compared to those who sell very little insurance?" An alternative question might be, "What variables have a causal effect on the amount of insurance sold by agents?" Upon reflection, you might realize that the insurance company really only wants to know what things management can do to help agents to sell more. This might eliminate from consideration certain personality traits or demographic variables that are not under management's control, and substantially narrow the focus of the research program. Upon further refinement, a more specific statement of the research question might be, "What variables under the control of management have a causal effect on the amount of insurance sold by agents?" Once you define the research question(s) clearly, you are in a better position to develop a good hypothesis that provides an answer to the question(s).

The Hypothesis

A **hypothesis** is a statement about the predicted relationships among observed events or factors. A good hypothesis in the present case might identify which *specific variables* will have a causal effect on the amount of insurance sold by agents. For example, a hypothesis might predict that agents' level of training will have a positive effect on the amount of insurance sold. Or, it might predict that agents' level of motivation will positively affect sales.

In developing the hypothesis, you might be influenced by any number of sources: an existing theory; some related research; or even personal experience. Let's assume that you have been influenced by **goal-setting theory** that states, among other things, that higher levels of work performance are achieved when employees have difficult work-related goals. Drawing on goal-setting theory, you now state the following hypothesis: "The difficulty of goals that agents set for themselves is positively related to the amount of insurance they sell."

Notice how this statement satisfies our definition for a hypothesis as it is a statement about the assumed causal relationship between two variables. The first variable can be labeled *Goal Difficulty*, and the second can be labeled *Amount of Insurance Sold*. This relationship is illustrated in Figure 1.1:

Figure 1.1 Hypothesized Relationship between Goal Difficulty and Amount of Insurance Sold

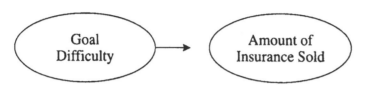

The same hypothesis could be restated in a number of other ways. For example, the following hypothesis makes the same basic prediction: "Agents who set difficult goals for themselves sell greater amounts of insurance than agents who do not set difficult goals."

Notice that these hypotheses are stated in the present tense. It is also acceptable to state hypotheses in the past tense. For example, the preceding could have been stated: "Agents who set difficult goals for themselves sold greater amounts of insurance than agents who did not set difficult goals." The verb tense of the hypothesis depends on whether the researcher will be examining data already collected or will undertake data collection at some future point.

You should also note that these two hypotheses are quite broad in nature. In many research situations, it is helpful to state hypotheses that are more specific in the predictions they make. A more specific hypothesis for the present study might be: "Agents who score above 60 on the Smith Goal Difficulty Scale will sell greater amounts of insurance than agents who score below 40 on the Smith Goal Difficulty Scale."

Defining the Instrument, Gathering Data, Analyzing Data, and Drawing Conclusions

With the hypothesis stated, you can now test it by conducting a study in which you gather and analyze relevant data. *Data* can be defined as a collection of scores obtained when participants' characteristics and/or performance are assessed. For instance, you might decide to test your hypothesis by conducting a simple correlational study. As an example, you might identify a group of 100 agents and determine:

- the difficulty of the goals that have been set for each agent;
- the amount of insurance sold by each agent.

Different types of instruments are used to obtain different types of data. For example, you can use a questionnaire to assess goal difficulty, but rely on company records for measures of insurance sold. Once the data are gathered, each agent will have a score indicating the difficulty of his or her goals and a second score indicating the amount of insurance that he or she has sold.

With the data in hand, you analyze these data to determine if the agents with the more difficult goals did, as hypothesized, sell more insurance. If yes, the study provides support for your hypothesis; if no, it fails to provide support. In either case, you draw conclusions regarding the tenability of your hypotheses. This information is then considered with respect to your research question. These findings might stimulate new questions and hypotheses for subsequent research and the cycle would repeat. For example, if you found support for your hypothesis with the current correlational study, you might choose to follow up with a study using a different method, such as an experimental study. (The difference between these methods is described later.) Over time, a body of research evidence would accumulate, and researchers would be able to review study findings to draw conclusions about the determinants of insurance sales.

Variables, Values, and Observations

Variables

When discussing data, social scientists often speak in terms of variables, values, and observations. For the type of research discussed here, a **variable** refers to some specific participant characteristic that can assume different values (i.e., the values vary). For the participants in the study just described, Amount of Insurance Sold is an example of a variable: some participants had sold a lot of insurance whereas others had sold less. A second variable was Goal Difficulty: some participants set more difficult goals while others set less difficult goals. Participant Age and Sex (male or female) were other variables, though not considered in our initial hypothesis.

Values

A **value** is a specific quantity or category of a variable. For example, Amount of Insurance Sold is a variable that can assume a range of values. One agent might sell $2,000,000 worth of insurance in one year, one might sell $100,000, whereas another may sell nothing (i.e., $0). Age is another variable that can assume a variety of values. In our example, these values range from a low of 22 years (the youngest agent) to a high of 64 years (the oldest agent). In other words, age is a variable that is comprised of a wide array of specific values.

Quantitative Variables versus Classification Variables

For both types of these variables, a given value is the specific score that indicates participants' standing with respect to variable of interest. Amount of Insurance Sold and Age are **quantitative variables** since numbers serve as values. The word "score" is an appropriate substitute for the word value in these cases.

A different type of variable is a **classification variable** or, alternatively, **qualitative variable** or **categorical variable**. With classification variables, different values represent different groups to which participants can belong. Sex is a good example of a classification variable as it (generally) assumes only one of two values (i.e., participants are classified as either male or female). Race is an example of a classification variable that can assume a larger number of values: participants can be classified as Caucasian; African American; Asian American; or as belonging to a large number of other groups. Notice why these variables are classification variables and not quantitative variables. The values represent only group membership; they do not represent a characteristic that some participants possess in greater quantity than others.

Observational Units

In discussing data, researchers often make reference to **observational units** that can be defined as individual participants (or objects) that serve as the source of the data. (Observational units are also referred to as *units of analysis*.) Within the social sciences, a person usually serves as the observational unit under study (although it is also possible to use some other entity such as an individual school or organization as the observational unit). In this text, the person is used as the observational unit in all examples. Researchers often refer to the number of **observations** (or **cases**) included in their datasets. This simply refers to the number of participants studied.

For a more concrete illustration of the concepts discussed so far, consider the following dataset:

Table 1.1

Insurance Sales Data

Observation	Name	Sex	Age	Scores	Goal Difficulty Rank	Sales
1	Bob	M	34	97	2	$598,243
2	Pietro	M	56	80	1	$367,342
3	LeMont	M	36	67	4	$254,998
4	Susan	F	24	40	3	$80,344
5	Saleem	M	22	37	5	$40,172
6	Mack	M	44	24	6	$0

The preceding table reports information about six research participants: Bob; Pietro; LeMont; Susan; Saleem; and Mack. Therefore, the dataset includes six observations. Information about a given observation (participant) appears as a **row** running left to right across the table. The first **column** of the dataset (running vertically) indicates the observation number, and the second column reports the name of the participant. The remaining five columns report information on the five research variables under study. The "Sex" column reports participant sex, which can assume one of two values: "M" for male and "F" for female. The "Age" column reports participants' age in years. The "Goal Difficulty Scores" column reports participants' scores on a fictitious goal difficulty scale. (Assume that each participant completed a 20-item questionnaire that assessed the difficulty of his or her work goals.)

Depending on their responses, participants receive a score that can range from a low of 0 (meaning that work goals are quite easy) to a high of 100 (meaning that they are quite difficult). The "Rank" column shows how participants were ranked by their supervisor according to their overall effectiveness as agents when compared to one another. A rank of 1 represents the most effective agent whereas a rank of 6 represents the least effective. Finally, the "Sales" column indicates the amount of insurance sold by each agent (in dollars) during the past year.

This example illustrates a very small dataset with six observations and five research variables (i.e., Sex, Age, Goal Difficulty, Rank and Sales). One variable is a classification variable (Sex), while the remainder are quantitative variables. The numbers or letters that appear within a column represent some of the values that can be assumed by that variable.

Scales of Measurement

One of the most important schemes for classifying a variable involves its **scale of measurement**. Data generally fall within four different scales of measurement: nominal; ordinal; ratio; and interval. Before analyzing a dataset, it is important to determine which scales of measurement were used because certain types of statistical procedures require specific scales of measurement. In this text, each chapter that deals with a specific statistical procedure specifies what scale of measurement is required. The researcher must determine scales of measurement for study variables before selecting which statistical procedures to use.

Nominal Scales

A **nominal scale** is a classification system that places people, objects, or other entities into mutually exclusive categories. A variable measured along a nominal scale is a *classification variable*; it simply indicates the group to which each participant belongs. The examples of classification variables provided earlier (e.g., Sex and Ethnicity) also are examples of nominal-level variables; they tell us to which group a participant belongs but they do not provide any quantitative information. That is, the Sex variable identifies participants as either male or female but it does not tell us that participants possess more or less of a specific characteristic relative to others. However, the remaining three scales of measurement—ordinal, interval, and ratio—provide some quantitative information.

Ordinal Scales

Values on an **ordinal scale** represent the rank order of participants with respect to the variable being assessed. For example, the preceding table includes one variable called Rank that represents the rank ordering of participants according to their overall effectiveness as agents. Values for this ordinal scale represent a *hierarchy of levels* with respect to the construct of "effectiveness." We know that the agent ranked 1 was perceived as being more effective than the agent ranked 2, that the agent ranked 2 was more effective than the one ranked 3, and so forth.

The information conveyed by an ordinal scale is limited because equal differences in scale values do not necessarily have equal quantitative meaning. For example, notice the following rankings:

Rank	Name
1	Bob
2	Pietro
3	Susan
4	LeMont
5	Saleem
6	Mack

Notice that Bob was ranked 1 while Pietro was ranked 2. The difference between these two rankings is 1 (because $2 - 1 = 1$), so there is one unit of difference between Bob and Pietro. Now notice that Saleen was ranked 5 while Mack was ranked 6. The difference between these two rankings is also 1 (because $6 - 5 = 1$), so there is also 1 unit of difference between Saleem and Mack. Putting the two together, the difference in ranking between Bob and Pietro is equal to the difference in ranking between Saleem and Mack.

But, does this mean that the difference in overall effectiveness between Bob and Pietro is equal to the difference in overall effectiveness between Saleem and Mack? Not necessarily. It is possible that Bob was significantly superior to Pietro in effectiveness, while Saleem might have been only slightly superior to Mack. In fact, this appears to be the case. Whereas Bob had sold policies totaling $598,243, Pietro had sold $367,342 for a difference of $230,830 between the two. In contrast, the difference in sales between Saleem ($40,170) and Mack ($0) was only a faction of the difference between Bob and Pietro (i.e., $40,170 vs. $230,830). This example indicates that these rankings reveal very little about the quantitative differences between participants with regard to the underlying construct (effectiveness, in this case). An ordinal scale simply provides a rank order. Other scales of measurement are required to provide this added level of measurement.

Interval Scales

With an **interval scale**, equal differences between values have equal quantitative meaning. For this reason, it can be seen that an interval scale provides more quantitative information than an ordinal scale. A good example of interval measurement is the Fahrenheit scale used to measure temperature. With the Fahrenheit scale, the difference between 70 degrees and 75 degrees is equal to the difference between 80 degrees and 85 degrees. In other words, the units of measurement are equal throughout the full range of the scale.

However, the interval scale also has a limitation; it does not have a true zero point. A **true zero point** means that a value of zero on the scale represents zero quantity of the variable being assessed. It should be obvious that the Fahrenheit scale does not have a true zero point; when the thermometer reads 0 degrees Fahrenheit, that does not mean that there is absolutely no heat present in the environment.

Social scientists often assume that many of their man-made variables are measured on an interval scale. In the preceding study involving insurance agents, for example, you would probably assume that scores from the goal difficulty questionnaire constitute an interval-level scale (i.e., you would likely assume that the difference between a score of 50 and 60 is approximately equal to the difference between a score of 70 and 80). Many researchers would also assume that scores from an instrument such as an intelligence test are also measured at the interval level of measurement.

On the other hand, some researchers are skeptical that instruments such as these have true equal-interval properties and prefer to refer to them as **quasi-interval** scales (e.g., Likert-type scales to which respondents indicate their degree of agreement to a series of statements with a fixed number of response alternatives such as strongly disagree, disagree, neutral, agree, and strongly agree). Disagreements concerning the level of measurement with such instruments continue to be a controversial topic within the social sciences (i.e., whether scale responses ranging from *strongly disagree* to *strongly agree* constitute ordinal- or interval-level measurement).

It is clear that there is no true zero point with either of the preceding instruments. A score of 0 on the goal difficulty scale does not indicate the complete absence of goal difficulty, and a score of 0 on an intelligence test does not indicate the complete absence of intelligence. A true zero point can be found only with variables measured on a ratio scale.

Ratio Scales

Ratio scales are similar to interval scales in that equal differences between scale values have equal quantitative meaning. However, ratio scales also have a true zero point which gives them an additional property. With ratio scales, it is possible to make meaningful statements about the ratios between scale values. For example, the system of inches used with a common ruler is an example of a ratio scale. There is a true zero point with this system in which zero inches does, in fact, indicate a complete absence of length. With this scale, therefore, it is possible to make meaningful statements about ratios. It is appropriate to say that an object four inches long is twice as long as an object two inches long. Age, as measured in years, is also on a ratio scale as a 10-year old house is twice as old as a 5-year old house. Notice that it is not possible to make these statements about ratios with the interval-level variables discussed above. One would not say that a person with an IQ of 160 is twice as intelligent as a person with an IQ of 80.

Although ratio-level scales might be easiest to find when one considers the physical properties of objects (e.g., height and weight), they are also common in the type of research discussed in this text. For example, the study discussed previously included the variables for age and amount of insurance sold (in dollars). Both of these have true zero points and are measured as ratio scales.

Basic Approaches to Research

Nonexperimental Research

Much research can be categorized as being either experimental or nonexperimental in nature. In **nonexperimental research** (also called **nonmanipulative** or **correlational** research), the investigator simply studies the association between two or more naturally occurring variables. A **naturally occurring variable** is one that is not manipulated or controlled by the researcher; it is simply observed and measured (e.g., the age of insurance salespersons).

The insurance study described previously is a good example of nonexperimental research since you simply measured two naturally occurring variables (i.e., goal difficulty and amount of insurance sold) to determine whether they were related. If, in a different study, you investigated the relationship between IQ and college grade point average (GPA), this would also be an example of nonexperimental research.

With nonexperimental research designs, social scientists often refer to criterion variables and predictor variables. A **criterion variable** is an outcome variable that might be predicted by one or more other variables. The criterion variable is generally the main focus of the study; it is the outcome variable mentioned in the statement of the research problem. In our example, the criterion variable is Amount of Insurance Sold.

The **predictor variable**, on the other hand, is that variable used to predict or explain values of the criterion. In some studies, you might even believe that the predictor variable has a causal effect on the criterion. In the insurance study, for example, the predictor variable was Goal Difficulty. Because you believed that Goal Difficulty positively affects insurance sales, you conducted a study in which Goal Difficulty is identified as the predictor and Sales as the criterion. You do not necessarily have to believe that there is a causal relationship between Goal Difficulty and Sales to conduct this study. You might simply be interested in determining whether there is an association between these two variables (i.e., as the values for the predictor change, a corresponding change in the criterion variable is observed).

You should note that nonexperimental research that examines the relationship between just two variables generally provides little evidence concerning cause-and-effect relationships. The reasons for this can be seen by reviewing the study on insurance sales. If the social scientist conducts this study and finds that the agents with the more difficult goals also tend to sell more insurance, does that mean that having difficult goals *caused* them to sell more insurance? Not necessarily. You can argue that selling a lot of insurance increases the agents' self-confidence and that this, in turn, causes them to set higher work goals for themselves. Under this second scenario, it was actually the insurance sales that had a causal effect on Goal Difficulty.

As this example shows, with nonexperimental research it is often possible to obtain a result consistent with a range of causal explanations. Hence, a strong inference that "variable A had a causal effect on variable B" is seldom possible when you conduct simple correlational research with just two variables. To obtain stronger evidence of cause and effect, researchers generally either analyze the relationships between a larger number of variables using sophisticated statistical procedures that are beyond the scope of this text, or drop the nonexperimental approach entirely and, instead, use experimental research methods. The nature of experimental research is discussed in the following section.

Experimental Research

Most experimental research can be identified by three important characteristics:

- Participants are randomly assigned to experimental and control conditions.
- The researcher manipulates one or more variables.
- Participants in different experimental conditions are treated similarly with regard to all variables except the manipulated variable.

To illustrate these concepts, assume that you conduct an experiment to test the hypothesis that goal difficulty positively affects insurance sales. Assume that you identify a group of 100 agents who will serve as study participants. You randomly assign 50 agents to a "difficult-goal" condition. Participants in this group are told by their superiors to make at least 25 cold calls (sales calls) to potential policyholders per week. The other 50 agents assigned to the "easy-goal" condition have been told to make just five cold calls to potential policyholders per week. The design of this experiment is illustrated in Figure 1.2.

Figure 1.2 Design of the Experiment Used to Assess the Effects of Goal Difficulty

Group	Treatment Conditions Under the Independent Variable (Goal Difficulty)	Results Obtained with the Dependent Variable (Amount of Insurance Sold)
Group 1 ⟶ ($n = 50$)	Difficult-Goal ⟶ Condition	\$156,000 in Sales
Group 2 ⟶ ($n = 50$)	Easy-Goal ⟶ Condition	\$121,000 in Sales

After 12 months, you determine how much new insurance each agent has sold that year. Assume that the average agent in the difficult-goal condition sold \$156,000 worth of new policies while the average agent in the easy-goal condition sold just \$121,000 worth.

It is possible to use some of the terminology associated with nonexperimental research when discussing this experiment. For example, it would be appropriate to continue to refer to Amount of Insurance Sold as being a criterion variable because this is the outcome variable of central interest. You could also continue to refer to Goal Difficulty as the predictor variable because you believe that this variable will predict sales to some extent.

Notice, however, that Goal Difficulty is now a somewhat different variable. In the nonexperimental study, Goal Difficulty was a naturally occurring variable that could take on a variety of values (whatever score participants received on the goal difficulty questionnaire). In the present experiment, however, Goal Difficulty is a **manipulated variable**, which means that you (as the researcher) determined what value of the variable would be assigned to both participant groups. In this experiment, Goal Difficulty could assume only one of two values. Therefore, Goal Difficulty is now a classification variable, assessed on a nominal scale.

Although it is acceptable to speak of predictor and criterion variables within the context of experimental research, it is more common to speak in terms of independent and dependent variables. The independent variable (IV) is that variable whose values (or levels) are selected by the experimenter to determine what effect the independent variable has on the dependent variable. The **independent variable** is the experimental counterpart to a predictor variable. A dependent variable (DV) is some aspect of the study participant's behavior that is assessed to reflect the effects of the independent variable. The **dependent variable** is the experimental counterpart to a criterion variable. In the present experiment, Goal Difficulty is the independent variable while Sales is the dependent variable. Remember that the terms predictor variable and criterion variable can be used with almost any type of research, but that the terms independent variable and dependent variable should be used only with experimental research.

Researchers often refer to the different **levels of the independent variable**. These levels are also referred to as **experimental conditions** or **treatment conditions** and correspond to the different groups to which participants can be assigned. The present example includes two experimental conditions: a difficult-goal condition and an easy-goal condition.

With respect to the independent variable, you can speak in terms of the experimental group versus the control group. Generally speaking, the **experimental group** receives the experimental treatment of interest while the **control group** is an equivalent group of participants who do not receive this treatment. The simplest type of experiment consists of just one experimental group and one control group. For example, the present study could have been redesigned so that it consisted of an experimental group that was assigned the goal of making 25 cold calls (the difficult-goal condition) and a control group in which no goals were assigned (the no-goal condition). Obviously, you can expand the study by creating more than one experimental group. You could do this in the present case by assigning one experimental group the difficult goal of 25 cold calls and the second experimental group the easy goal of just 5 cold calls.

Descriptive versus Inferential Statistical Analysis

To understand the difference between descriptive and inferential statistics, you must first understand the difference between populations and samples. A **population** is the *entire collection* of a carefully defined set of people, objects, or events. For example, if the insurance company in question employed 10,000 insurance agents in the European Union, then those 10,000 agents would constitute the population of agents hired by that company. A **sample**, on the other hand, is a subset of the people, objects, or events selected from a population. For example, the 100 agents used in the experiment described earlier constitute a sample.

Descriptive Analyses

A **parameter** is a descriptive characteristic of a population. For example, if you assessed the average amount of insurance sold by all 10,000 agents in this company, the resulting average would be a parameter. To obtain this average, of course, you would first need to tabulate the amount of insurance sold by each and every agent. In calculating this average, you are engaging in descriptive analysis. Descriptive analyses organize, summarize, and identify major characteristics of the population.

Most people think of populations as being very large groups, such as all of the people in the United Kingdom. However, a group does not have to be large to be a population, it only has to be the entire collection of the people or things being studied. For example, a teacher can define all twelfth-grade students in a single school as a population and then calculate the average score of these students on a measure of class satisfaction. The resulting average would be a population parameter.

Inferential Analyses

A **statistic**, on the other hand, is a numerical value that is computed from a sample and either describes some characteristic of that sample such as the average value, or is used to make inferences about the population from which the sample is drawn. For example, if you were to compute the average amount of insurance sold by your sample of 100 agents, that average would be a statistic because it summarizes a specific characteristic of the sample. Remember that the word "statistic" is generally associated with samples while "parameter" is generally associated with populations.

In contrast to descriptive analyses, **inferential statistics** involve information from a sample to make inferences, or estimates, about the population (i.e., infer from the sample to the larger population). For example, assume that you need to know how much insurance is sold by the average agent in the company. It might not be possible to obtain the necessary information from all 10,000 agents and then determine the average. An alternative would be to draw a random (and ideally representative) sample of 100 agents and determine the average amount sold by this subset. A **random sample** is a subset of the population in which each member of that population has an equal chance of selection. If this group of 100 sold an average of $179,322 worth of policies last year, then your best guess of the amount of insurance sold by all 10,000 agents would likewise be $179,322 on average. Here, you have used characteristics of the sample to make inferences about characteristics of the population. This is the real value of inferential statistical procedures; they allow you to examine information obtained from a relatively small sample and then make inferences about the overall population. For example, pollsters conduct telephone surveys to ascertain the voting preferences of Canadians leading up to, and between, federal elections. From randomly selected samples of approximately 1,200 participants, these pollsters can extrapolate their findings to the population of 20 million eligible voters with considerable accuracy (i.e., within relatively narrow limits).

Hypothesis Testing

Most of the procedures described in this text are inferential procedures that allow you to test specific hypotheses about the characteristics of populations. As an illustration, consider the simple experiment described earlier in which 50 agents were assigned to a difficult-goal condition and 50 other agents to an easy-goal condition. After one year, the agents with difficult goals had sold an average of $156,000 worth of insurance while the agents with easy goals had sold $121,000 worth. On the surface, this would seem to support your hypothesis that difficult goals cause agents to sell more insurance. But can you be sure of this? Even if goal setting had no effect at all, you would not really expect the two groups of 50 agents to sell exactly the same amount of insurance; one group would sell somewhat more than the other due to chance alone. The difficult-goal group did sell more insurance, but did it sell a sufficiently greater amount of insurance to suggest that the difference was due to your manipulation (i.e., random assignment to the experimental group)?

What's more, it could easily be argued that you don't even care about the amount of insurance sold by these two relatively small samples. What really matters is the amount of insurance sold by the larger populations that they represent. The first population could be defined as "the population of agents who are assigned difficult goals" and the second would be "the population of agents who are assigned easy goals." Your real research question involves the issue of whether the first population sells more than the second. This is where hypothesis testing comes in.

Types of Inferential Tests

Generally speaking, there are two types of tests conducted when using inferential procedures: tests of group differences and tests of association. With a **test of group differences,** you typically want to know whether populations differ with respect to their scores on some criterion variable. The present experiment would lead to a test of group differences because you want to know whether the average amount of insurance sold in the population of difficult-goal agents is different from the average amount sold in the population of easy-goal agents. A different example of a test of group differences might involve a study in which the researcher wants to know whether Caucasian-Americans, African-Americans, and Asian-Americans differ with respect to their scores on an academic achievement scale. Notice that in both cases, two or more distinct populations are being compared with respect to their scores on a single criterion variable.

With a **test of association** on the other hand, you are working with a single group of individuals and want to know whether or not there is a relationship between two or more variables. Perhaps the best-known test of association involves testing the significance of a correlation coefficient. Assume that you have conducted a simple correlational study in which you asked 100 agents to complete the 20-item goal-difficulty questionnaire. Remember that with this questionnaire, participants could receive a score ranging from a low of 0 to a high of 100 (interval measurement). You could then correlate these goal-difficulty scores with the amount of insurance sold by agents that year. Here, the goal-difficulty scores constitute the predictor variable while the amount of insurance sold serves as the criterion. Obtaining a strong positive correlation between these two variables would mean that the more difficult the agents' goals, the more insurance they tended to sell. Why would this be called a test of association? That is because you are determining whether there is an association, or *relationship*, between the predictor and criterion variables. Notice also that only one group is studied (i.e., there is no random assignment or experimental manipulation that creates a difficult-goal sample versus an easy-goal sample).

To be thorough, it is worth mentioning that there are some relatively sophisticated procedures that also allow you to perform a third type of test: whether the association between variables is the same across multiple groups. Analysis of covariance (ANCOVA) is one procedure that enables such a test. For example, you might hypothesize that the association between self-reported Goal Difficulty and insurance sales is stronger in the population of agents assigned difficult goals than it is in the population assigned easy goals. To test this assertion, you might randomly assign a group of insurance agents to either an easy-goal condition or a difficult-goal condition (as described earlier). Each agent could complete the 20-item self-report goal difficulty scale and then be exposed to the appropriate treatment. Subsequently, you would record each agent's sales. Analysis of covariance would allow you to determine whether the relationship between questionnaire scores and sales is stronger in the difficult-goal population than it is in the easy-goal population. (ANCOVA would also allow you to test a number of additional hypotheses.)

Types of Hypotheses

Two different types of hypotheses are relevant to most statistical tests. The first is called the null hypothesis, which is generally abbreviated as H_0. The **null hypothesis** is a statement that, in the population(s) being studied, there are either (a) *no difference* between the groups or; (b) *no relationship* between the measured variables. For a given statistical test, either (a) or (b) will apply, depending on whether one is conducting a test of group differences or a test of association, respectively.

With a **test of group differences**, the null hypothesis states that, in the population, there are no differences between groups studied with respect to their mean scores on the criterion variable. In the experiment in which a difficult-goal condition is being compared to an easy-goal condition, the following null hypothesis might be used:

H_0: In the population, individuals assigned difficult goals do not differ from individuals assigned easy goals with respect to the mean amount of insurance sold.

This null hypothesis can also be expressed mathematically with symbols in the following way:

H_0: $M_1 = M_2$

where:

H_0 represents null hypothesis

M_1 represents mean sales for the difficult-goal population

M_2 represents mean sales for the easy-goal population

In contrast to the null hypothesis, you will also form an alternative hypothesis (H_1) that states the opposite of the null. The **alternative hypothesis** is a statement that there *is a difference* between groups, or that there *is a relationship* between the variables in the population(s) studied.

Perhaps the most common alternative hypothesis is a **nondirectional alternative hypothesis** (often referred to as a 2-sided hypothesis). With a test of group differences, a no-direction alternative hypothesis predicts that the various populations will differ, but makes no specific prediction as to how they will differ (e.g., one outperforming the other). In the preceding experiment, the following nondirectional null hypothesis might be used:

> H_1: In the population, individuals assigned difficult goals differ from individuals assigned easy goals with respect to the amount of insurance sold.

This alternative hypothesis can also be expressed with symbols in the following way:

> H_1: $M_1 \neq M_2$

In contrast, a directional or 1-sided alternative hypothesis makes a more specific statement regarding the expected outcome of the analysis. With a test of group differences, a directional alternative hypothesis not only predicts that the populations differ, but also contends which will be relatively high and which will be relatively low. Here is a directional alternative hypothesis for the preceding experiment:

> H_1: The amount of insurance sold is higher in the population of individuals assigned difficult goals than in the population of individuals assigned easy goals.

This hypothesis can be symbolically represented in the following way:

> H_1: $M_1 > M_2$

Had you believed that the easy-goal population would sell more insurance, you would have replaced the "greater than" symbol ($>$) with the "less than" symbol ($<$), as follows:

> H_1: $M_1 < M_2$

Null and alternative hypotheses are also used with tests of association. For the study in which you correlated goal-difficulty questionnaire scores with the amount of insurance sold, you might have used the following null hypothesis:

> H_0: In the population, the correlation between goal-difficulty scores and the amount of insurance sold is zero.

You could state a nondirectional alternative hypothesis that corresponds to this null hypothesis in this way:

> H_1: In the population, the correlation between goal-difficulty scores and the amount of insurance sold is not equal to zero.

Notice that the preceding is an example of a nondirectional alternative hypothesis because it does not specifically predict whether the correlation is positive or negative, only that it is not zero. On the other hand, a directional alternative hypothesis might predict a positive correlation between the two variables. You could state such a prediction as follows:

> H_1: In the population, the correlation between goal-difficulty scores and the amount of insurance sold is greater than zero.

There is an important advantage associated with the use of directional alternative hypotheses compared to nondirectional hypotheses. Directional hypotheses allow researchers to perform 1-sided statistical tests (also called *1-tail tests*), which are relatively powerful. Here, "powerful" means that *one-sided* tests are more likely to find statistically significant differences between groups when differences really do exist. In contrast, nondirectional hypotheses allow only *2-sided* statistical tests (also called *2-tail tests*) that are less powerful.

Because they lead to more powerful tests, directional hypotheses are generally preferred over nondirectional hypotheses. However, directional hypotheses should be stated only when they can be justified on the basis of theory, prior research, or some other acceptable reason. For example, you should state the directional hypothesis that "the amount of insurance sold is higher in the population of individuals assigned difficult goals than in the population of individuals assigned easy goals" only if there are theoretical or empirical reasons to believe that the difficult-goal group will indeed score higher on insurance sales. The same should be true when you specifically predict a positive correlation rather than a negative correlation (or vice versa).

The *p* or Significance Value

Hypothesis testing, in essence, is a process of determining whether you can reject your null hypothesis with an acceptable level of confidence. When analyzing data with SAS, you will review the output for two pieces of information that are critical for this purpose: the obtained statistic and the probability (*p*) or significance value associated with that statistic. For example, consider the experiment in which you compared the difficult-goal group to the easy-goal group. One way to test the null hypothesis associated with this study would be to perform an independent samples *t* test (described in detail in Chapter 8, "*t* Tests: Independent Samples and Paired Samples"). When the data analysis for this study has been completed, you would review a *t* statistic and its corresponding *p* value. If the *p* value is very small (e.g., *p* < .05), you will reject the null hypothesis.

For example, assume that you obtain a *t* statistic of 0.14 and a corresponding *p* value of .90. This *p* value indicates that there are 90 chances in 100 that you would obtain a *t* statistic of 0.14 (or larger) if the null hypothesis were true. Because this probability is high, you would report that there is very little evidence to refute the null hypothesis. In other words, you would fail to reject your null hypothesis and would, instead, conclude that there is not sufficient evidence to find a statistically significant difference between groups (i.e., between group differences might well be due only to chance).

On the other hand, assume that the research project instead produces a *t* value of 3.45 and a corresponding *p* value of .01. The *p* value of .01 indicates that there is only one chance in 100 that you would obtain a *t* statistic of 3.45 (or larger) if the null hypothesis were true. This is so unlikely that you can be fairly confident that the null hypothesis is not true. You would therefore reject the null hypothesis and conclude that the two populations do, in fact, appear to differ. In rejecting the null hypothesis, you have tentatively accepted the alternative hypothesis.

Technically, the *p* value does not really provide the probability that the null hypothesis is true. Instead, it provides the probability that you would obtain the present results (the present *t* statistic, in this case) if the null hypothesis were true. This might seem like a trivial difference, but it is important that you not be confused by the meaning of the *p* value.

Notice that you were able to reject the null hypothesis only when the *p* value was a fairly small number (.01, in the above example). But how small must a *p* value be before you can reject the null hypothesis? A *p* value of .05 seems to be the most commonly accepted cutoff. Typically, when researchers obtain a *p* value *larger* than .05 (such as .13 or .37), they will fail to reject the null hypothesis and will instead conclude that the differences or relationships being studied were not statistically significant (i.e., differences can occur as a matter of chance alone). When they obtain a *p* value *smaller* than .05 (such as .04 or .02 or .01), they will reject the null hypothesis and conclude that differences or relationships being studied are statistically significant. The .05 level of significance is not an absolute rule that must be followed in all cases, but it should be serviceable for most types of investigations likely to be conducted in the social sciences.

Fixed Effects versus Random Effects

Experimental designs can be represented as mathematical models and these models can be described as fixed-effects models, random-effects models, or mixed-effects models. The use of these terms refers to the way that the levels of the independent (or predictor) variable were selected.

When the researcher arbitrarily selects the levels of the independent variable, the independent variable is called a **fixed-effects factor** and the resulting model is a **fixed-effects model**. For example, assume that, in the current study, you arbitrarily decide that participants in your easy-goal condition would be told to make just 5 cold calls per week and that participants in the difficult-goal condition would be told to make 25 cold calls per week. In this case, you have *fixed* (i.e., arbitrarily selected) the levels of the independent variable. Your experiment therefore represents a fixed-effects model.

In contrast, when the researcher randomly selects levels of the independent variable from a population of possible levels, the independent variable is called a **random-effects factor**, and the model is a **random-effects model**. For example, assume that you have determined that the number of cold calls that an insurance agent could possibly place in one week ranges from 0 to 45. This range represents the population of cold calls that you could possibly research. Assume that you use some random procedure to select two values from this population (perhaps by drawing numbers from a hat). Following this procedure, the values 12 and 32 are drawn. When conducting your study, one group of participants is assigned to make at least 12 cold calls per week, while the second is assigned to make 32 calls. In this instance, your study represents a random-effects model because the levels of the independent variable were randomly selected.

Most research in the social sciences involves **fixed-effects models**. As an illustration, assume that you are conducting research on the effectiveness of hypnosis in reducing anxiety among participants who suffer from test anxiety. Specifically, you could perform an experiment that compares the effectiveness of 10 sessions of relaxation training versus 10 sessions of relaxation training plus hypnosis. In this study, the independent variable might be labeled something like Type of Therapy. Notice that you did not randomly select these two treatment conditions from the population of all possible treatment conditions; you knew which treatments you wished to compare and designed the study accordingly. Therefore, your study represents a fixed-effects model.

To provide a nonexperimental example, assume that you were to conduct a study to determine whether Hispanic-Americans score significantly higher than Korean-Americans on academic achievement. The predictor variable in your study would be Ethnicity while the criterion variable would be scores on some index of academic achievement. In all likelihood, you would not have arbitrarily chosen "Hispanic American" versus "Korean American" because you are particularly interested in these two ethnic groups; you did not randomly select these groups from all possible ethnic groups. Therefore, the study is again an example of a fixed-effects model.

Of course, random-effects factors do sometimes appear in social science research. For example, in a repeated-measures investigation (in which repeated measures on the criterion variable are taken from each participant), *participant group* is viewed as a random-effects factor (assuming that they have been randomly selected). Some studies include both fixed-effects factors and random-effects factors. The resulting models are called **mixed-effects models**.

This distinction between fixed versus random effects has important implications for the types of inferences that can be drawn from statistical tests. When analyzing a fixed-effects model, you can generalize the results of the analysis only to the specific levels of the independent variable that were manipulated in that study. This means that if you arbitrarily selected 5 cold calls versus 25 cold calls for your two treatment conditions, once the data are analyzed you can draw conclusions only about the population of agents assigned 5 cold calls versus the population assigned 25 cold calls.

If, on the other hand, you randomly selected two values for your treatment conditions (say 12 versus 32 cold calls) from the population of possible values, your model is a random-effects model. This means that you can draw conclusions about the entire population of possible values that your independent variable could assume; these inferences would not be restricted to just the two treatment conditions investigated in the study. In other words, you could draw inferences about the relationship between the population of the possible number of cold calls to which agents might be assigned and the criterion variable (insurance sales).

Conclusion

Regardless of discipline, researchers need a common language when discussing their work. This chapter has reviewed the basic concepts and terminology of research that will be referred to throughout this text. Now that you can speak the language, you are ready to move on to Chapter 2 where you will learn how to prepare a simple SAS program.

Introduction to SAS® Programs, SAS® Logs, and SAS® Output

> **Overview.** This chapter describes the three types of files that you will work with while using SAS: the SAS program; the SAS log; and the SAS output file. You are presented with a very simple SAS program, along with the log and output files produced by that program. This chapter provides the big picture regarding the steps you need to follow when performing data analyses with SAS.

Introduction: What Is SAS?

SAS is a modular, integrated, and hardware-independent system of statistical software. It is a particularly powerful tool for social scientists because it allows them to easily perform a myriad of statistical analyses that might be required in the course of conducting research. SAS is comprehensive enough to perform the most sophisticated multivariate analyses (i.e., multiple dependent variables), but is also easy to use so that undergraduates can also perform basic analyses after only a short period of instruction.

In a sense, SAS can be viewed as a library of prewritten statistical algorithms. By submitting a short SAS program, you can access a prewritten procedure and use it to analyze a set of data. For example, below are the SAS statements used to call up the algorithm that calculates Pearson correlation coefficients:

```
PROC CORR    DATA=D1;
   RUN;
```

The preceding statements cause SAS to compute correlation coefficients for all numeric variables in your dataset. The ability to call up complex procedures with such a simple statement makes this system powerful and easy to use. By contrast, if you had to prepare your own programs to compute correlations by using a programming language such as FORTRAN or BASIC, it would require many statements, and there would be several opportunities for error. By using SAS instead, most of the work is completed and allows you to focus on the *results* of the analysis rather than on the *mechanics* for obtaining those results.

> **Where is SAS installed?** SAS computer software products are installed at over 40,000 business, government, and university sites in 105 countries. More than 2,000 customer sites are universities.

Three Types of SAS Files

Subsequent chapters of this manual provide details on how to write a SAS program: how to create and manage data; how to request specific statistical procedures; and so forth. This chapter presents a short SAS program and discusses the resulting output. Little elaboration is offered.

The purpose of this chapter is to provide a very general sense of what it entails to submit a SAS program and interpret the results. You are encouraged to copy the program that appears in the following example, submit it for analysis, and verify that the resulting output matches the output reproduced here. This exercise will provide you with the SAS big picture, and this perspective will facilitate learning the programming details presented in subsequent chapters.

You work with three types of "files" when using SAS: one file contains the SAS program; one contains the SAS log; and one contains the SAS output. The following sections discuss the differences among these files.

The SAS Program

A SAS program consists of a set of statements written by the researcher or programmer. These statements provide SAS with the data to be analyzed, tell it about the nature of these data, and indicate which statistical analyses should be performed on the data.

This section illustrates a simple SAS program by analyzing some fictitious data from a fictitious study. Assume that six high school students have taken the *Graduate Record Examinations* (GRE). This test provides two scores for each student: a score on the GRE verbal test; and a score on the GRE math test. With both tests, scores can range from 200 to 800 with higher scores indicating higher achievement levels.

Assume that you now want to obtain simple descriptive statistics regarding six students' scores on these two tests. For example, what is their *average* score on the GRE verbal test or on the GRE math test? What is the *standard deviation* of the scores on the two tests?

To perform these analyses, you prepare the following SAS program:

```
DATA D1;
INPUT PARTICIPANT GREVERBAL GREMATH;

DATALINES;
1 520 490
2 610 590
3 470 450
4 410 390
5 510 460
6 580 350
;
RUN;
```

DATA step

```
                        ⎧    PROC MEANS   DATA=D1
    PROC step           ⎨        VAR  GREVERBAL  GREMATH;
                        ⎩    RUN;
```

The preceding code shows that a SAS program consists of two parts: a **DATA step** which is used to read data and create a SAS dataset; and a **PROC step** which is used to process or analyze the data. The differences between these steps are described in the next two sections.

The DATA Step

In the DATA step, programming statements create and/or modify a SAS dataset. Among other things, these statements can:

- provide a name for the dataset;
- provide a name for the variables to be included in the dataset;
- provide the actual data to be analyzed.

In the preceding program, the DATA step begins with the DATA statement and ends with the semicolon and RUN statement. The RUN statement immediately precedes the PROC MEANS statement.

The first statement of the preceding program begins with the word DATA and specifies that SAS should create a dataset to be called D1. The next line contains the INPUT statement, which indicates that three variables will be contained in this dataset. The first variable will be called PARTICIPANT, and this variable will simply provide a participant number for each student. The second variable will be called GREVERBAL (for the GRE verbal test), and the third will be called GREMATH (for the GRE math test).

The DATALINES statement indicates that data lines containing your data will appear on the following lines. The first line after the DATALINES statement contains the data (test scores) for participant 1. You can see that this first data line contains the numbers 520 and 490 meaning that participant 1 received a score of 520 on the GRE verbal test and a score of 490 on the GRE math test. The next data line shows that participant 2 received a score of 610 for the GRE verbal and a score of 590 for the GRE math. The semicolon and RUN statement after the last data line signal the end of the data.

The PROC Step

In contrast to the DATA step, the PROC step includes programming statements that request specific statistical analyses of the data. For example, the PROC step might request that correlations be performed between all quantitative variables or might request that a *t* test be performed. (For more information, see Chapter 8, "*t* Tests: Independent Samples and Paired Samples.) In the preceding example, the PROC step consists of the last three lines of the program.

The first line after the DATA step is the PROC MEANS statement. This requests that SAS use a procedure called MEANS to analyze the data. The MEANS procedure computes means, standard deviations, and other descriptive statistics for numeric variables in the dataset. Immediately after the words PROC MEANS are the words

DATA=D1. This tells the system that the data to be analyzed are in a dataset named D1. (Remember that D1 is the name of the dataset just created.)

Following the PROC MEANS statement is the VAR statement, which includes the names of two variables: GREVERBAL and GREMATH. This requests that the descriptive statistics be performed on GREVERBAL (GRE verbal test scores) and GREMATH (GRE math test scores).

Finally, the last line of the program is the RUN statement that signals the end of the PROC step. If a SAS program requests multiple PROCs (procedures), you have two options for using the RUN statement:

- You can place a separate RUN statement following each PROC statement.

- You can place a single RUN statement following the last PROC statement.

> **What is the single most common programming error?** For new SAS users, the single most common error involves leaving off a required semicolon. Remember that every SAS statement must end with a semicolon. In the preceding program, notice that the DATA statement ends with a semicolon as does the INPUT statement, the DATALINES statement, the PROC MEANS statement, and the RUN statement. When you obtain an error in running a SAS program, one of the first things you should do is look over the program for missing semicolons.

Once you submit the preceding program for analysis, SAS creates two types of files reporting the results of the analysis. One file is called the **SAS log** or **log file** in this text. This file contains notes, warnings, error messages, and other information related to the execution of the SAS program. The other file is referred to as the **SAS output file**. The SAS output file contains the results of the requested statistical analyses.

The SAS Log

The SAS log is a listing of notes and messages that help you verify that your SAS program was executed successfully. Specifically, the log provides the following:

- a reprinting of the SAS program that was submitted;

- a listing of notes indicating how many variables and observations are contained in the dataset;

- a listing of any errors made in the execution of the SAS program.

Log 2.1 provides a reproduction of the SAS log for the preceding program:

Log 2.1 SAS Log for the Preceding Program

```
NOTE: Copyright (c) 2002-2003 by SAS Institute Inc., Cary, NC, USA.
NOTE: SAS (r) 9.1 (TS1M0)
      Licensed to NORM O'ROURKE, Site 0042223001.
NOTE: This session is executing on the WIN_PRO  platform.



NOTE: SAS initialization used:
      real time              23.83 seconds
      cpu time               4.47 seconds

1     DATA D1;
2     INPUT PARTICIPANT GREVERBAL GREMATH;
3
4     DATALINES;

NOTE: The data set WORK.D1 has 6 observations and 3 variables.
NOTE: DATA statement used (Total process time):
      real time              1.39 seconds
      cpu time               0.27 seconds



11    ;
12    RUN;
13    PROC MEANS    DATA=D1;
14       VAR  GREVERBAL  GREMATH;
15    RUN;

NOTE: There were 6 observations read from the data set WORK.D1.
NOTE: PROCEDURE MEANS used (Total process time):
      real time              1.09 seconds
      cpu time               0.25 seconds
```

Notice that the statements constituting the SAS program are assigned line numbers and are reproduced in the SAS log. The data lines are not normally reproduced as part of the SAS log unless they are specifically requested.

About halfway down the log, a note indicates that the dataset contains six observations and three variables. You would check this note to verify that the dataset contains all of the variables that you intended to input (in this case, three) and that it contains data from all of your participants (in this case, six). So far, everything appears to be correct.

If you made any errors in writing the SAS program, there would also have been ERROR messages in the SAS log. Often, these error messages help you determine what was wrong with the program. For example, a message might indicate that SAS was expecting a program statement that was not included. Whenever you encounter an error message, read it carefully and review all of the program statements that preceded it. Often, the error appears in the program statements that immediately precede the error message; in other cases, the error might be hidden much earlier in the program.

If more than one error message is listed, do not panic; there still might be only one error. Sometimes, a single error causes a large number of subsequent error messages.

Once the error or errors are identified, you must revise the original SAS program and resubmit it for analysis. Review the new SAS log to see if the errors have been eliminated. If the log indicates that the program ran correctly, you are free to review the results of the analyses in the SAS output file.

The SAS Output File

The SAS output file contains the results of the statistical analyses requested in the SAS program. Because the program in the previous example requested the MEANS procedure, the corresponding output file contains means and other descriptive statistics for the variables analyzed. In this text, the SAS output file is sometimes referred to as the **lst file**. "Lst" is an abbreviation for "listing of results."

The following is a reproduction of the SAS output file that would be produced from the preceding SAS program:

Output 2.1 Results of the MEANS Procedure

```
                               The SAS System

                             The MEANS Procedure

Variable     N          Mean           Std Dev         Minimum          Maximum
-------------------------------------------------------------------------------
GREVERBAL    6      516.6666667       72.5718035      410.0000000      610.0000000
GREMATH      6      455.0000000       83.3666600      350.0000000      590.0000000
-------------------------------------------------------------------------------
```

Below the heading "Variable," SAS prints the names of each of the variables being analyzed. In this case, the variables are called GREVERBAL and GREMATH. To the right of the heading GREVERBAL, descriptive statistics for the GRE verbal test may be found. Figures for the GRE math test appear to the right of GREMATH.

Below the heading "N," the number of observations analyzed is reported. The average score on each variable is reproduced under "Mean" and standard deviations appear in the column "Std Dev." Minimum and maximum scores for the two variables appear in the remaining two columns. You can see that the mean score on the GRE verbal test was 516.67, and the standard deviation of these scores was 72.57. For the GRE math test, the mean was 455.00, and the standard deviation was 83.37.

 SAS Output generally reports findings to several decimal places. In the body of this text, however, numbers will be reported to only two decimal places—rounded where necessary—in keeping with the publication manual of the American Psychological Association (APA, 2001).

The statistics included in the preceding output are printed by default (i.e., without asking for them specifically). Later in this text, you will learn that there are many additional statistics that you can request as options with PROC MEANS.

SAS Customer Support Center

Although the context of this text provides examples of many of the analyses commonly performed in the social sciences, specific questions to certain problems might arise. One alternative for registered SAS users (or their institutions) is use of the SAS Customer Support Center (http://support.sas.com). SAS maintains a comprehensive Web site with up-to-date information. For one option that is particularly useful for novice (and not-so-novice) SAS users, follow this path: **Documentation → Sample and Technical Tips → Search Samples and Tips for**. Here, you can search for information regarding specific statistical procedures covered, and not covered, in this text.

It is also possible to pose specific questions (**Technical Support → Submit a Problem**) via the **SAS System** or **SAS inSchool** menu options. To use this, you need to provide an e-mail address to which replies are sent, identify your school or institution, and provide a customer site number or license information. This latter information can be found in any SAS Log file. (See Log 2.1 where the release version and license number are specified in the first section.) Whenever possible, SAS strives to respond to all inquiries within one business day.

Conclusion

Regardless of the computing environment in which you work, the basics of using SAS remain the same: you prepare the SAS program; submit it for analysis; review the resulting log for any errors; and examine the output files to view the results of your analyses. This chapter has provided a brief introduction to SAS. You are now ready to move on to Chapter 3, "Data Input," where the fundamentals of creating SAS datasets are introduced.

Reference

American Psychological Association (2001). *Publication manual of the American Psychological Association.* (5th ed.). Washington, DC: Author.

Data Input

Overview. This section shows how to create a SAS dataset. A SAS dataset contains the information (e.g., independent variables, dependent variables, survey responses) to be analyzed by SAS procedures such as PROC MEANS and PROC GLM. The chapter begins with a simple illustrative example in which a SAS dataset is created using the DATALINES statement. In subsequent sections, additional guidelines show how to input the different types of data that are most frequently encountered in social science research.

Introduction: Inputting Questionnaire Data versus Other Types of Data

This chapter shows how to create SAS datasets in a number of different ways, and it does this by illustrating how to input the types of data that are often obtained through questionnaire research. Questionnaire research generally involves distributing standardized instruments to a sample of participants, and asking them to respond by circling or checking fixed responses. For example, participants might be asked to indicate the extent to which they agree or disagree with a set of items by selecting a response along a 7-point Likert-type scale where 1 represents "strongly disagree" and 7 represents "strongly agree."

Because this chapter (and much of the entire book, for that matter) focuses on questionnaire research, some readers might be concerned that it is not useful for analyzing data that are obtained using different methods. This concern is understandable, because the social sciences are so diverse and so many different types of variables are investigated. These variables might be as different as "the number of aggressive acts performed by a child," "rated preferences for laundry detergents," or "levels of serotonin in the frontal lobes of chimpanzees."

However, because of the generality and flexibility of the basic principles of this discussion, you can expect to input virtually any type of data obtained in social science research upon completing this chapter. The same can be said for the remaining chapters of this book; although it emphasizes the analysis of questionnaire data, the concepts taught here can be readily applied to many types of data. This fact should become clear as the mechanics of using SAS are presented.

This book emphasizes the analysis of questionnaire data for two reasons. First, for better or for worse, many social scientists rely on questionnaire data when conducting their research. By focusing on this method, this text provides examples that are meaningful to the single largest subgroup of readers. Second, questionnaire data often create special entry and analysis problems that are not generally encountered with other research methods (e.g., large numbers of variables, "check all that apply" variables). This text addresses some of the most common of these difficulties.

Entering Data: An Illustrative Example

Before data can be entered and analyzed by SAS, they must be *entered* in some systematic way. There are a number of different approaches to entering data; to keep things simple, this chapter presents only the fixed format approach. With the **fixed format** method, each variable is assigned to a specific column (or set of columns) in the dataset. The fixed format method has the advantage of being very general: you can use it for almost any type of research problem. An additional advantage is that researchers are probably less likely to make errors when entering data if they adhere to this format.

In the following example, you actually enter some fictitious data from a fictitious study. Assume that you have developed a survey to measure attitudes toward *volunteerism*. A copy of the survey appears here:

```
                        Volunteerism Survey

Please indicate the extent to which you agree or disagree with each of
the following statements.  You will do this by circling the appropriate
number to the left of that statement.  The following format shows what
each response alternative represents:

        5 = Agree Strongly
        4 = Agree Somewhat
        3 = Neither Agree nor Disagree
        2 = Disagree Somewhat
        1 = Disagree Strongly

For example, if you "Disagree Strongly" with the first question, circle
the "1" to the left of that statement.  If you "Agree Somewhat," circle
the "4," and so on.

-------------
 Circle Your
  Response
-------------
1  2  3  4  5      1.      I feel a personal responsibility to
                           help needy people in my community.

1  2  3  4  5      2.      I feel I am personally obligated to
                           help homeless families.

1  2  3  4  5      3.      I feel no personal responsibility to
                           work with poor people in my community.

1  2  3  4  5      4.      Most of the people in my community are
                           willing to help the needy.

1  2  3  4  5      5.      A lot of people around here are willing
                           to help homeless families.

1  2  3  4  5      6.      The people in my community feel no personal
                           responsibility to work with poor people.

1  2  3  4  5      7.      Everyone should feel the responsibility to
                           perform volunteer work in his/her community.

What is your age in years? _____
```

Further assume that you administer this survey to 10 participants. For each of these individuals, you also obtain their intelligence quotient or IQ scores.

You then enter your data as a file in a computer. All of the survey responses and information about participant 1 appear on the first line of this file. All of the responses and information about participant 2 appear on the second line of this file, and so forth. You keep the data aligned so that responses to question 1 appear in column 1 for *all* participants, responses to question 2 appear in column 2 for all participants, and so forth. When you enter data in this fashion, your dataset should look similar to this:

```
2234243 22   98   1
3424325 20  105   2
3242424 32   90   3
3242323  9  119   4
3232143  8  101   5
3242242 24  104   6
4343525 16  110   7
3232324 12   95   8
1322424 41   85   9
5433224 19  107  10
```

You can think of the preceding dataset as a matrix consisting of 10 rows and 17 columns. The rows run horizontally (from left to right), and each row represents data for a different participant. The columns run vertically (up and down). For the most part, a given column represents a different variable that you measured or created. (Though, in some cases, a given variable is more than one column wide, but more on this later.)

For example, look at the last column in the matrix: the vertical column on the right side that goes from 1 (at the top) to 10 (at the bottom). This column codes the Participant Number variable. In other words, this variable simply tells us *which* participant's data are included on that line. For the top line, the assigned value of Participant Number is 1, so you know that the top line includes data for participant 1. The second line down has the value 2 in the participant number column, so this second line down includes data for participant 2, and so forth.

The *first* column of data includes participant responses to survey question 1. It can be seen that participant 1 selected "2" in response to this item, while participant 2 selected "3." The *second* column of data includes participants' responses to survey question 2, the *third* column codes question 3, and so forth. After entering responses to question 7, you left column 8 blank. Then, in columns 9 and 10, you enter each participant's age. We can see that participant 1 was 22 years old, while participant 2 was 20 years old. You left column 11 blank, and then entered the participants' IQs in columns 12, 13, and 14. (IQ can be a three-digit number, so it required three columns to enter it.) You left column 15 blank, and entered participant numbers in columns 16 and 17.

The following table presents a brief coding guide to summarize how you entered your data.

Column	Variable Name	Explanation
1	Q1	Responses to survey question 1
2	Q2	Responses to survey question 2
3	Q3	Responses to survey question 3
4	Q4	Responses to survey question 4
5	Q5	Responses to survey question 5
6	Q6	Responses to survey question 6
7	Q7	Responses to survey question 7
8	blank	
9-10	AGE	Participant's age in years
11	blank	
12-14	IQ	Participant's IQ score
15	blank	
16-17	NUMBER	Participant's number

Guides similar to this are used throughout this text to explain how datasets are arranged, so a few words of explanation are in order. This table identifies the specific columns in which variable values are assigned. For example, the first line of the preceding table indicates that in column 1 of the dataset, the values of a variable called Q1 are stored, and this variable includes responses to question 1. The next line shows that the values of variable Q2 are stored in column 2, and this variable includes responses to question 2. The remaining lines of the guide are interpreted in the same way. You can see, therefore, that it is necessary to read down the lines of this table to learn what is in each column of the dataset.

A few important notes about how you should enter data to be analyzed by SAS:

- Make sure that you enter variables in the correct column. For example, make sure that the data are lined up so that responses to question 6 always appear in column 6. If a participant happened to leave question 6 blank, then you should leave column 6 blank when you are entering your data. (Leave this column blank by pressing the space bar on your keyboard.) Then, go on to type the participant's response to question 7 in column 7. Do not enter a zero if the participant didn't answer a question; leave the space blank.

- It is also acceptable to enter a period (.) instead of a blank space to represent missing data. When using this convention, if a participant has a missing value on a variable, enter a single period in place of that missing value. If this variable happens to be more than one column wide, you should still enter just one period. For example, if the variable occupies columns 12 to 14 (as does IQ in the table), enter just one period in column 14; do not enter three periods in columns 12, 13, and 14.

- **Right-justify numeric data**. You should align numeric variables to the right side of columns in which they appear. For example, IQ is a three-digit variable (it could assume values such as 112 or 150). However, the IQ score for many individuals is a two-digit number (such as 99 or 87). Therefore, the two-digit IQ scores should appear to the right side of this three-digit column of values. A correct example of how to right-justify your data follows:

```
 99
109
100
 87
118
```

The following is *not* right-justified and is less preferable:

```
99
109
100
87
118
```

There are exceptions to this rule. For example, if numeric data contain decimal points, it is generally preferable to align the decimal points when entering the data so that the decimals appear in the same column. If there are no values to the right of the decimal point for a given participant, you can enter zeros to the right of the decimal point. Here is an example of this approach:

```
  3.450
 12.000
  0.133
144.751
  0.000
```

The preceding dataset includes scores for five participants for just one variable. Assume that possible scores for this variable range from 0.00 to 200.00. Participant 1 had a score of 3.45, participant 2 had a score of 12, and so forth. Notice that the scores have been entered so that the decimal points are aligned in the same vertical column.

Notice also that if a given participant's score does not include any digits to the right of the decimal point, zeros have been added. For example, participant 2 has a score of 12. However, this participant's score is entered as 12.000 so that it is aligned with the other scores.

Technically, it is not always necessary to align participant data in this way in order to include it in a SAS dataset: however, arranging data in an orderly fashion generally decreases the likelihood of making errors when entering data.

- **Left-justify character data**. Character variables can include letters of the alphabet. In contrast to numeric variables, you typically should left-justify character variables. This means that you align entries to the left, rather than to the right.

For example, imagine that you are going to enter two character variables for each participant. The first variable will be called FIRST, and this variable will include each participant's first name. You will enter this variable in columns 1 to 15. The second variable will be called LAST and will include each participant's surname. You will enter this variable in columns 16 to 25. Data for four participants are reproduced here:

```
Francis        Smith
Ishmael        Khmali
Michel         Hébert
Jose           Lopez
```

The preceding shows that the first participant is named Francis Smith, the second is named Ishmael Khmali, and so forth. Notice that the value "Francis" is moved to the left side of the column that include the FIRST variable (columns 1 to 15). The same is true for "Ishmael," as well as the remaining first names. In the same way, "Smith" is moved over to the left side of the columns that include the LAST variable (columns 16 to 25). The same is true for the remaining surnames.

- **Use of blank columns can be helpful but is not necessary**. Recall that when you entered your data, you left a blank column between Q7 and the AGE variable, and another blank column between AGE and IQ. Leaving blank columns between variables can be helpful because it makes it easier to look at your data and see if something has been entered out of place. However, leaving blank columns is not necessary for SAS to accurately read your data, so this approach is optional (though recommended).

Inputting Data Using the DATALINES Statement

Now that you know how to enter your data, you are ready to learn about the SAS statements that actually allow the computer to *read* the data and put them into a SAS dataset. There are several ways that you can input data, but this book focuses on two: use of the **DATALINES statement** that allows you to include the data within the SAS program itself; and the **INFILE statement** that allows you to include the data lines within an external file.

There are also several ways in which data can be read by SAS with regard to the instructions you provide concerning the location and format of your variables. Although SAS allows for list input, column input and formatted input, this text presents only formatted input because of its ability to easily handle many different types of data.

Here is the general form for inputting data using the DATALINES statement and the formatted input style:

```
DATA dataset-name;
    INPUT  #line-number   @column-number   variable-name  column-width.
                          @column-number   variable-name  column-width.
                          @column-number   variable-name  column-width. ;
DATALINES;
entered data are placed here
;
RUN;

PROC name-of-desired-statistical-procedure     DATA=dataset-name ;
RUN;
```

The following example shows a SAS program to analyze the preceding dataset. In the following example, the numbers on the far-left side are not actually part of the program. Instead, they are provided so that it will be easy to refer to specific lines of the program when explaining the meaning of the program in subsequent sections.

```
1            DATA D1;
2                INPUT     #1   @1   Q1       1.
3                               @2   Q2       1.
4                               @3   Q3       1.
5                               @4   Q4       1.
6                               @5   Q5       1.
7                               @6   Q6       1.
8                               @7   Q7       1.
9                               @9   AGE      2.
10                              @12  IQ       3.
11                              @16  NUMBER   2.   ;
12           DATALINES;
13           2234243 22   98  1
14           3424325 20  105  2
15           3242424 32   90  3
16           3242323  9  119  4
17           3232143  8  101  5
18           3242242 24  104  6
19           4343525 16  110  7
20           3232324 12   95  8
21           1322424 41   85  9
22           5433224 19  107 10
23           ;
24           RUN;
25
26           PROC MEANS   DATA=D1;
27           RUN;
```

A few important notes about these data input statements:

- **The DATA statement**. Line 1 from the preceding program includes the DATA statement, where the general form is:

  ```
  DATA dataset-name;
  ```

  In this case, you gave your dataset the name D1, so the statement reads

  ```
  DATA D1;
  ```

- **Dataset names and variable names**. The preceding paragraph stated that your dataset was assigned the name D1 on line 1 of the program. In lines 2 to 11 of the program, the dataset's variables are assigned names such as Q1, Q2, AGE, and IQ.

  You are free to assign a dataset or variable any name you like so long as it conforms to the following rules:

  - It must begin with a letter (rather than a number).

  - It contains no special characters such as "*" or "#".

  - It contains no blank spaces.

  Although the preceding dataset is named D1, it could have been given any of an almost infinite number of other names. Below are examples of other acceptable names for SAS datasets:

  ```
  SURVEY
  PARTICIPANT
  RESEARCH
  VOLUNTEER
  ```

- **The INPUT statement**. The INPUT statement has the following general form:

  ```
  INPUT   #line-number @column-number  variable-name  column-width.
                       @column-number  variable name  column-width.
                       @column-number  variable-name  column-width. ;
  ```

  Compare this general form to the actual INPUT statement that appears on lines 2 to 11 of the preceding SAS program, and note the values that were filled in to read your data. In the actual program, the word INPUT appears on line 2 and tells SAS that the INPUT statement has begun. SAS assumes that all of the instructions that follow are data input directions *until* it encounters a semicolon (;). At that semicolon, the INPUT statement ends. In this example, the semicolon appears at the end of line 11.

- **Line number controls**. To the right of the word INPUT is the following:

  ```
  #line-number
  ```

  This tells SAS what line it should read from in order to find specific variables. In some cases, there can be two or more lines of data for each participant. There is more information on this type of situation in a later section. For the present example, however, the situation is fairly simple: there is only one line of data for each participant so your program includes the following line number control (from line 2 of the program example):

  ```
  INPUT    #1
  ```

Technically, it is not necessary to include line number controls when there is only one line of data for each participant (as in the present case). In this text, however, line number controls appear for the sake of consistency.

- **Column location, variable name, and column width directions**. To the right of the line number directions, you place the column location, variable name, and column width directions. The syntax for this is as follows:

      @column-number   variable-name   column-width.

Where `column-number` appears above, you enter the number of the column in which a specific variable appears. If the variable occupies more than one column (such as IQ in columns 12, 13, and 14), you should enter the number of the column in which it begins (e.g., column 12). Where `variable-name` appears, you enter the name that you have given to that variable. And where `column width` appears, you enter how many columns are occupied by that variable. In the case of the preceding data, the first variable is Q1, which appears in column 1 and is only one column wide. This program example, therefore, provides the following column location controls (from line 2):

      @1   Q1   1.

The preceding line tells SAS to go to column 1. In that column, you find a variable called Q1. It is a number and it is one column wide.

You must follow the column width with a period. For column 1, the width is 1. It is important that you include this period; later, you will learn how the period provides information about decimal places.

Now that variable Q1 has been read, you must give SAS the directions required to read the remaining variables in the dataset. The completed INPUT statement appears as follows. Note that the line number controls are given only once because all of these variables come from the same line (for a given participant). However, there are different column controls for the different variables. Note also how column widths are different for AGE, IQ, and NUMBER:

```
INPUT   #1   @1    Q1       1.
             @2    Q2       1.
             @3    Q3       1.
             @4    Q4       1.
             @5    Q5       1.
             @6    Q6       1.
             @7    Q7       1.
             @9    AGE      2.
             @12   IQ       3.
             @16   NUMBER   2.    ;
```

Notice the semicolon that appears after the column width entry for the last variable (NUMBER). You must always end your input statement with a semicolon. It is easy to omit the semicolon, so always check for this semicolon if you get an error message following the INPUT statement. (More is said about error statements in later chapters.)

- **The DATALINES statement.** The DATALINES statement goes after the INPUT statement and tells SAS that raw data are to follow. Don't forget the semicolon after the word DATALINES. In the preceding program example, the DATALINES statement appears on line 12.

- **The data lines.** The data lines, of course, are the lines that contain the participants' values for the numeric and/or character variables. In the preceding program example, these appear on lines 13 to 22.

  The data lines should begin on the very next line after the DATALINES statement; there should be no blank lines. These data lines begin on line 13 in the preceding program example. On the very first line after the last of the data lines (line 23, in this case), you should add another semicolon to let SAS know that the data have ended. Do *not* place this semicolon at the end of the last line of data (i.e., on the *same line* as the data) as this might cause an error. After this semicolon, a RUN statement should appear at the end of the data lines. In the preceding program example, this statement appears on line 24.

  With respect to the data lines, the most important thing to remember is that you must enter a given variable in the column specified by the INPUT statement. For example, if your input statement contains the following line:

  ```
  @9   AGE   2.
  ```

  then make sure that the variable AGE really is a two-digit number found in columns 9 and 10.

- **PROC and RUN statements.** There is little to say about PROC and RUN statements at this point because most of the remaining text is concerned with using such SAS procedures. Suffice to say that a PROC (procedure) statement asks SAS to perform some statistical analysis. To keep things simple, this section uses a procedure called PROC MEANS. PROC MEANS asks SAS to calculate means, standard deviations, and other descriptive statistics for numeric variables. The preceding program includes the PROC MEANS statement on line 26.

  In most cases, your program ends with a RUN statement. In the preceding program example, a second RUN statement appears on line 27. A RUN statement executes any previously entered SAS statements; RUN statements are typically placed after every PROC statement. If your program includes a number of PROC statements in sequence, it is acceptable to place just one RUN statement after the final PROC statement.

If you submitted the preceding program for analysis, PROC MEANS would produce the results presented in Output 3.1:

Output 3.1 Results of the MEANS Procedure

```
The MEANS Procedure

     Variable     N          Mean       Std Dev       Minimum        Maximum
     ------------------------------------------------------------------------
     Q1          10     3.0000000     1.0540926     1.0000000      5.0000000
     Q2          10     2.6000000     0.8432740     2.0000000      4.0000000
     Q3          10     3.2000000     0.7888106     2.0000000      4.0000000
     Q4          10     2.6000000     0.8432740     2.0000000      4.0000000
     Q5          10     2.9000000     1.1972190     1.0000000      5.0000000
     Q6          10     2.6000000     0.9660918     2.0000000      4.0000000
     Q7          10     3.7000000     0.9486833     2.0000000      5.0000000
     AGE         10    20.3000000    10.2745641     8.0000000     41.0000000
     IQ          10   101.4000000     9.9241568    85.0000000    119.0000000
     NUMBER      10     5.5000000     3.0276504     1.0000000     10.0000000
     ------------------------------------------------------------------------
```

Additional Guidelines

Inputting String Variables with the Same Prefix and Different Numeric Suffixes

In this section, **prefix** refers to the first part of a variable's name, while **suffix** refers to the last part. For example, think about our variables Q1, Q2, Q3, Q4, Q5, Q6, and Q7. These are multiple variables with the same prefix (Q) and different numeric suffixes (i.e., 1, 2, 3, 4, 5, 6, and 7). Variables such as this are sometimes referred to as *string variables*. Earlier, this chapter provided one way of inputting these variables; the original INPUT statement is repeated here:

```
INPUT    #1    @1     Q1       1.
               @2     Q2       1.
               @3     Q3       1.
               @4     Q4       1.
               @5     Q5       1.
               @6     Q6       1.
               @7     Q7       1.
               @9     AGE      2.
               @12    IQ       3.
               @16    NUMBER   2.    ;
```

However, with string variables named in this way, there is an easier way of writing the INPUT statement. You could have written it this way:

```
INPUT    #1    @1     Q1-Q7    1.
               @9     AGE      2.
               @12    IQ       3.
               @16    NUMBER   2.    ;
```

The first line of this INPUT statement gives SAS the following directions: "Go to line #1. Once there, go to column 1. Beginning in column 1, you find variables Q1 through Q7. Each of these numeric variables is one column wide." With this second INPUT statement, SAS reads the data in exactly the same way that it would have using the original input statement.

As an additional example, imagine you had a 50-item survey instead of a 7-item survey. You called your variables Q1, Q2, Q3, and so forth. You entered your data in the following way:

Column	Variable Name	Explanation
1-50	Q1-Q50	Responses to survey questions 1-50
51	blank	.
52-53	AGE	Participant's age in years
54	blank	
55-57	IQ	Participant's IQ score
58	blank	
59-60	NUMBER	Participant's number

You could use the following INPUT to read these data:

```
INPUT    #1    @1    Q1-Q50   1.
               @52   AGE      2.
               @55   IQ       3.
               @59   NUMBER   2.   ;
```

Inputting Character Variables

This text deals with two types of basic variables: numeric and character variables. A **numeric variable** consists entirely of numbers, and it contains no letters. For example, all of your variables from the preceding dataset were numeric variables: Q1 could assume only the values of 1, 2, 3, 4, or 5. Similarly, AGE could take on only numeric values. On the other hand, a **character variable** can consist of either numbers, alphabetic characters (letters), or both.

Remember that responses to the seven questions of the Volunteerism Survey are entered in columns 1 to 7 in this dataset, AGE is entered in columns 9 to 10, IQ is entered in columns 12 to 14, and participant number is in columns 16 to 17.

You could include the sex of each participant and create a new variable called SEX. If a participant is male, SEX would assume the value "M." If a participant is female, SEX would assume the value "F." In the following, the new SEX variable appears in column 19 (the last column):

```
2234243 22  98  1 M
3424325 20 105  2 M
3242424 32  90  3 F
3242323  9 119  4 F
3232143  8 101  5 F
3242242 24 104  6 M
4343525 16 110  7 F
3232324 12  95  8 M
1322424 41  85  9 M
5433224 19 107 10 F
```

You can see that participants 1 and 2 are males whereas participants 3, 4, and 5 are females, and so forth.

You must use a special command within the INPUT statement to input a character variable. Specifically, in the column width region for the character variable, precede the column width with a dollar sign ($). For the preceding dataset, you would use the following INPUT statement. Note the dollar sign in the column width region for the SEX variable:

```
INPUT    #1    @1    Q1-Q7    1.
               @9    AGE      2.
               @12   IQ       3.
               @16   NUMBER   2.
               @19   SEX      $1.  ;
```

Using Multiple Lines of Data for Each Participant

Often, a researcher obtains so much data from each participant that it is impractical to enter all data on just one line. For example, imagine that you administer a 100-item questionnaire to a sample, and that you plan to enter responses to question 1 in column 1, responses to question 2 in column 2, and so forth. Following this process, you are likely to run into difficulty because you will need 100 columns to enter all responses from a given participant. Many computer monitors, however, allow no more than 79 columns. If you continue entering data past column 79, your data are likely to *wrap around* or appear in some way that makes it difficult to verify that you are entering a given value in the correct column.

In situations in which you require a large number of columns for your data, it is often best to divide each participant's data so that they appear on more than one line. (In other words, it is best to have multiple lines of data for each participant.) To do this, it is necessary to modify your INPUT statement.

To illustrate, assume that you obtained two additional variables for each participant in your study: their GRE verbal test scores and GRE math test scores. You decide to enter your data so that there are two lines of data for each participant. On line 1 for a given participant, you enter Q1 through Q7, AGE, IQ, NUMBER, and SEX (as above). On line

2 for that participant, you enter GREVERBAL (the GRE verbal test score) in columns 1 through 3, and you enter GREMATH (the GRE math test score) in columns 5 through 7:

```
2234243 22  98  1 M
520 490
3424325 20 105  2 M
440 410
3242424 32  90  3 F
390 420
3242323  9 119  4 F

3232143  8 101  5 F

3242242 24 104  6 M
330 340
4343525 16 110  7 F

3232324 12  95  8 M

1322424 41  85  9 M
380 410
5433224 19 107 10 F
640 590
```

GREVERBAL score for participant 1 is 520, and the GREMATH score is 490.

When a participant has no data for a variable which would normally appear on a given line, your dataset must still include a line for that participant, even if it is blank. For example, participant 4 is only 9 years old, so she has not yet taken the GRE and obviously does not have GRE scores. Nonetheless, you still need to include a second line for participant 4 even though it is blank. Notice that blank lines also appear for participants 5, 7, and 8, who are also too young to take the GRE.

Be warned that, with some text editors, it is necessary to create these blank lines by pressing the ENTER key, thus creating a *hard carriage return*. With these editors, using the directional arrows on the keypad might not create the necessary hard return. Problems in reading the data are also likely to occur if tabs are used; it is generally best to avoid the use of tabs or other hidden codes when entering data.

The following coding guide tells us where each variable appears. Notice that this guide indicates the line on which a variable is located, as well as the column where it is located.

Line	Column	Variable Name	Explanation
1	1-7	Q1-Q7	Survey questions 1-7
	8	blank	
	9-10	AGE	Participant's age in years
	11	blank	
	12-14	IQ	Participant's IQ score
	15	blank	
	16-17	NUMBER	Participant's number
	18	blank	
	19	SEX	Participant's sex
2	1-3	GREVERBAL	GRE-Verbal test score
	5-7	GREMATH	GRE-Math test score

When there are multiple lines of data for each participant, the INPUT statement must indicate on which line a given variable is located. This is done with the line number command (#) that was introduced earlier. You could use the following INPUT statement to read the preceding dataset:

```
INPUT   #1   @1    Q1-Q7         1.
             @9    AGE           2.
             @12   IQ            3.
             @16   NUMBER        2.
             @19   SEX           $1.
        #2   @1    GREVERBAL     3.
             @5    GREMATH       3.   ;
```

This INPUT statement tells SAS to begin at line #1 for a given participant, to go to column 1, and find variables Q1 through Q7. It continues to tell SAS where it will find each of the other variables located on line #1. After reading the SEX variable, SAS is told to move to line #2. There, it is to go to column 1 and find the variable GREVERBAL that is three columns wide. The variable GREMATH begins in column 5 and is also three columns wide. In theory, it is possible to have any number of lines of data for each participant so long as you use the line number command correctly.

Creating Decimal Places for Numeric Variables

Assume that you have obtained the high school grade point averages (GPAs) for a sample of five participants. You could create a SAS dataset containing these GPAs using the following program:

```
1          DATA D1;
2             INPUT   #1   @1   GPA   4.  ;
3          DATALINES;
4          3.56
5          2.20
6          2.11
7          3.25
8          4.00
9          ;
10         RUN;
11
12         PROC MEANS   DATA=D1;
13         RUN;
```

The INPUT statement tells SAS to go to line 1, column 1, to find a variable called GPA that is four columns wide. Within the dataset itself, values of GPA were entered using a period as a decimal point, with two digits to the right of the decimal point.

This same dataset could have been entered in a slightly different way. For example, what if the data had been entered without a decimal point, as follows?

```
356
220
211
325
400
```

It is still possible to have SAS insert a decimal point where it belongs, in front of the last two digits in each number. You do this in the column width command of the INPUT statement. With this column width command, you indicate how many columns the variable occupies, enter a period, and then indicate how many columns of data should appear to the right of the decimal place. In the present example, the GPA variable is three columns wide and two columns of data should appear to the right of the decimal place. So you would modify the SAS program in the following way. Notice the column width command:

```
1          DATA D1;
2             INPUT   #1   @1   GPA   3.2  ;
3          DATALINES;
4          356
5          220
6          211
7          325
8          400
9          ;
10         RUN;
11
12         PROC MEANS   DATA=D1;
13         RUN;
```

Inputting "Check All That Apply" Questions as Multiple Variables

A "check all that apply" question is a special type of questionnaire item that is often used in social science research. These items generate data that must be input in a special way. The following is an example of a "check all that apply" item that could have appeared on your volunteerism survey:

```
Below is a list of activities.  Please place a check mark next to
any activity in which you have engaged in the past six months.

Check here
-----

_____  1. Did volunteer work at a shelter for the homeless.
_____  2. Did volunteer work at a shelter for battered women.
_____  3. Did volunteer work at a hospital or hospice.
_____  4. Did volunteer work for any other community agency or
           organization.
_____  5. Donated money to the United Way.
_____  6. Donated money to a congregation-sponsored charity.
_____  7. Donated money to any other charitable cause.
```

An inexperienced researcher might think of the preceding as a single question with seven possible responses and try to enter the data in a single column in the dataset (e.g., in column 1). But this would lead to big problems. What would you enter in column 1 if a participant checked more than one category?

One way around this difficulty is to treat the seven possible responses as seven different questions. When entered, each of these questions is treated as a separate variable and appears in a separate column. For example, whether or not a participant checked activity 1 can be coded in column 1, whether the participant checked activity 2 can be coded in column 2, and so forth.

Researchers can code these variables by placing any values they like in these columns, but you should enter a two (2) if the participant did not check that activity and a one (1) if the participant did check it. Why code the variables using 1s and 2s? The reason is that this makes it easier to perform some types of analyses that you might later want to perform. A variable that can assume only two values is called a **dichotomous variable**, and the process of coding dichotomous variables with 1s and 2s is known as **dummy coding**. When dummy coding, we recommend that you do not use zeros to avoid the possibility that these might be confused with missing values.

Once a dichotomous variable is dummy coded, it can be analyzed using a variety of SAS procedures such as PROC REG to perform **multiple regression**, a procedure that allows you to assess the nature of the relationship between a single criterion variable and multiple predictor variables. If a dichotomous variable has been dummy coded properly, it can be used as a predictor variable in a multiple regression analysis. For these and other reasons, it is good practice to code dichotomous variables using 1s and 2s.

The following coding guide summarizes how you could enter responses to the preceding question:

Line	Column	Variable Name	Explanation
1	1-7	ACT1-ACT7	Responses regarding activities 1 through 7. For each activity, a 2 was recorded if the participant did not check the activity, and a 1 was recorded if the participant did check the activity.

When participants have responded to a "check all that apply" item, it is often best to analyze the resulting data with the FREQ (frequency) procedure. PROC FREQ indicates the actual number of people who appear in each category. In this case, PROC FREQ indicates the number of people who did not check a given activity versus the number who did. It also indicates the percentage of people who appear in each category, along with some additional information.

The following program inputs some fictitious data and requests frequency tables for each activity using PROC FREQ:

```
1          DATA D1;
2              INPUT   #1   @1   ACT1-ACT   1.  ;
3
4          DATALINES;
5          2212222
6          1211111
7          2221221
8          2212222
9          1122222
10         ;
11         RUN;
12
13         PROC FREQ      DATA=D1;
14             TABLES ACT1-ACT7;
15         RUN;
```

Data for the first participant appears on line 5 of the program. Notice that a 1 is entered in column 3 for this participant, indicating that he or she did perform activity 3 ("did volunteer work at a hospital or hospice") and that 1s are recorded for the remaining six activities, meaning that the participant did not perform those activities. The data entered for participant 2 on line 6 shows that this participant performed all of the activities except for activity 2.

Inputting a Correlation or Covariance Matrix

There are times when, for reasons of either necessity or convenience, you might choose to analyze a correlation matrix or covariance matrix rather than raw data (e.g., very large datasets). SAS allows you to input such a matrix as data and some (but not all) SAS procedures can then be used to analyze the dataset. For example, a correlation or covariance matrix can be analyzed using PROC REG, PROC FACTOR, or PROC CALIS, and other procedures.

Inputting a Correlation Matrix

This type of data input is sometimes necessary when a researcher obtains a correlation matrix from an earlier study (perhaps from an article published in a research journal) and wishes to perform further analyses on the data. You could input the published correlation matrix as a dataset and analyze it in the same way you would analyze raw data.

For example, imagine that you have read an article that tested a social psychology theory called the *investment model* (Rusbult, 1980). The investment model identifies a number of variables that are believed to influence a person's satisfaction with, and commitment to, a romantic relationship. The following are short definitions for the variables that constitute the investment model:

Commitment: the person's intention to remain in the relationship;

Satisfaction: the person's affective (emotional) response to the relationship;

Rewards: the number of good things or benefits associated with the relationship;

Costs: the number of bad things or hardships associated with the relationship;

Investment size: the amount of time, energy, and personal resources put into the relationship;

Alternative value: the attractiveness of alternatives to the relationship (e.g., attractiveness of alternative romantic partners).

One interpretation of the investment model predicts that commitment to the relationship is determined by satisfaction, investment size, and alternative value, while satisfaction with the relationship is determined by rewards and costs. The predicted relationships among these variables are shown in Figure 3.1:

Figure 3.1 Predicted Relationships between Investment Model Variables

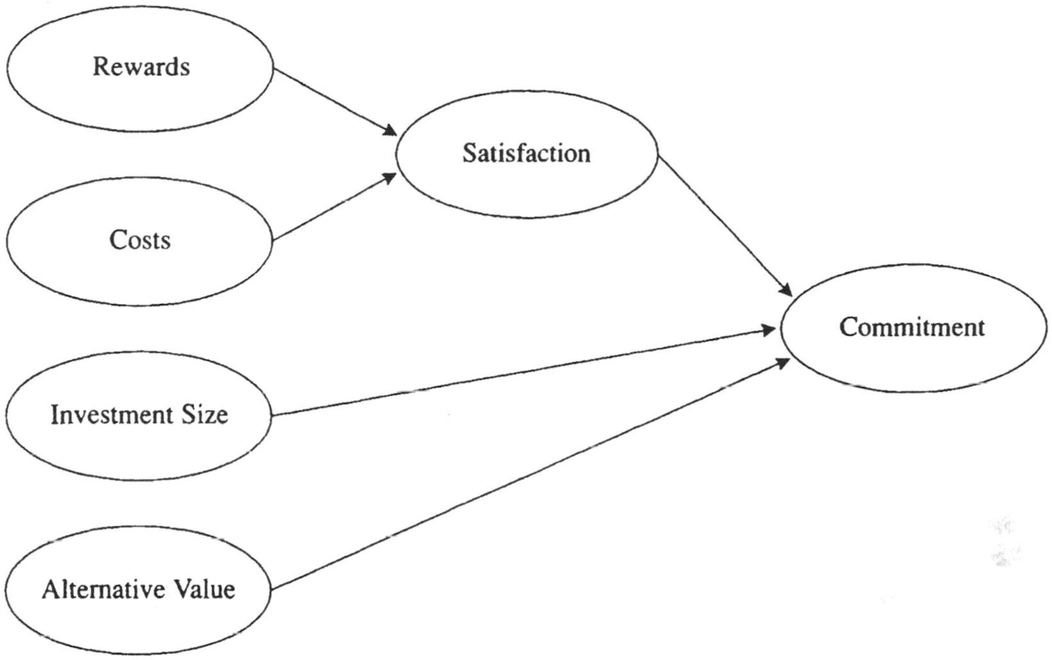

Assume that you have read an article that reports an investigation of the investment model and that the article included the following fictitious table:

Table 3.1

Standard Deviations and Intercorrelations for All Variables

		Intercorrelations					
Variable	SD	1	2	3	4	5	6
1. Commitment	2.3192	1.0000					
2. Satisfaction	1.7744	.6742	1.0000				
3. Rewards	1.2525	.5501	.6721	1.0000			
4. Costs	1.4086	-.3499	- .5717	-.4405	1.0000		
5. Investments	1.5575	.6444	.5234	.5346	-.1854	1.0000	
6. Alternatives	1.8701	-.6929	-.4952	-.4061	.3525	-.3934	1.0000

Note: N = 240.

Supplied with this information, you can now create a SAS dataset that includes just these correlation coefficients and standard deviations. Here are the necessary data input statements:

```
1       DATA D1(TYPE=CORR) ;
2         INPUT _TYPE_ $ _NAME_ $ V1-V6 ;
3         LABEL
4            V1 ='COMMITMENT'
5            V2 ='SATISFACTION'
6            V3 ='REWARDS'
7            V4 ='COSTS'
8            V5 ='INVESTMENTS'
9            V6 ='ALTERNATIVES' ;
10        DATALINES;
11      N       .     240     240     240     240     240     240
12      STD     .    2.3192  1.7744  1.2525  1.4086  1.5575  1.8701
13      CORR  V1   1.0000    .       .       .       .       .
14      CORR  V2    .6742   1.0000   .       .       .       .
15      CORR  V3    .5501    .6721  1.0000   .       .       .
16      CORR  V4   -.3499   -.5717  -.4405  1.0000   .       .
17      CORR  V5    .6444    .5234   .5346  -.1854  1.0000   .
18      CORR  V6   -.6929   -.4952  -.4061   .3525  -.3934  1.0000
19        ;
20      RUN;
```

The following shows the general form for this DATA step in which six variables are to be analyzed. The program would, of course, be modified if the analysis involved a different number of variables.

```
1       DATA dataset-name(TYPE=CORR) ;
2         INPUT _TYPE_ $ _NAME_ $ variable-list ;
3         LABEL
4            V1 ='long-name'
5            V2 ='long-name'
6            V3 ='long-name'
7            V4 ='long-name'
8            V5 ='long-name'
9            V6 ='long-name' ;
10        DATALINES;
11      N       .     n       n       n       n       n       n
12      STD     .    std     std     std     std     std     std
13      CORR  V1   1.0000    .       .       .       .       .
14      CORR  V2    r       1.0000   .       .       .       .
15      CORR  V3    r        r      1.0000   .       .       .
16      CORR  V4    r        r       r      1.0000   .       .
17      CORR  V5    r        r       r       r      1.0000   .
18      CORR  V6    r        r       r       r       r      1.0000
19        ;
20      RUN;
```

where:

variable-list =   List of variables (e.g., V1, V2,).

long-name =   Full name for the given variable. This is used to label the variable when it appears in the SAS output. If this is not desired, you can omit the entire LABEL statement.

n =   Number of observations contributing to the correlation matrix. Each correlation in this matrix should be based on the same observations and hence the same number of observations. (This is automatically the case if the matrix is created using the NOMISS option with PROC CORR, as discussed in Chapter 6.)

std =   Standard deviation obtained for each variable. These standard deviations are needed if you are performing an analysis on the correlation matrix so that SAS can convert the correlation matrix into a variance-covariance matrix. Instead, if you wish to perform an analysis on a variance-covariance matrix, then standard deviations are not required.

r =   Correlation coefficients between pairs of variables.

The observations that appear on lines 11 to 18 in the preceding program are easier to understand if you think of the observations as a matrix with eight rows and eight columns. The first column in this matrix (running vertically) contains the _TYPE_ variable. (Notice that the INPUT statement tells SAS that the first variable it will read is a character variable named "_TYPE_".) If an "N" appears as a value in this _TYPE_ column, then SAS knows that sample sizes will appear on that line. If "STD" appears as a value in the _TYPE_ column, then the system knows that standard deviations appear on that line. Finally, if "CORR" appears as a value in the _TYPE_ column, then SAS knows that correlation coefficients appear on that line.

The second column in this matrix contains short names for the observed variables. These names should appear only on the CORR lines. Periods (for missing data) should appear where the N and STD lines intersect with this column (i.e., above the diagonal).

Looking at the matrix from the other direction, you see eight rows running horizontally. The first row is the N row (or "line"); it should contain the following:

- the N symbol;
- a period for the missing variable name;
- the sample sizes for the variables, each separated by at least one blank space. The preceding program shows that the sample size was 240 for each variable.

The STD row (or line) should contain the following:

- the STD symbol;
- the period for the missing variable name;

- the standard deviations for the variables, each separated by at least one blank space. If the STD line is omitted, the analysis can be performed only on covariances, not correlation coefficients.

Finally, where rows 3 to 8 intersect with columns 3 to 8, the correlation coefficients should appear. These coefficients appear below the diagonal, ones should appear on the diagonal (i.e., the correlation coefficient of a number with itself is always equal to 1.0) and periods appear above the diagonal (where redundant correlation coefficients would again appear if this were a full matrix). Be very careful in entering these correlations; one missing period can cause an error in reading the data.

You can see that the columns of data in this matrix are lined up in an organized fashion. Technically, neatness is not required as this INPUT statement is in free format. You should try to be equally organized when preparing your matrix, however, as this will minimize the chance of leaving out an entry and causing an error.

Inputting a Covariance Matrix

The procedure for inputting a covariance matrix is similar to that used with a correlation matrix. An example is presented here:

```
1          DATA D1(TYPE=COV) ;
2            INPUT _TYPE_ $ _NAME_ $ V1-V6 ;
3            LABEL
4              V1 ='COMMITMENT'
5              V2 ='SATISFACTION'
6              V3 ='REWARDS'
7              V4 ='COSTS'
8              V5 ='INVESTMENTS'
9              V6 ='ALTERNATIVES' ;
10         DATALINES;
11         N      .    240     240     240     240     240     240
12         COV   V1 11.1284     .       .       .       .       .
13         COV   V2  5.6742   9.0054    .       .       .       .
14         COV   V3  4.5501   3.6721  6.8773    .       .       .
15         COV   V4 -3.3499  -5.5717 -2.4405 10.9936    .       .
16         COV   V5  7.6444   2.5234  3.5346 -4.1854  7.1185    .
17         COV   V6 -8.6329  -3.4952 -6.4061  4.3525 -5.3934  9.2144
18         ;
19         RUN;
```

Notice that the DATA statement now specifies TYPE=COV rather than TYPE=CORR. The line providing standard deviations is no longer needed and has been removed. The matrix itself now provides variances on the diagonal and covariances below the diagonal; the beginning of each line now specifies COV to indicate that this is a covariance matrix. The remaining sections are identical to those used to input a correlation matrix.

Inputting Data Using the INFILE Statement Rather than the DATALINES Statement

When working with a very large dataset, it might be more convenient to input data using the INFILE statement rather than the DATALINES statement. This involves:

- adding an INFILE statement to your program;

- placing your data lines in a second computer file, rather than in the file that contains your SAS program;

- deleting the DATALINES statement from your SAS program.

Your INFILE statement should appear *after* the DATA statement but *before* the INPUT statement. The general form for a SAS program using the INFILE statement is as follows:

```
DATA dataset-name;
    INFILE  'name-of-data-file' ;
    INPUT  #line-number    @column-number   variable-name  column-width.
                           @column-number   variable-name  column-width.
                           @column-number   variable-name  column-width. ;

PROC name-of-desired-statistical-procedure     DATA=dataset-name;
RUN;
```

Notice that the above is identical to the general form for a SAS program presented earlier except that an INFILE statement is added, and the DATALINES statement and data lines are deleted.

To illustrate the use of the INFILE statement, consider Dr. Lafleur's volunteerism study. The dataset itself is reproduced here:

```
2234243 22   98   1 M
3424325 20  105   2 M
3242424 32   90   3 F
3242323  9  119   4 F
3232143  8  101   5 F
3242242 24  104   6 M
4343525 16  110   7 F
3232324 12   95   8 M
1322424 41   85   9 M
5433224 19  107  10 F
```

If you were to input these data using the INFILE statement, you would enter the data in a separate computer file, giving it any name you like. Assume, in this case, that the preceding data file is named VOLUNTEER.DAT.

You must enter these data lines beginning on line 1 of the computer file; do *not* leave any blank lines at the top of the file. Similarly, there should be no blank lines at the end of the file (unless a blank line is appropriate because of missing data for the last participant).

Once the data are entered and saved in the file called VOLUNTEER.DAT, you could enter the SAS program itself in a separate file. Perhaps you would give this file a name such as SURVEY.SAS. A SAS program which would input the preceding data and calculate means for the variables appears here:

```
1          DATA D1;
2              INFILE 'A:/VOLUNTEER.DAT';
3              INPUT   #1    @1    Q1-Q7    1.
4                            @9    AGE      2.
5                            @12   IQ       3.
6                            @16   NUMBER   2.
7                            @19   SEX      $1.  ;
8
9          PROC MEANS    DATA=D1;
10         RUN;
```

Controlling the Output Size and Log Pages with the OPTIONS Statement

Although it is not really related to the topic of data input, the OPTIONS statement is introduced now so that you can modify the size of your output and log pages, if necessary. For example, when printing your output on a 132-character printer, you might want to modify your output so that each line can be up to 120 characters long. When working with a printer with a smaller platen width, however, you might want to produce output that is less than 120 characters in length. The OPTIONS statement allows you to do this.

This is the general form of the OPTIONS statement that allows you to control the maximum number of characters and lines that appear on each page in output and log files:

```
OPTIONS    LINESIZE=x    PAGESIZE=y ;
```

With the preceding general form, x = the maximum number of characters that you wish to appear on each line, and y = the maximum number of lines that you wish to appear on each page.

For example, to request output and log files in which each line can be up to 80 characters long and each page can contain up to 60 lines, use the following OPTIONS statement as the first line of your program:

```
OPTIONS    LINESIZE=80    PAGESIZE=60;
```

To request output and log files with lines that are up to 120 characters long, use the following:

```
OPTIONS    LINESIZE=120    PAGESIZE=60;
```

Conclusion

The material presented in this chapter has prepared you to input most types of data commonly encountered in social science research. Even when the data have been entered successfully, however, they are not necessarily ready to be analyzed. Perhaps you have inputted raw data, and need to transform the data in some way before they can be analyzed.

This is often the case with questionnaire data, as responses to multiple questions are often summed or averaged to create new variables to be analyzed. Or perhaps you have data from a large, heterogeneous sample and you want to perform analyses on only a subgroup of that sample (e.g., female, but not male, respondents). In these situations, some form of *data manipulation* or data subsetting is required. The following chapter shows how to do this.

Reference

Rusbult, C. E. (1980). Commitment and satisfaction in romantic associations: A test of the investment model. *Journal of Experimental Social Psychology 16*:172–186.

56

Working with Variables and Observations in SAS® Datasets

> **Overview**. This chapter shows how to modify a dataset so that existing variables are transformed or recoded so that new variables are created. The chapter shows how to eliminate unwanted observations from a dataset so that analyses are performed only on a specified subgroup or on participants that have no missing data. This chapter also shows the correct use of arithmetic operators, IF-THEN control statements, and comparison operators. Finally, the chapter shows how to concatenate and merge existing datasets to create new datasets.

Introduction: Manipulating, Subsetting, Concatenating, and Merging Data

Very often, researchers obtain a dataset in which the data are not yet in a form appropriate for analysis. For example, imagine that you are conducting research on job satisfaction. Perhaps you want to compute the correlation between participant age and a single index of job satisfaction. You administer a 10-item questionnaire to 200 employees to assess job satisfaction, and you enter their responses to the 10 individual questionnaire items.

You now need to add together each participant's response to those 10 items to arrive at a single composite score that reflects that participant's overall level of satisfaction. This computation is easy to perform by including a number of data-manipulation statements in the SAS program. **Data-manipulation statements** are SAS statements that transform the dataset in some way. They can be used to recode negatively keyed variables, create new variables from existing variables, and perform a wide range of other tasks.

At the same time, your original dataset might contain observations that you do not wish to include in your analyses. Perhaps you administered the questionnaire to hourly as well as salaried employees, and you want to analyze only data from the former. In addition, you might want to analyze data only from participants who have usable data on all of the study's variables. In these situations, you can include data-subsetting statements to eliminate unwanted responses from the sample. **Data-subsetting statements** are SAS statements that eliminate unwanted observations from a sample so that only a specified subgroup is included in the resulting dataset.

In other situations, it might be necessary to concatenate or merge datasets before you can perform the analyses you desire. When you **concatenate** datasets, you combine two previously existing datasets that contain data on the same variables but from different participants. The resulting concatenated dataset contains aggregate data from all participants. In contrast, when you **merge** datasets, you combine two datasets that involve the same participants but contain different variables. For example, assume that dataset D1 contains variables V1 and V2, while dataset D2 contains variables V3 and V4. Further assume that both datasets have a variable called ID (identification number) that is used to merge data from the same participants. Once D1 and D2 are merged, the resulting dataset (D3) contains V1, V2, V3, and V4 as well as ID.

The SAS programming language is so comprehensive and flexible that it can perform virtually any type of manipulation, subsetting, concatenating, or merging task. A complete treatment of these capabilities would easily fill a book. However, this chapter reviews some basic statements that can be used to solve a wide variety of problems that are commonly encountered in social science research (particularly in research that involves the analysis of questionnaire data).

Placement of Data Manipulation and Data Subsetting Statements

The use of data manipulation and data subsetting statements are illustrated here with reference to the fictitious study described in the preceding chapter. In that chapter, you were asked to imagine that you had developed a 7-item questionnaire dealing with volunteerism, as shown in the following example.

```
Volunteerism Survey

Please indicate the extent to which you agree or disagree with each
of the following statements.  You will do this by circling the
appropriate number to the left of that statement.  The following
format shows what each response alternative represents:

     5 = Agree Strongly
     4 = Agree Somewhat
     3 = Neither Agree nor Disagree
     2 = Disagree Somewhat
     1 = Disagree Strongly

For example, if you "Disagree Strongly" with the first question,
circle the "1" to the left of that statement.  If you "Agree
Somewhat," circle the "4," and so on.


 -------------
  Circle Your
   Response
 -------------
 1  2  3  4  5     1.     I feel a personal responsibility to help
                          needy people in my community.

 1  2  3  4  5     2.     I feel I am personally obligated to help
                          homeless families.

 1  2  3  4  5     3.     I feel no personal responsibility to work
                          with poor people in my community.

 1  2  3  4  5     4.     Most of the people in my community are
                          willing to help the needy.

 1  2  3  4  5     5.     A lot of people around here are willing to
                          help homeless families.
```

(continued on the next page)

(continued)

```
1   2   3   4   5      6.    The people in my community feel no personal
                             responsibility to work with poor people.

1   2   3   4   5      7.    Everyone should feel the responsibility to
                             perform volunteer work in his/her community.


What is your age in years?  _____
```

Assume that you administer this survey to a number of participants and you also obtain information concerning sex, IQ scores, GRE verbal test scores, and GRE math test scores for each participant. Once the data are entered, you might want to write a SAS program that includes some data-manipulation or data-subsetting statements to transform the raw data. But where within the SAS program should these statements appear?

In general, these statements should appear only within the DATA step. Remember that the DATA step begins with the DATA statement and ends as soon as SAS encounters a procedure. This means that if you prepare the DATA step, end the DATA step with a procedure, and then place some manipulation or subsetting statement immediately after the procedure, you will receive an error.

To avoid this error (and keep things simple), place your data-manipulation and data-subsetting statements in one of two locations within a SAS program:

- immediately following the INPUT statement;

- or immediately following the creation of a new dataset.

Immediately Following the INPUT Statement

The first of the two preceding guidelines indicates that the statements may be placed immediately following the INPUT statement. This guideline is illustrated again by referring to the volunteerism study. Assume that you prepare the following SAS program to analyze data obtained in your study. In the following program, lines 11 and 12 indicate where you can place data-manipulation or data-subsetting statements in that program. (To conserve space, only some of the data lines are reproduced in the program.)

```
1          DATA D1;
2               INPUT   #1    @1    Q1-Q7       1.
3                             @9    AGE         2.
4                             @12   IQ          3.
5                             @16   NUMBER      2.
6                             @19   SEX         $1.
7                       #2    @1    GREVERBAL   3.
8                             @5    GREMATH     3.  ;
9
10
11         place data-manipulation statements and
12         data-subsetting statements here
13
14         DATALINES;
```

```
15        2234243 22  98   1 M
16        520 490
17        3424325 20 105   2 M
18        440 410
19           .
20           .
21
22        5433224 19 107 10 F
23        640 590
24        ;
25        RUN;
26
27        PROC MEANS  DATA=D1;
28        RUN;
```

Immediately after Creating a New Dataset

The second guideline for placement provides another option regarding where you can place data-manipulation or data-subsetting statements; they can also be placed immediately following program statements that create a new dataset. A new dataset can be created at virtually any point in a SAS program (even after procedures are requested).

At times, you might want to create a new dataset so that, initially, it is identical to an existing dataset (perhaps the one created with a preceding INPUT statement). If data-manipulation or data-subsetting statements follow the creation of this new dataset, the new set displays the modifications requested by those statements.

To create a new dataset that is identical to an existing dataset, the general form is

```
DATA  new-dataset-name;
    SET  existing-dataset-name;
```

To create such a dataset, use the following statements:

```
DATA D2;
    SET D1;
```

These lines told SAS to create a new dataset called D2 and to make this new dataset identical to D1. Now that a new set has been created, you can write as many manipulation and subsetting statements as you like. However, once you write a procedure, that effectively ends the DATA step and you cannot write any more manipulation or subsetting statements beyond that point (unless you create another dataset later in the program).

The following is an example of how you might write your program so that the manipulation and subsetting statements follow the creation of the new dataset:

```
1        DATA D1;
2            INPUT   #1    @1    Q1-Q7        1.
3                          @9    AGE          2.
4                          @12   IQ           3.
5                          @16   NUMBER       2.
6                          @19   SEX          $1.
7                    #2    @1    GREVERBAL    3.
8                          @5    GREMATH      3.  ;
9
10       DATALINES;
11       2234243 22   98  1 M
12       520 490
13       3424325 20 105   2 M
14       440 410
15          .
16          .
17
18       5433224 19 107 10 F
19       640 590
20       ;
21       RUN;
22
23       DATA D2;
24           SET D1;
25
26       place data manipulation statements and
27       data subsetting statements here
28
29       PROC MEANS   DATA=D2;
30       RUN;
```

SAS creates two datasets according to the preceding program: D1 contains the original data; and D2 is identical to D1 except for modifications requested by the data-manipulation and data-subsetting statements.

Notice that the MEANS procedure in line 29 requests the computation of some simple descriptive statistics. It is clear that these statistics are performed on the data from dataset D2 because DATA=D2 appears in the PROC MEANS statement. If the statement, instead, specified DATA=D1, the analyses would have been performed on the original dataset.

The INFILE Statement versus the DATALINES Statement

The preceding program illustrates the use of the DATALINES statement rather than the INFILE statement. The guidelines regarding the placement of data-modifying statements are the same regardless of which approach is followed. The data-manipulation or data-subsetting statement should either immediately follow the INPUT statement or the creation of a new dataset. When a program is written using the INFILE statement rather than the DATALINES statement, data-manipulation and data-subsetting statements should appear *after* the INPUT statement but *before* the first procedure. For example, if your data are entered into an external file called VOLUNTEER.DAT, you can write the following program. (Notice where the manipulation and subsetting statements are placed.)

```
1    DATA D1;
2        INFILE 'A:/VOLUNTEER.DAT';
3        INPUT   #1   @1    Q1-Q7        1.
4                     @9    AGE          2.
5                     @12   IQ           3.
6                     @16   NUMBER       2.
7                     @19   SEX          $1.
8                #2   @1    GREVERBAL    3.
9                     @5    GREMATH      3.   ;
10
11       place data manipulation statements and
12       data subsetting statements here
13
14       PROC MEANS    DATA=D1;
15       RUN;
```

In the preceding program, the data-modifying statements again come immediately after the INPUT statement but before the first procedure, consistent with earlier recommendations.

Data Manipulation

Data manipulation involves performing a transformation on one or more variables in the DATA step. This section discusses several types of transformations that are frequently required in social science research. These include creation of duplicate variables with new variable names, creation of new variables from existing variables, recoding reversed items, and using IF-THEN/ELSE statements as well as other related procedures.

Creating Duplicate Variables with New Variable Names

Suppose that you give a variable a certain name when it is inputted, but then you want the variable to have a different, perhaps a more meaningful, name when it appears later in the SAS program or in the SAS output. This can easily be accomplished with a statement written according to the following syntax:

```
new-variable-name  =  existing-variable-name;
```

For example, in the preceding dataset, the first seven questions are given variable names of Q1 through Q7. Item 1 in the questionnaire reads, "I feel a personal responsibility to help needy people in my community." In the INPUT statement, this item was given a SAS variable name Q1, which is not very meaningful. RESNEEDY, which stands for "responsible for the needy," is a more meaningful name. Similarly, RESHOME is more meaningful than Q2, and NORES is more meaningful than Q3.

One way to rename an existing variable is to create a new variable that is identical to the existing variable and assign a new, more meaningful name to this variable. The following program renames Q1, Q2, and Q3 in this way.

This and later examples show only a portion of the entire program. However, enough of the program appears to illustrate where the remaining statements should be placed.

```
15          .
16          .
17
18          5433224 19 107 10 F
19          640 590
20          ;
21          RUN;
21
22          DATA D2;
23             SET D1;
24
25          RESNEEDY  = Q1;
26          RESHOME   = Q2;
27          NORES     = Q3;
28
29          PROC MEANS     DATA=D2;
30          RUN;
```

Line 25 tells SAS to create a new variable called RESNEEDY and for it to be identical to the existing variable, Q1. Variables RESNEEDY and Q1 now have identical data but RESNEEDY has a more meaningful name to facilitate the reading of printouts when statistical analyses are later performed.

When creating a new variable name, conform to the rules for naming SAS variables discussed in Chapter 3, "Data Input" (e.g., begins with a letter). Also, note that each statement that creates a duplicate of an existing variable must end with a semicolon.

Duplicating Variables versus Renaming Variables

Technically, the previous program did not really rename variables Q1, Q2, and Q3. Rather, the program created duplicates of these variables and assigned new names to these duplicate variables. Therefore, the resulting dataset contains both the original variables under their old names (Q1, Q2, and Q3) as well as the duplicate variables under their new names (RESNEEDY, RESHOME, and NORES). If, for some reason, you literally need to rename the existing variables so that the old variable names no longer exist in the dataset, consider using the RENAME statement.

Creating New Variables from Existing Variables

It is often necessary to perform mathematical operations on existing variables and use the results to create a new variable. With SAS, the following symbols can be used in arithmetic operations:

+   (addition)
-   (subtraction)
*   (multiplication)
/   (division)
=   (equals)

When writing formulae, you should make extensive use of parentheses. Remember that operations enclosed within parentheses are performed first, and operations outside of the parentheses are performed later. To create a new variable by performing a mathematical operation on an existing variable, use the following general form:

```
new-variable-name  =  formula-including-existing-variables;
```

For example, two existing variables in your dataset are GREVERBAL (GRE verbal test scores) and GREMATH (GRE math test scores). Suppose you wanted to create a new variable called GRECOMB. This variable includes each participant's combined GRE score. For each participant, you need to add together GREVERBAL and GREMATH scores; therefore, the GRECOMB value is the sum of the values for GREVERBAL and GREMATH. The program repeats this operation for each participant in the sample, using just one statement:

```
GRECOMB = (GREVERBAL + GREMATH);
```

The preceding statement tells SAS to create a new variable called GRECOMB and set it equal to the sum of GREVERBAL and GREMATH.

Suppose that you want to calculate the *average* of GREVERBAL and GREMATH scores. The new variable might be called GREAVG. The program repeats this operation for each participant in the sample using the following statement:

```
GREAVG = (GREVERBAL + GREMATH) / 2;
```

The preceding statement tells SAS to create a new variable called GREAVG by first adding together the values of GREVERBAL and GREMATH, then dividing this sum by 2. The resulting quotient is labeled GREAVG. You can also arrive at the same result by using two statements instead of one, as shown here:

```
GRECOMB = (GREVERBAL + GREMATH);
GREAVG  = GRECOMB/2;
```

Very often, researchers need to calculate the average of several items on a questionnaire. For example, look at items 1 and 2 in the questionnaire shown previously. Both items seem to be measuring participants' sense of personal responsibility to help the needy.

Rather than analyze responses to the items separately, it might be more useful to calculate the average of responses to those items. This average could then serve as participants' scores on some "personal responsibility" variable. For example, consider the following:

```
RESPONSE = (Q1 + Q2) / 2;
```

The preceding statement tells SAS to create a new variable called RESPONSE by adding together participants' scores for Q1 and Q2 and then dividing the resulting sum by 2. The resulting quotient creates the new RESPONSE variable.

When creating new variables in this manner, be sure that all variables on the right side of the equals sign are *existing* variables. This means that they already exist in the dataset, either because they are listed in the INPUT statement or because they were created with earlier data-manipulation statements.

Priority of Operators in Compound Expressions

A SAS expression (e.g., a formula) that contains just one operator is known as a **simple expression**. The following statement contains a simple expression. Notice that there is only one operator (+ sign) to the right of the = sign:

```
RESPONSE = Q1 + Q2;
```

In contrast, a **compound expression** contains more than one operator. A compound expression is illustrated in the following example. Notice that several different operators appear to the right of the = sign:

```
RESPONS = Q1 + Q2 - Q3 / Q4 * Q5;
```

When an expression contains more than one operator, SAS follows a set of rules that determine which operations are performed first, which are performed second, and so forth. The rules that pertain to mathematical operators (+, -, /, and *) are summarized here:

- Multiplication and division operators (* and /) have equal priority, and they are performed first.

- Addition and subtraction operators (+ and -) have equal priority, and they are performed second.

One point made in the preceding rules is that multiplication and division are performed prior to addition or subtraction. For example, consider the following statement:

```
RESPONS = Q1 + Q2 / Q3;
```

Since division has priority over addition, the operations in the preceding statement would be executed in this sequence:

- Q2 would first be divided by Q3.
- The resulting quotient would then be added to Q1.

Notice that division is performed first, even though the addition appears earlier in the formula (reading from left to right).

But what if multiple operators having equal priority appear in the same statement? In this situation, SAS reads the formula from left to right, and performs the operations in that sequence. For example, consider the following:

```
RESPONS = Q1 + Q2 - Q3;
```

The preceding expression contains only addition and subtraction operations that have equal priority. SAS therefore reads the statement from left to right: first Q1 is added to Q2; then Q3 is subtracted from the resulting sum.

Because different priority is given to different operators, it is unfortunately very easy to write a statement that results in operations being performed in a sequence other than that intended. For example, imagine that you want to create a new variable called RESPONSE. Each participant's score for RESPONSE is created by adding responses to Q1, Q2, and Q3 and by dividing this sum by 3. Imagine further that you try to achieve this with the following statement:

```
RESPONSE = Q1 + Q2 + Q3 / 3;
```

The preceding statement will not create the RESPONSE variable as you had intended. Because division has priority over addition, SAS performs the operations in the following order:

1. Q3 is divided by 3.
2. The resulting quotient is then added to Q1 and Q2.

Obviously, this is not what you intended.

To avoid mistakes such as this, it is important to use *parentheses* when writing formulae. Because operations inside parentheses are performed first, the use of parentheses gives you control over the sequence in which operations are executed. For example, the following statement creates the RESPONSE variable in the way originally intended because the lower priority operations (adding together Q1 plus Q2 plus Q3) are now included within parentheses:

```
RESPONSE = (Q1 + Q2 + Q3) / 3;
```

This statement tells SAS to add together Q1 plus Q2 plus Q3; the sum of these operations is then divided by 3.

This section has provided a brief introduction to the priority of a few operators that can be performed with SAS.

Recoding Reversed Variables

Very often, a questionnaire contains a number of reversed items. A reversed item is a question stated so that its meaning is opposite the meaning of other items in that group. For example, consider the meaning of the following items from the volunteerism survey:

```
1  2  3  4  5     1.   I feel a personal responsibility to help
                       needy people in my community.

1  2  3  4  5     2.   I feel I am personally obligated to help
                       homeless families.

1  2  3  4  5     3.   I feel no personal responsibility to work
                       with poor people in my community.
```

In a sense, all of these questions are measuring the same thing (i.e., whether the participant feels some sense of personal responsibility to help the needy). Items 1 and 2 are stated so that the more strongly you agree with these statements, the greater your sense of personal responsibility. This means that scores of 5 indicate a strong sense of responsibility and scores of 1 indicate a weak sense of responsibility. However, item 3 is a *reversed* or negatively keyed item. It is stated so that the more strongly you agree, the weaker your sense of personal responsibility. Here, a response of 1 indicates a strong sense of responsibility whereas a response of 5 indicates a weak sense of responsibility (which is just the reverse of items 1 and 2).

For later analyses, all three items must be consistent so that scores of 5 always indicate a strong sense of responsibility whereas scores of 1 always indicate a weak sense of responsibility. This requires that you recode item 3 so that those who actually select 5, instead, are given a score of 1; those who actually circle 4 are given, instead, a score of 2; those who actually select 2 are given a score of 4; and those who select 1 are given a score of 5. This can be done very well with the following statement:

```
Q3 = 6 - Q3;
```

The preceding statement tells SAS to create a new version of the variable Q3, then take the number 6 and subtract from it participants' existing (old) scores for Q3. The result is a new score for Q3. Notice that with this statement, if an initial score for Q3 was 5, the new score becomes 1; and if the initial score was 1, the new score is 5.

The syntax for this recoding statement is as follows:

```
existing-variable  =  constant  -  existing-variable;
```

The constant is always equal to the number of response points on your survey plus 1. For example, the volunteerism survey included 5 response points: participants could circle "1" for "Disagree Strongly" all the way through "5" for "Agree Strongly." It was a 5-point scale, so the constant is $5 + 1 = 6$. What would the constant be if the following 7-point scale had been used instead?

```
7 = Agree Very Strongly
6 = Agree Strongly
5 = Agree Somewhat
4 = Neither Agree nor Disagree
3 = Disagree Somewhat
2 = Disagree Strongly
1 = Disagree Very Strongly
```

It would be 8 because $7 + 1 = 8$, and the recoding statement would read as follows:

```
Q3 = 8 - Q3;
```

> *Where* **should the recoding statements go?** In most cases, reversed items should be recoded before other data manipulations are performed. For example, assume that you want to create a new variable called RESPONSE, which stands for "personal responsibility." With this scale, higher scores indicate higher levels of perceived personal responsibility. Scores on this scale are the average of participant responses to items 1, 2, and 3 from the survey. Because item 3 is a reversed item, it is important that it be recoded before it is added to items 1 and 2 when calculating the overall scale score. Therefore, the correct sequence of statements is as follows:
>
> ```
> Q3 = 6 - Q3;
> RESPONSE = (Q1 + Q2 + Q3) / 3;
> ```
>
> The following sequence is *not* correct:
>
> ```
> RESPONSE = (Q1 + Q2 + Q3) / 3;
> Q3 = 6 - Q3;
> ```

Using IF-THEN Control Statements

An IF-THEN control statement allows you to make sure that operations are performed on data only if certain conditions are true. The following comparison operators can be used with IF-THEN statements:

```
      = (equal to)
     NE (not equal to)
GT or > (greater than)
     GE (greater than or equal to)
LT or < (less than)
     LE (less than or equal to)
```

The general form for an IF-THEN statement is as follows:

```
IF  expression  THEN  statement ;
```

The expression usually consists of some comparison involving existing variables. The statement usually involves some operation performed on existing variables or new variables. For example, assume that you want to create a new variable called GREVGRP for "GRE-verbal group." This variable will be created so that:

- If you do not know participants' GRE verbal test scores, they will be assigned a score of "." (for "missing data").

- If participants' scores are less than 500 on the GRE verbal test, they will be assigned a score of 1 for GREVGRP.

- If the participant's score is 500 or greater on the GRE verbal test, the participant will have a score of 2 for GREVGRP.

Assume that the variable GREVERBAL already exists in your dataset and that it contains each participant's score for the GRE verbal test. You can use it to create the new variable GREVGRP by writing the following statements:

```
GREVGRP = .;
IF GREVERBAL LT 500 THEN GREVGRP = 1;
IF GREVERBAL GE 500 THEN GREVGRP = 2;
```

The preceding statements tell SAS to create a new variable called GREVGRP and begin by setting everyone's score as equal to "." (i.e., missing). If participants' scores for GREVERBAL are less than 500, then their score for GREVGRP will be equal to 1. If participants' scores for GREVERBAL are greater than or equal to 500, then their score for GREVGRP is equal to 2.

Using ELSE Statements

In reality, you can perform the preceding operations more efficiently by using the ELSE statement. The general form for using the ELSE statement, in conjunction with the IF-THEN statement, is presented as follows:

```
IF   expression  THEN   statement  ;
   ELSE  IF  expression  THEN  statement;
```

The ELSE statement provides alternative actions that SAS can take when the original IF expression is not true. For example, consider the following:

```
GREVGRP = .;
IF GREVERBAL LT 500 THEN GREVGRP = 1;
ELSE IF GREVERBAL GE 500 THEN GREVGRP = 2;
```

The preceding tells SAS to create a new variable called GREVGRP and initially assign all participants a value of "missing." If a given participant has a GREVERBAL score less than 500, the system assigns that participant a score of 1 for GREVGRP. Otherwise, if the participant has a GREVERBAL score greater than or equal to 500, then the system assigns that participant a score of 2 for GREVGRP.

Obviously, the preceding statements are identical to the earlier statements that created GREVGRP, except that the word ELSE is added to the beginning of the third line. In fact, these two approaches actually result in assigning exactly the same values for

GREVGRP to each participant. So, what is the advantage of including the ELSE statement? The answer has to do with *efficiency*. When an ELSE statement is included, the actions specified by that statement are executed only if the expression in the preceding IF statement is not true.

For example, consider the situation in which participant 1 has a GREVERBAL score less than 500. Line 2 in the preceding statements assigns that participant a score of 1 for GREVGRP. SAS then ignores line 3 (because it contains the ELSE statement), thus saving computing time. If line 3 did not contain the word ELSE, SAS would have executed the command, checking to see whether the GREVERBAL score for participant 1 is greater than or equal to 500 (which is actually unnecessary, given what was learned in line 2).

A word of caution regarding missing data is required at this point. Notice that line 2 of the preceding program assigns participants to group 1 (under GREVGRP) if their values for GREVERBAL are less than 0. Unfortunately, a value of "missing" (i.e., a value of ".") for GREVERBAL is viewed as being less than 500 (actually, it is viewed as being less than 0) by SAS. This means that participants with missing data for GREVERBAL are assigned to group 1 under GREVGRP by line 2 of the preceding program. This is not desirable.

To prevent this from happening, you can rewrite the program in the following way:

```
GREVGRP = .;
IF GREVERBAL GT 0 AND GREVERBAL LT 500 THEN GREVGRP = 1;
ELSE IF GREVERBAL GE 500 THEN GREVGRP = 2;
```

Line 2 of the program now tells SAS to assign participants to group 1 only if their values for GREVERBAL are both greater than 0 and less than 500. This modification involves the use of the conditional AND statement, which is discussed in greater detail in the following section.

Finally, remember that the ELSE statement should be used only in conjunction with a preceding IF statement. In addition, always remember to place the ELSE statement *immediately* following the relevant IF statement.

Using the Conditional Statements AND and OR

As the preceding section indicates, you can also use the conditional statement AND within an IF-THEN statement or an ELSE statement. For example, consider the following:

```
GREVGRP = .;
IF GREVERBAL GT 0 AND GREVERBAL LT 500 THEN GREVGRP = 1;
ELSE IF GREVERBAL GE 500 THEN GREVGRP = 2;
```

The second statement in the preceding program tells SAS that if GREVERBAL is greater than 0 and less than 500, then a score of 1 is given to participants for the GREVGRP variable. This means that all are given a value of 1 *only* if they are both over 0 and under 500. What happens to those who have a score of 0 or less for GREVERBAL? They are given a value of "." for GREVGRP. That is, they are classified as having a missing value for GREVGRP. This is because they (along with everyone else) were initially given a value of "." in the first statement, and neither of the later statements replaces that "." with 1 or 2. However, for those with GREVERBAL scores greater than 0, one of the subsequent statements replaces "." with either 1 or 2.

You can also use the conditional statement OR within an IF-THEN statement or an ELSE statement. For example, assume that you have a variable in your dataset called ETHNIC. With this variable, participants were assigned the value 5 if they are Caucasian, 6 if they are African American, or 7 if they are Asian American. Assume that you now wish to create a new variable called MAJORITY. Participants are assigned a value of 1 for this variable if they are in the majority group (i.e., if they are Caucasians), and they are assigned a value of 2 for this variable if they are in a minority group (i.e., if they are either African Americans or Asian Americans). This variable is created with the following statements:

```
MAJORITY=.;
IF ETHNIC = 5 THEN MAJORITY = 1;
ELSE IF ETHNIC = 6 OR ETHNIC = 7 THEN MAJORITY = 2;
```

In the preceding statements, all participants are first assigned a value of "missing" for MAJORITY. If their value for ETHNIC is 5, their value for MAJORITY changes to 1 and SAS ignores the following ELSE statement. If their value for ETHNIC is not 5, then SAS proceeds to the ELSE statement. There, if participants' value for ETHNIC is either 6 or 7, they are then assigned a value of 2 for MAJORITY.

Working with Character Variables

When working with character variables (i.e., variables in which the values consist of letters rather than numbers), you must enclose values within single quotation marks in the IF-THEN and ELSE statements. For example, suppose you want to create a new variable called SEXGRP. With this variable, males are given a score of 1 and females are given a score of 2. The variable SEX already exists in your dataset, and it is a character variable in which males are coded with the letter "M" and females are coded with the letter "F." You can create the new SEXGRP variable using the following statements:

```
SEXGRP = .;
IF SEX = 'M' THEN SEXGRP = 1;
ELSE IF SEX = 'F' THEN SEXGRP = 2;
```

Using the IN Operator

The IN operator makes it easy to determine whether a given value is among a specified list of values. Because of this, a single IF statement including the IN operator can perform comparisons that could otherwise require a large number of IF statements. The general form for using the IN operator is as follows:

```
IF   variable   IN   value-1,value-2, ...value-n   THEN   statement;
```

Notice that each value in the preceding list must be separated by a comma.

For example, assume that you have a variable in your dataset called MONTH. The values assumed by this variable are the numbers 1 through 12. With these values, 1 represents January, 2 represents February, 3 represents March, and so forth. Assume that these values for MONTH indicate the month in which a given participant was born, and that you have data for 100 participants.

Imagine that you now wish to create a new variable called SEASON. This variable will indicate the season in which each participant was born. Participants are assigned values for SEASON according to the following guidelines:

- Participants are assigned a value of 1 for SEASON if they were born in January, February, or March (months 1, 2, 3).

- Participants are assigned a value of 2 for SEASON if they were born in April, May, or June (months 4, 5, 6).

- Participants are assigned a value of 3 for SEASON if they were born in July, August, or September (months 7, 8, 9).

- Participants are assigned a value of 4 for SEASON if they were born in October, November, or December (months 10, 11, 12).

One way to create the new SEASON variable involves using four IF-THEN statements, as shown here:

```
SEASON = .;
IF MONTH = 1  OR MONTH = 2  OR MONTH = 3  THEN SEASON = 1;
IF MONTH = 4  OR MONTH = 5  OR MONTH = 6  THEN SEASON = 2;
IF MONTH = 7  OR MONTH = 8  OR MONTH = 9  THEN SEASON = 3;
IF MONTH = 10 OR MONTH = 11 OR MONTH = 12 THEN SEASON = 4;
```

However, the same results can be achieved somewhat more easily by using the IN operator within IF-THEN statements, as shown here:

```
SEASON = .;
IF MONTH IN (1,2,3)    THEN SEASON = 1;
IF MONTH IN (4,5,6)    THEN SEASON = 2;
IF MONTH IN (7,8,9)    THEN SEASON = 3;
IF MONTH IN (10,11,12) THEN SEASON = 4;
```

In the preceding example, all variable values are numbers. However, the IN operator can also be used with character variables. As always, it is necessary to enclose all character variable values within single quotation marks. For example, assume that MONTH is actually a character variable that assumes values such as "Jan," "Feb," "Mar," and so forth. Assume further that SEASON assumes the values "Winter," Spring," Summer," and "Fall." Under these circumstances, the preceding statements would be modified in the following way:

```
SEASON = '.';
IF MONTH IN ('Jan', 'Feb', 'Mar') THEN SEASON = 'Winter';
IF MONTH IN ('Apr', 'May', 'Jun') THEN SEASON = 'Spring';
IF MONTH IN ('Jul', 'Aug', 'Sep') THEN SEASON = 'Summer';
IF MONTH IN ('Oct', 'Nov', 'Dec') THEN SEASON = 'Fall';
```

Data Subsetting

Using a Simple Subsetting Statement

Often, it is necessary to perform an analysis on only a subset of the participants in the dataset. For example, you might want to review the mean survey responses provided by just the female participants. A subsetting IF statement can be used to accomplish this, and the general form is presented here:

```
DATA  new-dataset-name;
   SET  existing-dataset-name;
IF  comparison;

PROC  name-of-desired-statistical-procedure   DATA=new-dataset-name;
RUN;
```

The comparison described in the preceding statements generally includes an existing variable and at least one comparison operator. The following statements enable you to calculate the mean survey responses for only the female participants.

```
15        .
16        .
17
18        5433224 19 107 10 F
19        640 590
20        ;
21        RUN;
22
23        DATA D2;
24           SET D1;
25
26           IF SEX = 'F';
27
28        PROC MEANS    DATA=D2;
29        RUN;
```

The preceding statements tell SAS to create a new dataset called D2 and to make it identical to D1. However, the program keeps a participant's data only if her SEX has a value of F. Then the program executes the MEANS procedure for the data that are retained.

Using Comparison Operators

All of the comparison operators previously described can be used in a subsetting IF statement. For example, consider the following:

```
DATA D2;
   SET D1;

   IF SEX = 'F' AND AGE GE 65;

PROC MEANS    DATA=D2;
RUN;
```

The preceding statements analyze only data from women who are 65 or older.

Eliminating Observations with Missing Data for Some Variables

One of the most common difficulties encountered by researchers in the social sciences is the problem of missing data. Briefly, missing data involves not having scores for all variables for all participants in a dataset. This section discusses the problem of missing data, and shows how a subsetting IF statement can be used to deal with it.

Assume that you administer your volunteerism survey to 100 participants, and you use their scores to calculate a single volunteerism score for each. You also obtain a number of additional variables regarding the participants. The SAS names for the study's variables and their descriptions are as follows:

VOLUNTEER: participant scores on the volunteerism questionnaire, where higher scores reveal greater likelihood to engage in unpaid prosocial activities;

GREVERBAL: participant scores on the verbal subtest of the Graduate Record Examinations;

GREMATH: participant scores on the math subtest of the Graduate Record Examinations;

IQ: participant scores on a standard intelligence test (intelligence quotient).

Assume further that you obtained scores for VOLUNTEER, GREVERBAL, and GREMATH for all 100 participants. Due to a recordkeeping error, however, you were able to obtain IQ scores for only 75 participants.

You now want to analyze your data using a procedure called multiple regression. (This procedure is covered in a later chapter; you do not need to understand multiple regression to understand the points to be made here.) In Analysis #1, VOLUNTEER is the criterion variable, and GREVERBAL and GREMATH are the predictor variables. The multiple regression equation for Analysis #1 is represented in the following PROC REG statement:

Analysis #1:

```
PROC REG   DATA=D1;
   MODEL VOLUNTEER  =  GREVERBAL   GREMATH ;
RUN;
```

When you review the analysis results, note that the analysis is based on 100 participants. This makes sense because you had complete data on all of the variables included in this analysis.

In Analysis #2, VOLUNTEER is again the criterion variable; this time, however, the predictor variables include GREVERBAL and GREMATH as well as IQ. The equation for Analysis #2 is as follows:

Analysis #2:

```
PROC REG   DATA=D1;
   MODEL VOLUNTEER  =  GREVERBAL   GREMATH   IQ;
RUN;
```

When you review the results of analysis #2, you see that you have encountered a problem. The SAS output indicates that the analysis is based on only 75 participants. At first you might not understand this because you know that there are 100 participants in the dataset. But then you remember that you did not have complete data for one of the variables. You had values for the IQ variable for only 75 participants. The REG procedure (and many other SAS procedures) includes in the analysis only those participants who have complete data for *all* of the variables analyzed with that procedure. For Analysis #2, this means that any participant with missing data for IQ will be eliminated from the analysis. Twenty-five participants had missing data for IQ and were therefore eliminated.

Why were these 25 participants not eliminated from Analysis #1? Because that analysis did not involve the IQ variable. It involved only VOLUNTEER, GREVERBAL, and GREMATH; and all 100 participants had complete data for each of these three variables.

In a situation such as this, you have a number of options with respect to how you might perform these analyses and summarize the results. One option is to retain the results described previously. You could report that you performed one analysis on all 100 participants and a second analysis on just the 75 who had complete data for the IQ variable.

This approach might leave you open to criticism, however. The beginning of your research paper probably reported demographic characteristics for all 100 participants (e.g., how many were female, mean age). However, you might not have a section providing demographics for the subgroup of 75. This might lead readers to wonder if the subgroup differed in some important way from the aggregate group.

There are statistical reasons why this approach might cause problems as well. For example, you might wish to test the significance of the difference between the squared multiple correlation (R^2) value obtained from Analysis #1 and the R^2 value obtained from Analysis #2. (This test is described in Chapter 14, "Multiple Regression.") When performing this test, it is important that both R^2 values be based on exactly the same participants in both analyses. This is obviously not the case in your study as 25 of the participants used in Analysis #1 were not used in Analysis #2.

In situations such as this, you are usually better advised to ensure that every analysis is performed on exactly the same sample. This means that, in general, any participant who has missing data for variables to be included in any (reported) analysis should be deleted before the analyses are performed. In this instance, therefore, it is best to see to it that both Analysis #1 and Analysis #2 are performed on only those 75 participants who had complete data for all four variables (i.e., VOLUNTEER, GREVERBAL, GREMATH, and IQ). Fortunately, this can easily be done using a subsetting IF statement.

Recall that with SAS, a missing value is represented with a period ("."). You can take advantage of this to eliminate any participant with missing data for any analyzed variable. For example, consider the following subsetting IF statement:

```
DATA D2;
   SET D1;
IF VOLUNTEER NE  . AND GREVERBAL NE  . AND
   GREMATH    NE . AND IQ    NE . ;
```

The preceding statements tell the system to do the following:

1. Create a new dataset named D2, and make it an exact copy of D1.
2. Retain a participant in this new dataset only if (for that participant):
    *   VOLUNTEER is not equal to missing;
    *   GREVERBAL is not equal to missing;
    *   GREMATH is not equal to missing;
    *   IQ is not equal to missing.

In other words, the system creates a new dataset named D2; this new dataset contains only the 75 participants who have complete data for all four variables of interest. You can now specify DATA=D2 in all SAS procedures, with the result that all analyses will be performed on exactly the same 75 participants.

The following SAS program shows where these statements should be placed:

```
14          .
15          .
16          .
17
18          5433224 19 107 10 F
19          640 590
20          ;
21          RUN;
22
23          DATA D2;
24             SET D1;
25
26          IF VOLUNTEER NE . AND GREVERBAL NE . AND
27             GREMATH   NE . AND IQ   NE . ;
28
29          PROC REG   DATA=D2;
30             MODEL VOLUNTEER =  GREVERBAL  GREMATH ;
31          RUN;
32
33          PROC REG   DATA=D2;
34             MODEL VOLUNTEER =  GREVERBAL  GREMATH  IQ ;
35          RUN;
```

As evident above, the subsetting IF statement must appear in the program before the procedures that request the modified dataset (dataset D2, in this case).

How should I enter missing data? If you are entering data and come to a participant with a missing value for some variable, you do not need to record a "." to represent the missing data. So long as your data are being input using the DATALINES statement and the conventions discussed here, it is acceptable to simply leave that column (or those columns) blank by hitting the space bar on your keyboard. SAS will internally assign that participant a missing data value (".") for that variable. In some cases, however, it might be useful to enter a "." for variables with missing data, as this can make it easier to keep your place when entering information.

When using a subsetting IF statement to eliminate participants with missing data, exactly *which* variables should be included in that statement? In most cases, it should be those variables, and only those variables, that are ultimately discussed. This means that you might not know exactly which variables to include until you actually begin analyzing the data. For example, imagine that you conduct your study and obtain data for the following number of participants for each of the following variables:

Variable	Number of Participants with Valid Data for This Variable
VOLUNTEER	100
GREVERBAL	100
GREMATH	100
IQ	75
AGE	10

As before, you obtained complete data for all 100 participants for VOLUNTEER, GREVERBAL and GREMATH, and you obtained data for 75 participants on IQ. But notice the last variable. You obtained information regarding age for only 10 participants. What would happen if you included the variable AGE in the subsetting IF statement, as shown here?

```
IF VOLUNTEER NE . AND GREVERBAL NE . AND
   GREMATH  NE . AND IQ  NE . AND AGE  NE . ;
```

This IF statement causes the system to eliminate from the sample anyone who does not have complete data for all five variables. Since only 10 participants have values for the AGE variable, you know that the resulting dataset includes just these 10 participants. However, this sample is too small for virtually all statistical procedures. At this point, you have to decide whether to gather more data or forget about doing any analyses with the AGE variable.

In summary, one approach for identifying those variables to be included in the subsetting IF statement is to do the following:

- perform some initial analyses;
- decide which variables will be included in the final analyses (for your study);
- include all of those variables in the subsetting IF statement;
- perform all analyses on this reduced dataset so that all analyses reported in the study are performed on exactly the same sample.

Of course, there are circumstances in which it is neither necessary nor desirable that all analyses be performed on exactly the same group of participants. The purpose of the research, along with other considerations, should determine when this is appropriate.

A More Comprehensive Example

Often, a single SAS program contains a large number of data-manipulation and subsetting statements. Consider the following example which makes use of the INFILE statement rather than the DATALINES statement:

```
 1    DATA D1;
 2        INFILE 'A:/VOLUNTEER.DAT' ;
 3        INPUT    #1    @1    Q1-Q7        1.
 4                       @9    AGE          2.
 5                       @12   IQ           3.
 6                       @16   NUMBER       2.
 7                       @19   SEX          $1.
 8               #2      @1    GREVERBAL    3.
 9                       @5    GREMATH      3.   ;
10
11    DATA D2;
12        SET D1;
13
14    Q3 = 6 - Q3;
15    Q6 = 6 - Q6;
16    RESPONSE = (Q1 + Q2 + Q3) / 3;
17    TRUST    = (Q4 + Q5 + Q6) / 3;
18    SHOULD   = Q7;
19
20    PROC MEANS    DATA=D2;
21    RUN;
22
23    DATA D3;
24        SET D2;
25        IF SEX = 'F';
26
27    PROC MEANS    DATA=D3;
28    RUN;
29
30    DATA D4;
31        SET D2;
32        IF SEX = 'M';
33
34    PROC MEANS    DATA=D4;
35    RUN;
```

In the preceding program, lines 11 and 12 create a new dataset called D2 and set it identical to D1. All data-manipulation commands that appear between those lines and PROC MEANS on line 20 are performed on dataset D2. Notice that a new variable called TRUST is created on line 17. TRUST is the average of participants' responses to items 4, 5, and 6. Look over these items on the volunteerism survey to see why the name TRUST makes sense. On line 18, variable Q7 is duplicated, and the resulting new variable is called SHOULD. Why does this make sense? PROC MEANS appears on line 20, so the means and other descriptive statistics are calculated for all of the quantitative variables in the most recently created dataset, which is D2. This includes all variables inputted in dataset D1 as well as the new variables that were just created.

In lines 23 through 25, a new dataset called D3 is created; only responses from female participants are retained in this dataset. Notice that the SET statement sets D3 equal to D2 rather than D1. This enables the newly created variables such as TRUST and SHOULD to appear in this all-female dataset. In lines 30 through 32, a new dataset called D4 is created which is also set equal to D2 (not D3). This new dataset contains data from only males.

After this program is submitted for analysis, the SAS output contains three tables of means. The first table is based on lines 1 through 21, and gives the means based on all participants. The second table is based on lines 23 through 28 and gives the means based on the responses of females. The third table is based on lines 30 through 35, and is based on the responses of males.

Concatenating and Merging Datasets

The techniques described to this point in this chapter are designed to help you transform data within a single dataset (e.g., to recode a variable within a single dataset). However, you frequently need to perform transformations that involve combining more than one dataset to create a new dataset. For example, **concatenating** involves creating a new dataset by combining two or more previously existing datasets. With concatenation, the same variables typically appear in both of the previously existing datasets, but the two sets contain data from *different participants*. By concatenating the two previously existing sets, you create a new set that contains data from all participants.

In contrast, **merging** combines datasets in a different way. With merging, each previously existing dataset typically contains data from the same participants. However, the different, previously existing sets usually contain *different variables*. By merging these sets, you can create a new dataset that contains all variables found in the previously existing datasets. For example, assume that you conduct a study with 100 participants. Dataset A contains each participant's age, while dataset B contains questionnaire responses from the same 100 participants. By merging datasets A and B, you can create a new dataset called C that, again, contains just 100 observations. A given observation in dataset C contains a given participant's age as well as the questionnaire responses made by that participant. Now that the datasets are merged, it is possible to correlate participant age with responses to the questionnaire. These coefficients could not be calculated when AGE was in one dataset and the questionnaire responses were in another.

Concatenating Datasets

Imagine that you are conducting research that involves the Graduate Record Examinations (GRE). You obtain data from four participants: John; Sally; Miguel; and Mats. You enter information about these four participants into a SAS dataset called A. This dataset contains three variables:

- NAME, which contains the participant's first name;
- GREVERBAL, which contains the participant's score on the GRE verbal test;
- GREMATH, which contains the participant's score on the GRE math test.

The contents of dataset A appear below in Table 4.1. You can see that John has a score of 520 for GREVERBAL and a score of 500 for GREMATH, Sally had a score of 610 for GREVERBAL and 640 for GREMATH, and so forth.

Table 4.1

Contents of Dataset A

NAME	GREVERBAL	GREMATH
John	520	500
Sally	610	640
Miguel	490	470
Mats	550	560

Imagine that later you create a second dataset called B that contains data from four different participants: Susan; Jiri; Cheri; and Zdeno. Values for these participants for GREVERBAL and GREMATH appear in Table 4.2.

Table 4.2

Contents of Dataset B

NAME	GREVERBAL	GREMATH
Susan	710	650
Jiri	450	400
Cheri	570	600
Zdeno	680	700

Assume that you would like to perform some analyses on a single dataset that contains scores from all eight of these participants. But you encounter a problem; the values in dataset A were entered differently from the values of dataset B, making it impossible to read data from both sets with a single INPUT statement. For example, perhaps you entered GREVERBAL in columns 10 to 12 in dataset A, but entered it in columns 11 to 13 in dataset B. Because the variable was entered in different columns in the two datasets, it is not possible to write a single INPUT statement that will input this variable (assuming that you use a formatted input approach).

One way to deal with this problem is to input A and B as separate datasets and then concatenate them to create a single dataset that contains all eight observations. You can then perform analyses on the new dataset. The following is the general form for concatenating multiple datasets into a single dataset:

```
DATA   new-dataset-name;
      SET   dataset-1   dataset-2 ... dataset-n;
```

In the present situation, you want to concatenate two datasets (A and B) to create a new dataset named C. This could be done in the following statements:

```
DATA C;
    SET A B;
```

The entire program follows that places these statements in context. This program

- inputs dataset A;
- inputs dataset B;
- concatenates A and B to create C;
- uses PROC PRINT to print the contents of dataset C. (PROC PRINT is discussed in greater detail in Chapter 5, "Exploring Data with PROC MEANS, PROC FREQ, PROC PRINT, and PROC UNIVARIATE.")

```
1        DATA  A;
2           INPUT  #1    @1    NAME       $7.
3                        @10   GREVERBAL  3.
4                        @14   GREMATH    3.   ;
5
6        DATALINES;
7        John      520 500
8        Sally     610 640
9        Miguel    490 470
10       Mats      550 560
11       ;
12       RUN;
13
14       DATA  B;
15          INPUT  #1    @1    NAME       $7.
16                       @11   GREVERBAL  3.
17                       @15   GREMATH    3.   ;
18
19       DATALINES;
20       Susan      710 650
21       Jiri       450 400
22       Cheri      570 600
23       Zdeno      680 700
24       ;
25       RUN;
26
27       DATA  C;
28          SET   A   B;
29
30       PROC PRINT   DATA=C
31       RUN;
```

In the preceding program, dataset A is input in program lines 1 through 12, and dataset B is input in lines 14 through 25. In lines 27 and 28, the two datasets are concatenated to create dataset C. In lines 30 and 31, PROC PRINT is used to print the contents of dataset C, and the results of this procedure are reproduced as Output 4.1. The results of Output 4.1 show that dataset C contains eight observations: the four observations from dataset A along with the four observations from dataset B. To perform additional statistical analyses on this combined dataset, you would specify DATA=C in the PROC statement of your SAS program.

Output 4.1 Results of Performing PROC PRINT on Dataset C

```
           OBS    NAME     GREVERBAL    GREMATH

            1     John        520         500
            2     Sally       610         640
            3     Miguel      490         470
            4     Mats        550         560
            5     Susan       710         650
            6     Jiri        450         400
            7     Cheri       570         600
            8     Zdeno       680         700
```

Merging Datasets

As stated earlier, you would normally merge datasets when

- you are working with two datasets;
- both datasets contain information for the same participants, but one dataset contains one set of variables, while the other dataset contains a different set of variables.

Once these two datasets are merged, you have a single dataset that contains all variables. Having all dataset variables in one dataset allows you to assess the associations among variables, should you want to do so.

As an illustration, assume that your sample consists of just four participants: John; Sally; Miguel; and Mats. Assume that you have obtained the social security number for each participant, and that these numbers are included as a SAS variable named SOCSEC in both previously existing datasets. In dataset D, you have GRE verbal test scores and GRE math test scores for these participants (represented as SAS variables GREVERBAL and GREMATH, respectively). In dataset E, you have college cumulative grade point average for the same four participants (represented as GPA). Table 4.3 and Table 4.4 show the content of these two datasets.

Table 4.3

Contents of Dataset D

NAME	SOCSEC	GREVERBAL	GREMATH
John	232882121	520	500
Sally	222773454	610	640
Miguel	211447653	490	470
Mats	222671234	550	560

Table 4.4

Contents of Dataset E

NAME	SOCSEC	GPA
John	232882121	2.70
Sally	222773454	3.25
Miguel	211447653	2.20
Mats	222671234	2.50

Assume that, in conducting your research, you would like to compute the correlation coefficient between GREVERBAL and GPA. (Let's forget for the moment that you really shouldn't perform a correlation using such a small sample!) Computing this correlation should be possible because you do have values for these two variables for all four of your participants. However, you cannot compute this correlation until both variables appear in the same dataset. Therefore, it is necessary to merge the variables contained in datasets D and E.

There are actually two ways of merging datasets. Perhaps the simplest way is the one-to-one approach. With **one-to-one merging**, observations are merged according to their order of appearance in the datasets. For example, imagine that you were to merge datasets D and E using one-to-one merging. In doing this, SAS would take the first observation from dataset D and pair it with the first observation from dataset E. The result would become the first observation in the new dataset (dataset F). If the observations in datasets D and E were in exactly the same sequence, this method would work fine. Unfortunately, if any of the observations were out of sequence, or if one dataset contained more observations than another, then this approach could result in the incorrect pairing of observations. For this reason, we recommend a different strategy for merging: the match-merging approach next described.

Match-merging seems to be the method that is least likely to produce undesirable results and errors. With **match-merging**, both datasets must contain a common variable, so that values for this common variable can be used to combine observations from the two previously existing datasets into observations for the new dataset (often the participant identification number). For example, consider datasets D and E from Table 4.3 and Table 4.4. The variable SOCSEC appears in both of these datasets; thus, it is a common variable. When SAS uses match-merging to merge these two datasets according to values on SOCSEC, it will:

- read the social security number for the first participant in dataset D;

- look for a participant in dataset E who has the same social security number;

- merge the information from that participant's observation in dataset D with his or her information from dataset E (if it finds a participant in dataset E with the same social security number);

- combine the information into a single observation in the new dataset, F;

- repeat this process for all participants.

As the preceding description suggests, the variable that you use as your common variable must be chosen carefully. Ideally, each participant should be assigned a *unique value* for this common variable. This means that no two participants should have the same value for the common variable. This objective would be achieved when social security numbers are used as the common variable, because no two people have the same social security number (assuming that the data are entered correctly).

The SAS procedure for match-merging datasets is somewhat more complex than for concatenating datasets. In part, this is because both previously existing datasets must be sorted according to values for the common variable prior to merging. This means that the observations must be rearranged in a consistent order with respect to values for the common variable. Fortunately, this is easy to do with PROC SORT, a SAS procedure that allows you to sort variables. This section shows how PROC SORT can be used to achieve this.

The general form for match-merging two previously existing datasets is presented as follows:

```
PROC SORT  DATA=dataset-1;
   BY  common-variable;
RUN;

PROC SORT  DATA=dataset-2;
   BY  common-variable;
RUN;

DATA  new-dataset-name;
   MERGE  dataset-1  dataset-2;
   BY  common-variable;
RUN;
```

To illustrate, assume that you want to match-merge datasets D and E from Table 4.3 and Table 4.4; to do this, use SOCSEC as the common variable. In the following program, these two datasets are entered, sorted, and then merged using the match-merge approach:

```
1        DATA   D;
2           INPUT   #1    @1    NAME         $9.
3                         @10  SOCSEC        9.
4                         @20  GREVERBAL     4.
5                         @23  GREMATH       4.   ;
6
7        DATALINES;
8        John       232882121 520 500
9        Sally      222773454 610 640
10       Miguel     211447653 490 470
11       Mats       222671234 550 560
12       ;
13       RUN;
14
15
16       DATA   E;
17          INPUT   #1    @1    NAME         $9.
18                        @10   SOCSEC       9.
19                        @20   GPA          4.   ;
20
21       DATALINES;
22       John       232882121 2.70
23       Sally      222773454 3.25
24       Miguel     211447653 2.20
25       Mats       222671234 2.50
26       ;
27       RUN;
28
29       PROC SORT   DATA=D;
30          BY   SOCSEC;
31       RUN;
32
33       PROC SORT   DATA=E;
34          BY   SOCSEC;
35       RUN;
36
37       DATA   F;
38          MERGE   D   E;
39          BY   SOCSEC;
40       RUN;
41
42       PROC PRINT   DATA=F;
43       RUN;
```

In the preceding program, dataset D was input in lines 1 to 11, and dataset E was input in lines 16 through 27. In lines 29 through 35, both datasets were sorted according to values for SOCSEC, and the two datasets were merged according to values of SOCSEC in lines 37 through 40. Finally, the PROC PRINT on lines 42 and 43 requests a printout of the raw data contained in the new dataset.

Output 4.2 contains the results of PROC PRINT, which printed the raw data now contained in dataset F. You can see that each observation in this new dataset now contains the merged data from the two previous datasets D and E. For example, the line for the participant named Miguel now contains his scores on the verbal and math sections of the GRE (which came from dataset D), as well as his grade point average score (which came from dataset E). The same is true for the remaining participants. It now is possible to correlate GREVERBAL with GPA, if that analysis were desired.

Output 4.2 Results of Performing PROC PRINT on Dataset F

```
                              The SAS System

          Obs     NAME      SOCSEC      GREVERBAL     GREMATH      GPA
           1     Miguel    211447653       490          470       2.20
           2     Mats      222671234       550          560       2.50
           3     Sally     222773454       610          640       3.25
           4     John      232882121       520          500       2.70
```

Notice that the observations in Output 4.2 are not in the same order in which they appeared in Tables 4.3 and 4.4. This is because they have now been sorted according to values for SOCSEC by the PROC SORT statements in the preceding SAS program.

Conclusion

After completing this chapter, you should be prepared to modify datasets, isolate subgroups of participants for analysis, and perform other tasks that are often required when performing quantitative research in the social sciences. At this point, you should be prepared to proceed to the stage of analyzing data to determine what they mean. Some of the most basic statistics for this purpose (descriptive statistics and related procedures) are covered in the following chapter.

Exploring Data with
PROC MEANS, PROC FREQ,
PROC PRINT, and
PROC UNIVARIATE

> **Overview.** This chapter illustrates the use of four procedures: PROC MEANS, used to calculate means, standard deviations, and other descriptive statistics for quantitative variables; PROC FREQ, used to construct frequency distributions; PROC PRINT, used to create a printout of the raw dataset; and PROC UNIVARIATE, used to test for normality and produce stem-and-leaf plots. Once data are entered, these procedures can be used to screen for errors, test statistical assumptions, and obtain simple descriptive statistics.

Introduction: Why Perform Simple Descriptive Analyses?

The procedures discussed in this chapter are useful for (at least) three important purposes. The first involves the concept of data screening. **Data screening** is the process of carefully reviewing data to ensure that they were entered correctly and are being read correctly by the computer. Before conducting any of the more sophisticated analyses to be described in this text, you should carefully screen your data to avoid computational errors (e.g., numbers that were accidentally entered, out-of-range values, numbers that were entered in the wrong column). The process of data screening does not guarantee that your data are correct, but it does increase the likelihood by identifying and fixing obvious errors.

Second, these procedures are useful because they allow you to explore the shape of your data distribution. Among other things, understanding the shape of your data helps you choose the appropriate measure of central tendency (i.e., the mean, mode, or median). In addition, many statistical procedures require that sample data are normally distributed, or at least that the sample data do not display a *marked* departure from normality. You can use the procedures discussed herein to produce graphic plots of the data, as well as test the null hypothesis that the data are from a normal population.

Finally, the nature of the research question itself might require use of a procedure such as PROC MEANS or PROC FREQ to obtain a desired statistic. For example, if your research question is "What is the average age at which women married in 1991?" you could obtain data from a representative sample of women who married in that year, analyze their ages with PROC MEANS, and review the results to determine the mean age.

Similarly, in almost any research study it is desirable to report demographic information about the sample. For example, if a study is performed on a sample that includes participants from a variety of demographic groups, it is desirable to report the percent of participants of each sex, the percent of participants by race, the mean age, and so forth. You can also use PROC MEANS and PROC FREQ to obtain this information.

Example: An Abridged Volunteerism Survey

To help illustrate these procedures, assume that you conduct a scaled-down version of your study on volunteerism. You construct a new questionnaire which asks just one question related to helping behavior. The questionnaire also contains an item that assesses participant sex, and another that determines each participant's class in college (e.g., freshman, sophomore). See the questionnaire below:

```
Please indicate the extent to which you agree or disagree with the
following statement:

1.  I feel a personal responsibility to help needy people in my
    community. (please check your response below)

      (5) _____ Agree Strongly
      (4) _____ Agree Somewhat
      (3) _____ Neither Agree nor Disagree
      (2) _____ Disagree Somewhat
      (1) _____ Disagree Strongly

2.  Your sex (please check one):

      (F) _____ Female
      (M) _____ Male

3.  Your classification as a college student:

      (1) _____ Freshman
      (2) _____ Sophomore
      (3) _____ Junior
      (4) _____ Senior
      (5) _____ Other
```

Notice that this instrument is printed so that data entry will be relatively simple. With each variable, the value that will be entered appears to the left of the corresponding participant response. For example, with question 1 the value "5" appears to the left of "Agree Strongly." This means that the number "5" will be entered for any participant checking that response. For participants checking "Disagree Strongly," a "1" will be entered. Similarly, notice that, for question 2, the letter "F" appears to the left of "Female," so an "F" will be entered for participants checking this response.

The following format is used when entering the data:

COLUMN	VARIABLE NAME	EXPLANATION
1	RESNEEDY	Responses to question 1: Participant's perceived responsibility to help the needy
2	blank	
3	SEX	Responses to question 2: Participant's sex
4	blank	
5	CLASS	Responses to question 3: Participant's classification as a college student

You administer the questionnaire to 14 students. The following is the entire SAS program used to analyze the data, including the raw data:

```
1       DATA D1;
2          INPUT   #1   @1   RESNEEDY   1.
3                            @3   SEX        $1.
4                            @5   CLASS      1.    ;
5       DATALINES;
6       5 F 1
7       4 M 1
8       5 F 1
9         F 1
10      4 F 1
11      4 F 2
12      1 F 2
13      4 F 2
14      1 F 3
15      5 M
16      4 F 4
17      4 M 4
18      3 F
19      4 F 5
20      ;
21      RUN;
22
23      PROC MEANS    DATA=D1;
24         VAR RESNEEDY CLASS;
25      RUN;
26      PROC FREQ    DATA=D1;
27         TABLES SEX CLASS RESNEEDY;
28      RUN;
29      PROC PRINT    DATA=D1;
30         VAR RESNEEDY SEX CLASS;
31      RUN;
```

The data obtained from the first participant appears on line 6 of the preceding program. This participant has a value of "5" on the RESNEEDY variable (indicating that she checked "Agree Strongly"), has a value of "F" on the SEX variable (indicating that she is a female), and has a value of "1" on the CLASS variable (indicating that she is a freshman).

Notice that there are some missing data in this dataset. On line 9 in the program, you can see that this participant indicated that she was a female freshman, but did not answer question 1. That is why the corresponding space in column 1 is left blank. In addition, there appears to be missing data for the CLASS variable on lines 15 and 18. Unfortunately, missing data are common in questionnaire research.

Computing Descriptive Statistics with PROC MEANS

You can use PROC MEANS to analyze quantitative (numeric) variables. For each variable analyzed, it provides the following information:

- the number of observations on which calculations were performed (abbreviated "N" in the output);
- the mean;
- the standard deviation;
- the minimum (smallest) value observed;
- the maximum (largest) value observed.

These statistics are produced by default, and some additional statistics (to be described later) can also be requested as options.

Here is the general form for PROC MEANS:

```
PROC MEANS   DATA=dataset-name
             option-list
             statistic-keyword-list ;
   VAR   variable-list  ;
RUN;
```

The PROC MEANS Statement

The PROC MEANS statement begins with "PROC MEANS" and ends with a semicolon. It is recommended (on some platforms, it is required) that the statement should also specify the name of the dataset to be analyzed with the DATA= option.

The "option-list" appearing in the preceding program indicates that you can request a number of options with PROC MEANS. A complete list of these options appears in the *SAS/STAT User's Guide*. Some options especially useful for social science research are:

MAXDEC=N
> Specifies the maximum number of decimal places (digits to the right of the decimal point) to be used when printing results; the possible range is 0 to 8.

VARDEF=divisor
> Specifies the divisor to be used when calculating variances and covariances. Following are two possible divisors:

VARDEF=DF	Divisor is the degrees of freedom for the analysis: (n–1). This is the default.
VARDEF=N	Divisor is the number of observations, n.

The "statistic-keyword-list" appearing in the program indicates that you can request a number of statistics to replace the default output. Some statistics that can be of particular value in social science research include the following. See the *SAS/STAT User's Guide* for a more complete listing:

NMISS	The number of observations in the sample that displayed missing data for this variable.
RANGE	The range of values displayed in the sample.
SUM	The sum.
CSS	The corrected sum of squares.
USS	The uncorrected sum of squares.
VAR	The variance.
STDERR	The standard error of the mean.
SKEWNESS	The skewness displayed by the sample. **Skewness** refers to the extent to which the sample distribution departs from the normal curve because of a long "tail" on either side of the distribution. If the long tail appears on the right side of the sample distribution (where the higher values appear), it is described as being **positively skewed**. If the long tail appears on the left side of the distribution (where the lower values appear), it is described as being **negatively skewed.**
KURTOSIS	The kurtosis displayed by the sample. **Kurtosis** refers to the extent to which the sample distribution departs from the normal curve because it is either peaked or flat. If the sample distribution is relatively peaked (tall and skinny), it is described as being **leptokurtic**. If the distribution is relatively flat, it is described as being **platykurtic**.
T	The obtained value of Student's *t* test for testing the null hypothesis that the population mean is 0.
PRT	The *p* value for the preceding *t* test; that is, the probability of obtaining a *t* value this large or larger if the population mean were 0.

To illustrate the use of these options and statistic keywords, assume that you want to use the MAXDEC option to limit the printing of results to two decimal places, use the VAR keyword to request that the variances of all quantitative variables be printed, and use the

KURTOSIS keyword to request that the kurtosis of all quantitative variables be printed. You could do this with the following PROC MEANS statement:

```
PROC MEANS    DATA=D1    MAXDEC=2    VAR    KURTOSIS ;
```

The VAR Statement

Here again is the general form of the statements requesting the MEANS procedure, including the VAR statement:

```
PROC MEANS   DATA=dataset-name
             option-list
             statistic-keyword-list ;
   VAR   variable-list  ;
RUN;
```

In the place of "variable-list" in the preceding VAR statement, you can list the quantitative variables to be analyzed. Each variable name should be separated by at least one blank space. If no VAR statement is used, SAS performs PROC MEANS on all of the quantitative variables in the dataset. This is true for many other SAS procedures as well, as explained in the following note:

> **What happens if I do not include a VAR statement?** For many SAS procedures, failure to include a VAR statement causes the system to perform the requested analyses on *all* variables in the dataset. For datasets with a large number of variables, leaving off the VAR statement can unintentionally result in a very long output file.

The program used to analyze your dataset includes the following statements. RESNEEDY and CLASS are specified in the VAR statement so that descriptive statistics would be calculated for both variables:

```
PROC MEANS    DATA=D1;
   VAR RESNEEDY CLASS;
RUN;
```

Output 5.1 Results of the MEANS Procedure

Variable	N	Mean	Std Dev	Minimum	Maximum
RESNEEDY	13	3.6923077	1.3155870	1.0000000	5.0000000
CLASS	12	2.2500000	1.4222262	1.0000000	5.0000000

Reviewing the Output

Output 5.1 contains the results created by the preceding program. Before doing any more sophisticated analyses, you should always perform PROC MEANS on each quantitative variable and carefully review the output to ensure that everything looks right. Under the heading "Variable" is the name of each variable being analyzed. The descriptive statistics for that variable appear to the right of the variable name. Below the heading "N" is the number of valid cases, or observations, on which calculations were performed. Notice that in this instance, calculations were performed on only 13 observations for RESNEEDY. This might come as a surprise, because the dataset actually contains 14 cases. Recall, however, that one participant did not respond to this question (question 1 on the survey). It is for this reason that N is equal to 13 rather than 14 for RESNEEDY in these analyses.

You should next examine the mean for the variable, to verify that it is a reasonable number. Remember that, with question 1, responses could range from 1 "Disagree Strongly" to 5 "Agree Strongly." Therefore, the mean response should be somewhere between 1.00 and 5.00 for the RESNEEDY variable. If it is outside of this range, you know that an error has been made. In the present case, the mean for RESNEEDY is 3.69, which is within the predetermined range, so everything appears correct so far.

Using the same reasoning, it is prudent to next check the column headed "Minimum." Here, you will find the lowest value on RESNEEDY that appeared in the dataset. If this is less than 1.00, you will again know that an error was made, because 1 was the lowest value that could have been assigned to a participant. On the printout, the minimum value is 1.00, which indicates no problems. Under "Maximum," the largest value observed for that variable is reported. This should not exceed 5.00, because 5 was the largest score a participant could obtain on item 1. The reported maximum value is 5.00, so again it appears that there were no obvious errors in entering the data or writing the program.

Once you have reviewed the results for RESNEEDY, you should also inspect the results for CLASS. If any of the observed values are out of range, you should carefully review the program for programming errors, and the dataset for data entry error. In some cases, you might want to use PROC PRINT to print out the raw dataset because this makes the review easier. PROC PRINT is described later in this chapter.

Creating Frequency Tables with PROC FREQ

The FREQ procedure produces frequency distributions for quantitative variables as well as classification variables. For example, you can use PROC FREQ to determine the percentage of participants who "agreed strongly" with a statement on a questionnaire, the percentage who "agreed somewhat," and so forth.

The PROC FREQ and TABLES Statements

The general form for the procedure is as follows:

```
PROC FREQ   DATA=dataset-name;
    TABLES  variable-list  /  options;
RUN;
```

In the TABLES statement, you list the names of the variables to be analyzed, with each name separated by at least one blank space. Below are the PROC FREQ and TABLES statements from the program presented earlier in this chapter (analyzing data from the volunteerism survey):

```
PROC FREQ   DATA=D1;
    TABLES SEX CLASS RESNEEDY;
RUN;
```

Reviewing the Output

These statements will cause SAS to create three frequency distributions: one for the SEX variable; one for CLASS; and one for RESNEEDY. This output appears in Output 5.2.

Output 5.2 Results of the FREQ Procedure

```
                         The FREQ Procedure

                                      Cumulative    Cumulative
      SEX     Frequency    Percent    Frequency      Percent
      --------------------------------------------------------
      F           11        78.57        11          78.57
      M            3        21.43        14         100.00


                                      Cumulative    Cumulative
     CLASS    Frequency    Percent    Frequency      Percent
     ---------------------------------------------------------
       .           2        14.29         2          14.29
       1           5        35.71         7          50.00
       2           3        21.43        10          71.43
       3           1         7.14        11          78.57
       4           2        14.29        13          92.86
       5           1         7.14        14         100.00


                                      Cumulative    Cumulative
   RESNEEDY   Frequency    Percent    Frequency      Percent
   -----------------------------------------------------------
       .           1         7.14         1           7.14
       1           2        14.29         3          21.43
       3           1         7.14         4          28.57
       4           7        50.00        11          78.57
       5           3        21.43        14         100.00
```

Output 5.2 shows that the name for the variable being analyzed appears on the far left side of the frequency distribution, just above the dotted line. The values assumed by the variable appear below this variable name. The first distribution provides information about the SEX variable; below the word "SEX" appear the values "F" and "M." Information about female participants appears to the right of "F," and information about males appears to the right of "M." When reviewing a frequency distribution, it is useful to think of these different values as representing categories to which participants can belong.

Under the heading "Frequency," the output indicates the number of individuals in a given category. Here, you can see that 11 participants were female while 3 were male. Below "Percent," the percent of participants in each category appears. The table shows that 78.57% of the participants were female while 21.43% were male.

Under "Cumulative Frequency" is the number of observations that appear in the current category plus all of the preceding categories. For example, the first (top) category for SEX is "female." There were 11 participants in that category so the cumulative frequency is 11. The next category is "male," and there are 3 participants in that category. The cumulative frequency for the "male" category is therefore 14 (because 11 + 3 = 14). In the same way, the "Cumulative Percent" category provides the percent of observations in the current category plus all of the preceding categories.

The next table presents results for the CLASS variable. Notice that the first line of this table begins with a period (.) in the left-hand column (row 1). This indicates that there are missing data for two participants on the CLASS variable, a fact that can be verified by reviewing the data as they appear in the preceding program: CLASS values are blank for two participants.

If no participant appears in a given category, the value representing that category does not appear in the frequency distribution at all. This is demonstrated with the third table, which presents the frequency distribution for the RESNEEDY variable. Notice that, under the "RESNEEDY" heading, you can find only the values "1," "3," "4," and "5." The value "2" does not appear because none of the participants checked "Disagree Somewhat" for question 1.

Printing Raw Data with PROC PRINT

PROC PRINT can be used to create a printout of your raw data as they exist in the computer's internal memory. The output of PROC PRINT shows each participant's value on each of the requested variables. You can use this procedure with both quantitative variables and classification variables. The general form is:

```
PROC PRINT    DATA=dataset-name;
   VAR  variable-list  ;
RUN;
```

In the variable list, you can request any variable that is specified in the INPUT statement, as well as any new variable that is created from existing variables. If you do not include the VAR statement, then all existing variables will be printed. The program presented earlier in this chapter included the following PROC PRINT statements:

```
PROC PRINT    DATA=D1;
   VAR RESNEEDY SEX CLASS;
   RUN;
```

These statements produce Output 5.3.

Output 5.3 Results of the PRINT Procedure

Obs	RESNEEDY	SEX	CLASS
1	5	F	1
2	4	M	1
3	5	F	1
4	.	F	1
5	4	F	1
6	4	F	2
7	1	F	2
8	4	F	2
9	1	F	3
10	5	M	.
11	4	F	4
12	4	M	4
13	3	F	.
14	4	F	5

The first column of output is headed "Obs" for "observation." This variable is created by SAS to give an observation number to each participant. The second column provides the raw data for the RESNEEDY variable, the third column displays the SEX variable, and the last displays the CLASS variable. The output shows that observation 1 (participant 1) provided a value of 5 on RESNEEDY, was a female, and provided a value of 1 on the CLASS variable. Notice that SAS prints periods where there are missing values.

PROC PRINT is helpful for verifying that your data are entered correctly and that SAS is reading the data correctly. It is particularly useful for studies with a large number of variables; for instance, when you use a questionnaire with a large number of questions. In these situations, it is often difficult to visually inspect the data as they exist in the SAS program file. In the SAS program file, participant responses are often entered immediately adjacent to each other, making it difficult to determine just which number represents question 24 as opposed to question 25. When the data are printed using PROC PRINT, however, the variables are separated from each other and are clearly labeled with their variable names.

After entering questionnaire data, you should compare the results of PROC PRINT with several of the questionnaires as they were actually completed by participants. Verify that participants' responses on the PROC PRINT output correspond to the original responses on the questionnaire. If not, it is likely that mistakes were made in either entering the data or in writing the SAS program.

Testing for Normality with PROC UNIVARIATE

The normal distribution is a symmetrical, bell-shaped distribution of values. The shape of the normal distribution is shown in Figure 5.1.

Figure 5.1 The Normal Distribution

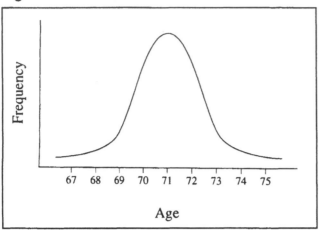

To understand the distribution in Figure 5.1, assume that you are interested in conducting research on people who live in retirement communities. Imagine for a moment that it is possible to assess the age of every person in this population. To summarize this distribution, you prepare a figure similar to Figure 5.1: the variable AGE is plotted on the horizontal axis, and the frequency of persons at each age is plotted on the vertical axis. Figure 5.1 suggests that many of your participants are around 71 years of age, since the distributions of ages "peaks" near the age of 71. This suggests that the mean of this distribution will likely be somewhere around 71. Notice also that most of your participants' ages are between 67 (near the lower end of the distribution) and 75 (near the upper end of the distribution). This is the approximate range of ages that we would expect for persons living in a retirement community.

Why Test for Normality?

Normality is an important concept in quantitative analyses because there are at least two problems that can result when data are not normally distributed. The first problem is that markedly non-normal data can lead to incorrect conclusions in inferential statistical analyses. Many inferential procedures are based on the assumption that the sample of observations is normally distributed. If this assumption is violated, the statistic might give misleading findings. For example, the independent groups *t* test assumes that both samples in the study were drawn from normally distributed populations. If this assumption is violated, then performing the analysis can cause you to incorrectly reject the null hypothesis (or incorrectly fail to reject the null hypothesis). Under these circumstances, you should instead analyze the data using a procedure that does not assume normality (e.g., some nonparametric procedure).

The second problem is that markedly non-normal data can have a biasing effect on correlation coefficients, as well as more sophisticated procedures that are based on correlation coefficients. For example, assume that you compute the Pearson correlation coefficient between two variables. If one or both of these variables are markedly non-normal, this can cause your obtained coefficient to be much larger (or much smaller) than the actual correlation between these variables in the population. Your obtained correlation is essentially misleading. To make matters worse, many sophisticated data analysis procedures (such as principal component analysis) are actually performed on a matrix of correlation coefficients. If some or all of these correlations are distorted due to departures from normality, then the results of the analyses can again be misleading. For this reason, many experts recommend that researchers routinely check their data for major departures from normality prior to performing sophisticated analyses such as principal component analysis (e.g., Rummel, 1970).

Departures from Normality

Assume that you draw a random sample of 18 participants from your population of persons living in retirement communities. There are several ways that your data can display a departure from normality.

Figure 5.2 shows the distribution of ages in two samples of participants drawn from the population of retirees. This figure is somewhat different from Figure 5.1 because the distributions have been "turned on their sides" so that age is now plotted on the vertical axis rather than on the horizontal axis. (This is so that these figures will be more similar to the stem-and-leaf plots produced by PROC UNIVARIATE, discussed in a later section.) Each small circle in Figure 5.2 represents one participant in a given distribution. For example, in the distribution for Sample A, you can see that there is one participant at age 75, one at age 74, two at age 73, three at age 72, and so forth. The ages of the 18 participants in Sample A range from a low of 67 to a high of 75.

Figure 5.2 Sample with an Approximately Normal Distribution and a Sample with an Outlier

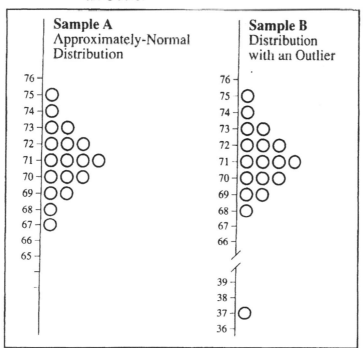

The data in Sample A form an approximately normal distribution (called *approximately normal* because it is difficult to form a perfectly normal distribution using a small sample of just 18 cases). An inferential test (discussed later) will show that Sample A does not demonstrate a significant departure from normality. Therefore, it probably is appropriate to include the data in Sample A in an independent samples *t* test, for example.

In contrast, there are problems with the data in Sample B. Notice that its distribution is very similar to that of Sample A, except that there is an outlier at the lower end of the distribution. An **outlier** is an extreme value that differs substantially from the other values in the distribution. In this case, the outlier represents a participant whose age is only 37. Obviously, this person's age is markedly different from that of the other participants in your study. Later, you will see that this outlier causes the dataset to demonstrate a significant departure from normality, making the data inappropriate for some statistical procedures. When you observe an outlier such as this, it is important to determine whether it should be either corrected or simply deleted from the dataset. Obviously, if the outlier exists because an error was made in entering the data, it should be corrected.

A sample can also depart from normality because it displays kurtosis. **Kurtosis** refers to the peakedness of the distribution. The two samples displayed in Figure 5.3 demonstrate different types of kurtosis:

Figure 5.3 Samples Displaying Positive versus Negative Kurtosis

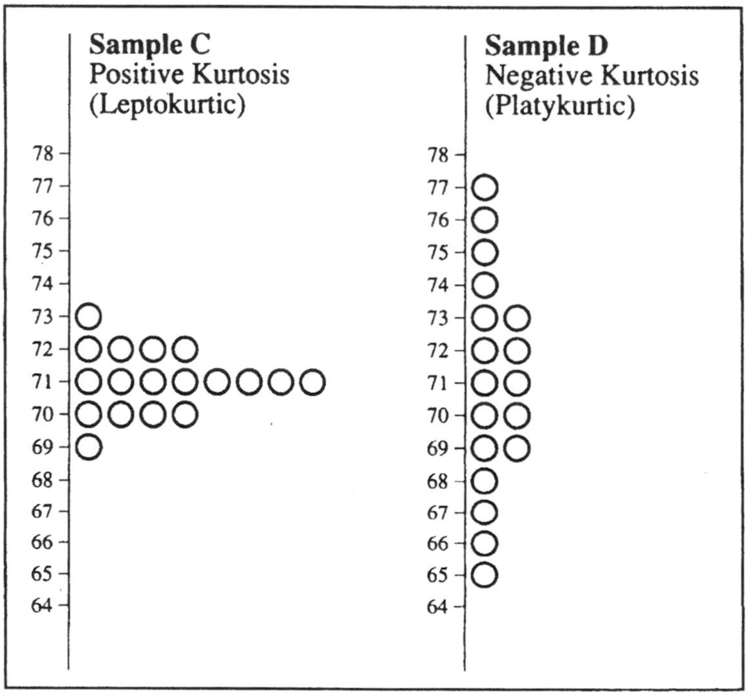

Sample C in Figure 5.3 displays **positive kurtosis**, which means that the distribution is relatively peaked (tall and skinny) rather than flat. Notice that, with Sample C, there are a relatively large number of participants who cluster around the central part of the distribution (around age 71). This is what makes the distribution peaked (relative to Sample A, for example). Distributions with positive kurtosis are also called **leptokurtic**. A mnemonic device to remember the meaning of this word is to think of the distribution *leaping* upward (i.e., a leptokurtic distribution leaped up).

In contrast, Sample D in the same figure displays **negative kurtosis**, which means that the distribution is relatively flat. Flat distributions are described as being **platykurtic**. A mnemonic device to remember the meaning of this word is to think of the distribution as flat as a *plate*.

In addition to kurtosis, distributions can also demonstrate varying degrees of **skewness**, or sidedness. A distribution is skewed if the tail on one side of the distribution is longer than the tail on the other side. The distributions in Figure 5.4 show two types of skewness:

Figure 5.4 Samples Displaying Positive versus Negative Skewness

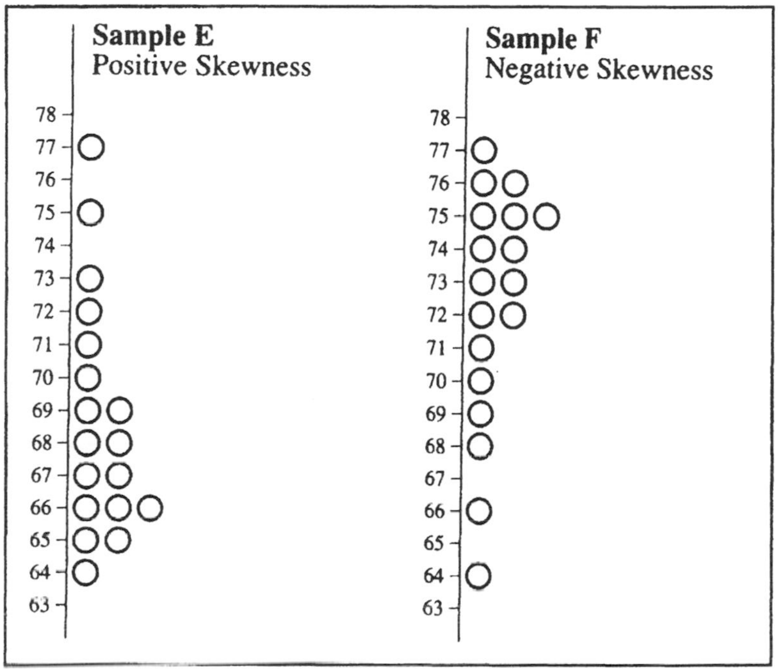

Consider Sample E in Figure 5.4. Notice that the largest number of participants in this distribution tends to cluster around the age of 66. The tail of the distribution that stretches above 66 (from 67 to 77) is relatively long, while the tail of the distribution that stretches below 66 (from 65 to 64) is relatively short. Clearly, this distribution is skewed. A distribution is said to be **positively skewed** if the longer tail of a distribution points in the direction of *higher* values. You can see that Sample E displays positive skewness, because its longer tail points toward larger numbers such as 75, 77, and so forth.

On the other hand, if the longer tail of a distribution points in the direction of lower values, the distribution is said to be **negatively skewed**. You can see that Sample F in Figure 5.4 displays negative skewness because in that sample the longer tail points downward, in the direction of lower values (such as 66 and 64).

General Form for PROC UNIVARIATE

Like the MEANS procedure, PROC UNIVARIATE provides a number of descriptive statistics for quantitative variables, including the mean, standard deviation, kurtosis, and skewness. However, PROC UNIVARIATE has the added advantage of printing a significance test for the null hypothesis that the data come from a normally distributed population. The procedure also provides plots that will help you understand the shape of your sample's distribution, along with additional information that will help you understand *why* your data depart from normality (if, indeed, they do). This text describes just a few of the features of PROC UNIVARIATE. See the *SAS/STAT User's Guide* for a more complete listing.

Here is the general form for the PROC UNIVARIATE statements that produce the output discussed in this chapter:

```
PROC UNIVARIATE   DATA=dataset-name   NORMAL   PLOT;
    VAR variable-list;
    ID identification-variable;
RUN;
```

In the preceding program, the NORMAL option requests a significance test for the null hypothesis that the sample data are from a normally distributed population. The Shapiro-Wilk statistic is printed for samples of 2000 or less; for larger samples, the Kolmogorov statistic is printed. See the *SAS/STAT User's Guide* for details.

The PLOT option in the preceding program produces a stem-and-leaf plot, a box plot, and a normal probability plot, each of which is useful for understanding the shape of the sample's distribution. This book shows how to interpret the stem-and-leaf plot.

The names of the variables to be analyzed should be listed in the VAR statement. The ID statement is optional but is useful (and recommended) for identifying outliers. PROC UNIVARIATE prints an "extremes" table that lists the five largest and five smallest values in the dataset. These values are identified by the identification variable listed in the ID statement. For example, assume that AGE (participant age) is listed in the VAR statement, and SOCSECURITY (for participant Social Security number) is listed in the ID statement. PROC UNIVARIATE will print the Social Security numbers for the participants with the five largest and five smallest values on AGE. This should make it easier to identify the specific participant who represents an outlier in your dataset. (This use of the extremes table is illustrated here.)

Results for an Approximately Normal Distribution

For purposes of illustration, assume that you want to analyze the data that are illustrated as Sample A of Figure 5.2 (the approximately normal distribution). You prepare a SAS program in which participant age is entered as a variable called AGE, and participant identification numbers are entered as a variable called PARTICIPANT. Here is the entire program that will input these data and analyze them using PROC UNIVARIATE:

```
1            DATA D1;
2               INPUT  #1   @1   PARTICIPANT  2.
3                           @4   AGE          2.   ;
4            DATALINES;
5             1 72
6             2 69
7             3 75
8             4 71
9             5 71
10            6 73
11            7 70
12            8 67
13            9 71
14           10 72
15           11 73
16           12 68
```

```
17              13  69
18              14  70
19              15  70
20              16  71
21              17  74
22              18  72
23              ;
24              RUN;
25              PROC UNIVARIATE    DATA=D1    NORMAL    PLOT;
26                 VAR AGE;
27                 ID PARTICIPANT;
28              RUN;
```

The preceding program requests that PROC UNIVARIATE be performed on the variable AGE. Values of the variable PARTICIPANT are used to identify outlying values of AGE in the extremes table.

This output would contain the following:

- a **moments table** that includes the mean, standard deviation, variance, skewness, kurtosis, and other statistics;

- a table of **basic statistical measures** that provide indices of central tendency and variability estimates;

- tests of normality such as the Shapiro-Wilk statistic;

- a **quantiles table** that provides the median, 25th percentile, 75th percentile, and related information;

- **extreme observations** that provides the five highest values and five lowest values on the variable being analyzed;

- a stem-and-leaf plot, box plot, and normal probability plot.

Output 5.4 includes the Moments table, basic statistical measures, tests for normality, quantiles table, extremes table, and a stem-and-leaf plot for Sample A.

Output 5.4 Tables from PROC UNIVARIATE for Sample A

```
                       The UNIVARIATE Procedure
                          Variable:  AGE

                             Moments

N                          18     Sum Weights                   18
Mean                       71     Sum Observations            1278
Std Deviation      2.05798302     Variance              4.23529412
Skewness                    0     Kurtosis              -0.1357639
Uncorrected SS          90810     Corrected SS                  72
Coeff Variation    2.89856764     Std Error Mean        0.48507125


                    Basic Statistical Measures

            Location                      Variability

      Mean      71.00000     Std Deviation            2.05798
      Median    71.00000     Variance                 4.23529
      Mode      71.00000     Range                    8.00000
                             Interquartile Range      2.00000


                       Tests for Normality

      Test                   --Statistic---    -----p Value------

      Shapiro-Wilk           W    0.983895     Pr < W      0.9812
      Kolmogorov-Smirnov     D    0.111111     Pr > D     >0.1500
      Cramer-von Mises       W-Sq 0.036122     Pr > W-Sq  >0.2500
      Anderson-Darling       A-Sq 0.196144     Pr > A-Sq  >0.2500


                     Quantiles (Definition 5)

            Quantile        Estimate

            100% Max           75
            99%                75
            95%                75
            90%                74
            75% Q3             72
            50% Median         71
            25% Q1             70
            10%                68
            5%                 67
            1%                 67
            0% Min             67
```

(continued on the next page)

Output 5.4 *(continued)*

```
                    Extreme Observations

       ----------Lowest----------        ----------Highest----------

   Value   PARTICIPANT    Obs        Value   PARTICIPANT    Obs

     67            8       8           72           18      18
     68           12      12           73            6       6
     69           13      13           73           11      11
     69            2       2           74           17      17
     70           15      15           75            3       3


          Stem Leaf                      #          Boxplot
            75 0                         1             |
            74 0                         1             |
            73 00                        2             |
            72 000                       3          +-----+
            71 0000                      4          *--+--*
            70 000                       3          +-----+
            69 00                        2             |
            68 0                         1             |
            67 0                         1             |
              ----+----+----+----+
```

At the top of Output 5.4, the note "Variable: AGE" indicates that AGE is the name of the variable being analyzed by PROC UNIVARIATE. The moments table is the first table reproduced in Output 5.4. On the upper-left side of the moments table is the heading "N"; to the right of this, you can see that the analysis is based on 18 observations. Below "N" are the headings "Mean" and "Std Deviation." To the right of these, you can see that the mean and standard deviation for AGE are 71 and 2.06 (rounded to two decimal places), respectively.

To the right of "Skewness," you can see that the skewness statistic for AGE is 0. In interpreting the skewness statistic, keep in mind the following:

- A skewness value of 0 means that the distribution is not skewed. In other words, this means that the distribution is symmetrical, that neither tail is longer than the other.

- A positive skewness value means that the distribution is positively skewed, that the longer tail points toward higher values in the distribution (as with Sample E in Figure 5.4).

- A negative skewness value means that the distribution is negatively skewed, that the longer tail points toward lower values in the distribution (as with Sample F in Figure 5.4).

Since the AGE variable of Sample A displays a skewness value of 0, we know that neither tail is longer than the other in this sample.

A closer look at the moments table in Output 5.4 shows that it actually consists of two columns of statistics. The column on the left provides statistics such as the sample size, the mean, the standard deviation, and so forth. The column on the right contains headings such as "Sum Weights," "Sum Observations," and "Variance." Notice that in this right-hand column, the fourth entry down has the heading "Kurtosis" (just below "Variance"). To the right of "Kurtosis," you can see that the kurtosis statistic for AGE is approximately –.14. When interpreting this kurtosis statistic, keep in mind the following:

- A kurtosis value of 0 means that the distribution displays no kurtosis. In other words, the distribution is neither relatively peaked nor is it relatively flat compared to the normal distribution.

- A positive kurtosis value means that the distribution is relatively peaked, or leptokurtic.

- A negative kurtosis value means that the distribution is relatively flat, or platykurtic.

The small negative kurtosis value of –.14 in Output 5.4 indicates that Sample A is slightly flat, or platykurtic.

Further down, in the third grouping of information, the Shapiro-Wilk statistic appears at the top of the left column. As you will recall from earlier in this chapter, this statistic tests the null hypothesis that the sample data are normally distributed. To the right of the "W," you can see that the value for the Shapiro-Wilk statistic is 0.98. To the immediate right of this statistic is its corresponding p value. This p value appears as the first value in the right column, to the right of the heading "Pr < W." In this instance, the p value is 0.98. Remember that this statistic tests the null hypothesis that the data are normally distributed. This p value is very large at .98, meaning that there are approximately 98 chances in 100 that you would obtain the present results if the data were drawn from a normal population. Because this statistic gives so little evidence to reject the null hypothesis, you can tentatively accept it. This makes sense when you review the shape of the distribution of Sample A in Figure 5.2 as the sample data clearly appear to be normally distributed. In general, you should reject the null hypothesis of normality when p values are less than .05.

Results for a Distribution with an Outlier

The data of Sample A in Figure 5.2 displayed an approximately normal distribution. For purposes of contrast, assume that you now use PROC UNIVARIATE to analyze the data of Sample B from Figure 5.2. You will remember that Sample B was similar in shape to Sample A except that Sample B contained an outlier. The lowest value in Sample B was 37, which was an extremely low score compared to the other values in the sample. (If necessary, turn back to Figure 5.2 at this time to verify this.)

The raw data from Sample B follow. Columns 1 to 2 contain values of PARTICIPANT, the participant identification number, and columns 4 to 5 contain AGE values. Notice that these data are identical to those of Sample A, except for participant 8. In Sample A, participant 8's age was listed as 67; in Sample B, it is listed as 37.

```
 1 72
 2 69
 3 75
 4 71
 5 71
 6 73
 7 70
 8 37
 9 71
10 72
11 73
12 68
13 69
14 70
15 70
16 71
17 74
18 72
```

When analyzed with PROC UNIVARIATE, the preceding data would again produce the following output. Some of the results of this analysis are presented in Output 5.5.

Output 5.5 Selected Tables from PROC UNIVARIATE for Sample B

```
                        The UNIVARIATE Procedure
                           Variable:  AGE

                                Moments

N                           18     Sum Weights               18
Mean                 69.3333333     Sum Observations        1248
Std Deviation        8.26758376     Variance          68.3529412
Skewness             -3.9049926     Kurtosis          16.0332475
Uncorrected SS            87690     Corrected SS            1162
Coeff Variation      11.9243996     Std Error Mean    1.94868818


                       Basic Statistical Measures

            Location                        Variability

     Mean      69.33333     Std Deviation              8.26758
     Median    71.00000     Variance                  68.35294
     Mode      71.00000     Range                     38.00000
                            Interquartile Range        2.00000


                          Tests for Normality

     Test                    --Statistic---      -----p Value------

     Shapiro-Wilk         W     0.458117      Pr < W      <0.0001
     Kolmogorov-Smirnov   D     0.380384      Pr > D      <0.0100
     Cramer-von Mises     W-Sq  0.696822      Pr > W-Sq   <0.0050
     Anderson-Darling     A-Sq  3.681039      Pr > A-Sq   <0.0050
```

(continued on the next page)

Output 5.5 *(continued)*

```
                      Quantiles (Definition 5)

                      Quantile      Estimate

                      100% Max          75
                      99%               75
                      95%               75
                      90%               74
                      75% Q3            72
                      50% Median        71
                      25% Q1            70


                      The UNIVARIATE Procedure
                            Variable:  AGE

                      Quantiles (Definition 5)

                      Quantile      Estimate

                      10%               68
                      5%                37
                      1%                37
                      0% Min            37


                        Extreme Observations

    -----------Lowest-----------        -----------Highest----------

   Value   PARTICIPANT      Obs        Value   PARTICIPANT      Obs

      37             8        8           72            18       18
      68            12       12           73             6        6
      69            13       13           73            11       11
      69             2        2           74            17       17
      70            15       15           75             3        3


         Stem Leaf                        #         Boxplot
            7 5                           1            |
            7 0001111222334              13         +-----+
            6 899                         3            +
            6
            5
            5
            4
            4
            3 7                           1            *
              ----+----+----+----+
         Multiply Stem.Leaf by 10**+1
```

By comparing the moments table in Output 5.5 (for Sample B) to that in Output 5.4 (for Sample A), you can see that the inclusion of the outlier has had a considerable effect on some of the descriptive statistics for AGE. The mean of Sample B is now 69.33, down from the mean of 71 found for Sample A. More dramatic is the effect that the outlier has had on the standard deviation. With the approximately normal distribution, the standard deviation is only 2.05. With the outlier included, the standard deviation is much larger at 8.27.

Output 5.5 shows that the skewness index for Sample B is –3.90. A negative skewness index such as this is just what you would expect. The outlier has, in essence, created a long tail that points toward the lower values in the AGE distribution. You will remember that this generally results in a negative skewness index.

Output 5.5 shows that the test for normality for Sample B results in a Shapiro-Wilk statistic of approximately .46 (to the right of "W") and a corresponding *p* value of less than .01 (to the right of "Pr < W"). Because this *p* value is below .05, you reject the null hypothesis and conclude that Sample B data are not normally distributed. In other words, you can conclude that Sample B displays a statistically significant departure from normality.

The extreme observations table for Sample B appears just below the quantiles table in Output 5.5. On the left side of the extremes table, below the heading "Lowest," PROC UNIVARIATE prints the lowest values observed for the variable specified in the VAR statement (AGE, in this case). Here, you can see that the lowest five values were 37, 68, 69, 69, and 70. To the immediate right of each value is the identification number for the participant who contributed that value to the dataset. The participant identification variable is specified in the ID statement (PARTICIPANT, in this case). Reviewing these values shows you that participant 8 contributed the AGE value of 37, participant 12 contributed the AGE value of 68, and so forth. Compare the results in this extremes table with the actual raw data (reproduced earlier) to verify that these are, in fact, the specific participants who provided these values on AGE.

On the right side of the extremes table, similar information is provided; though, in this case, it is provided for the five *highest* values observed in the dataset. Under the heading "Highest" (and reading from the bottom up), you can see that the highest value on age was 75, and it was provided by participant 3, the next highest value was 74, provided by participant 17, and so forth.

This extremes table is useful for quickly identifying the specific participants who might have contributed outliers to a dataset. For example, in the present case you were able to determine that it is participant 8 who contributed the low outlier on AGE. Using the extreme observations table might not be necessary when working with a very small dataset (as in the present situation), but it can be invaluable when dealing with a large dataset. For example, if you know that you have an outlier in a dataset with 1,000 observations, the extreme observations table can immediately identify outliers. This saves you the tedious chore of examining data lines for each of the 1,000 observations individually.

Understanding the Stem-and-Leaf Plot

A stem-and-leaf plot provides a visual representation of your data with conventions somewhat similar to those used with Figures 5.2, 5.3, and 5.4. Output 5.6 provides the stem-and-leaf plot for Sample A (the approximately normal distribution):

Output 5.6 Stem-and-Leaf Plot from PROC UNIVARIATE for Sample A
(Approximately Normal Distribution)

```
        Stem Leaf                          #           Boxplot
         75  0                             1              |
         74  0                             1              |
         73  00                            2              |
         72  000                           3           +------+
         71  0000                          4           *--+--*
         70  000                           3           +------+
         69  00                            2              |
         68  0                             1              |
         67  0                             1              |
            ----+----+----+----+
```

To understand a stem-and-leaf plot, it is necessary to think of a given participant's score on AGE as consisting of a "stem" and a "leaf." The **stem** is that part of the value that appears to the left of the decimal point, and the **leaf** consists of that part that appears to the right of the decimal point. For example, participant 8 in Sample A had a value on AGE of 67. For this participant, the stem is 67 (because it appears to the left of the decimal point), and the leaf is 0 (because it appears to the right). Participant 12 had a value on age of 68, so the stem for this value is 68, and the leaf is again 0.

In the stem-and-leaf plot in Output 5.6, the vertical axis (running up and down) plots the various stems that could be encountered in the dataset (these appear under the heading "Stem"). Reading from the top down, these stems are 75, 74, 73, and so forth. Notice that at the very bottom of the plot is the stem 67. To the right of this stem appears a single leaf (a single 0). This means that there was only one participant in Sample A with a stem-and-leaf of 67 (i.e., a value on AGE of 67). Move up one line, and you see the stem 68. To the right of this, again one leaf appears (i.e., one 0 appears), meaning that only one participant had a score on AGE of 68. Move up an additional line, and you see the stem 69. To the right of this, two leaves appear (that is, two 0s appear). This means that there were two participants with a stem-and-leaf of 69 (two participants with values on AGE of 69). Continuing up the plot in this fashion, you can see that there were three participants at age 70, four participants at age 71, three at age 72, two at age 73, one at 74, and one at 75.

On the right side of the stem-and-leaf plot appears a column headed "#". This column prints the number of observations that appear at each stem. Reading from the bottom up, this column again confirms that there was one participant with a score on AGE of 67, one with a score of 68, two with a score of 69, and so forth.

Reviewing the stem-and-leaf plot in Output 5.6 shows that its shape is very similar to the shape portrayed for Sample A in Figure 5.2. This is to be expected, since both figures apply similar conventions and both describe the data of Sample A. In Output 5.6, notice that the shape of the distribution is symmetrical (i.e., neither tail is longer than the other). This, too, is to be expected since Sample A demonstrated 0 skewness.

In some cases, the stem-and-leaf plot produced by UNIVARIATE will be somewhat more complex than the one reproduced in Output 5.6. For example, Output 5.7 includes the stem-and-leaf plot produced by Sample B from Figure 5.2 (the distribution with an outlier). Consider the stem-and-leaf at the very bottom of this plot. The stem for this entry is 3, and the leaf is 7, meaning that the stem-and-leaf is 3.7. Does this mean that some participant had a score on AGE of 3.7? Not at all.

Output 5.7 Stem-and-Leaf Plot from PROC UNIVARIATE for Sample B
(Distribution with Outlier)

```
     Stem Leaf                        #          Boxplot
       7 5                            1             |
       7 0001111222334               13          +-----+
       6 899                          3             +
       6
       5
       5
       4
       4
       3 7                            1             *
         ----+----+----+----+
     Multiply Stem.Leaf by 10**+1
```

Notice the note at the bottom of this plot, which says "Multiply Stem.Leaf by 10**+1." This means "Multiply the stem-and-leaf by 10 raised to the first power." Ten raised to the first power (or 10^1), of course, is merely 10. This means that to find a participant's *actual* value on AGE, you must multiply a stem-and-leaf for that participant by 10.

For example, consider what this means for the stem-and-leaf at the bottom of this plot. This stem-and-leaf was 3.7. To find the actual score that corresponds to this stem-and-leaf, you would perform the following multiplication:

```
3.7 X 10 = 37
```

This means that, for the participant who had a stem-and-leaf of 3.7, the actual value of AGE was 37.

Move up one line in the plot, and you come to the stem "4". Note, however, that there are no leaves for this stem, which means that there were no participants with a stem of 4.0. Reading up the plot, note that no leaves appear until you reach the stem "6." The leaves on this line suggest that there is one participant with a stem-and-leaf of 6.8, and two participants with a stem-and-leaf of 6.9. Multiply these values by 10 to determine their actual values on AGE:

```
6.8 X 10 = 68
6.9 X 10 = 69
```

Move up an additional line, and note that there are actually two stems for the value 7. The first stem (moving up the plot) includes stem-and-leaf values from 7.0 through 7.4, while the next stem includes stem-and-leaf values from 7.5 through 7.9. Reviewing values in these rows, you can see that there are three participants with a stem-and-leaf of 7.0, four with a stem-and-leaf of 7.1, and so forth.

The note at the bottom of the plot tells you to multiply each stem-and-leaf by 10 raised to the first power. However, sometimes this note will tell you to multiply by 10 raised to a different power. For example, consider the following note:

```
Multiply Stem.Leaf by 10**+2
```

This note tells you to multiply by 10 raised to the second power (i.e., 10^2), or 100. Notice what some of the actual values on AGE would have been if this note had appeared (needless to say, such large values would not have made sense for the AGE variable):

```
6.8 X 100 = 680
6.9 X 100 = 690
```

All of this multiplication probably seems somewhat tedious at this point, but there is a simple rule that you can use to ease the interpretation of the note that sometimes appears at the bottom of a stem-and-leaf plot. With respect to this note, remember that the power to which 10 is raised indicates the *number of decimal places* you should move the decimal point in the stem-and-leaf. Once you have moved the decimal point this number of spaces, your stem-and-leaf will represent the actual value of interest. For example, consider the following note:

```
Multiply Stem.Leaf by 10**+1
```

This note tells you to multiply the stem-and-leaf by 10 raised to the power of 1; in other words, move the decimal point *one space to the right*. Imagine that you start with a stem-and-leaf of 3.7. Moving the decimal point one space to the right results in an actual value on AGE of 37. If you begin with a stem-and-leaf of 6.8, this becomes 68.

On the other hand, consider if the plot had included this note:

```
Multiply Stem.Leaf by 10**+2
```

It would have been necessary to move the decimal point *two* decimal spaces to the right. In this case, a stem-and-leaf of 3.7 would become 370; 6.8 would become 680. (Again, these values do not make sense for the AGE variable; they are used only for purposes of demonstration.) Finally, remember that, if no note appears at the bottom of the plot, it is not necessary to move the decimal points in the stem-and-leaf values at all.

Results for Distributions Demonstrating Skewness

Output 5.8 provides some results from the PROC UNIVARIATE analysis of Sample E from Figure 5.4. You will recall that this sample demonstrated a positive skew.

Output 5.8 Tables and Stem-and-Leaf Plot from PROC UNIVARIATE for Sample E (Positive Skewness)

```
                        The UNIVARIATE Procedure
                            Variable:  AGE

                                Moments

N                          18     Sum Weights               18
Mean                68.7777778     Sum Observations        1238
Std Deviation        3.62273143     Variance           13.124183
Skewness            0.86982584     Kurtosis           0.11009602
Uncorrected SS          85370     Corrected SS       223.111111
Coeff Variation     5.26729933     Std Error Mean     0.85388599


                    Basic Statistical Measures

        Location                        Variability

    Mean     68.77778     Std Deviation            3.62273
    Median   68.00000     Variance                13.12418
    Mode     66.00000     Range                   13.00000
                          Interquartile Range      5.00000


                      Tests for Normality

    Test                   --Statistic--        -----p Value------

    Shapiro-Wilk         W     0.929575     Pr < W        0.1909
    Kolmogorov-Smirnov   D     0.14221      Pr > D       >0.1500
    Cramer-von Mises     W-Sq  0.074395     Pr > W-Sq     0.2355
    Anderson-Darling     A-Sq  0.465209     Pr > A-Sq     0.2304


                    Quantiles (Definition 5)

            Quantile        Estimate

            100% Max           77
            99%                77
            95%                77
            90%                75
            75% Q3             71
            50% Median         68
            25% Q1             66


                The UNIVARIATE Procedure
                    Variable:  AGE

            Quantiles (Definition 5)

            Quantile        Estimate

            10%                65
            5%                 64
            1%                 64
            0% Min             64
```

(continued on the next page)

Output 5.8 *(continued)*

```
                           Extreme Observations

       -----------Lowest-----------           ----------Highest----------

    Value    PARTICIPANT      Obs        Value    PARTICIPANT      Obs

      64           18          18          71           5            5
      65           17          17          72           4            4
      65           16          16          73           3            3
      66           15          15          75           2            2
      66           14          14          77           1            1


              Stem Leaf                      #           Boxplot
                76  0                        1              |
                74  0                        1              |
                72  00                       2              |
                70  00                       2           +------+
                68  0000                     4           *--+--*
                66  00000                    5           +------+
                64  000                      3              |
                    ----+----+----+----+
```

Remember that when the approximately normal distribution was analyzed, it displayed a skewness index of 0. In contrast, note that the skewness index for Sample E in Output 5.8 is approximately .87. This positive skewness index is what you would expect, given the positive skew of the data. The skew is also reflected in the stem-and-leaf plot that appears in Output 5.8. Notice the relatively long tail that points in the direction of higher values for AGE (such as 74 and 76).

Although this sample displays positive skewness, it does not display a significant departure from normality. In the moments table in Output 5.8, you can see that the Shapiro-Wilk statistic (to the right of "W") is .93; its corresponding p value (to the right of "Pr < W") is .19. Because this p value is greater than .05, you need not reject the null hypothesis. With small samples such as the one examined here, this test is not very powerful (i.e., not very sensitive). This is why the sample was not found to display a significant departure from normality, even though it was clearly skewed.

For purposes of contrast, Output 5.9 presents the results of an analysis of Sample F from Figure 5.4. Sample F displayed negative skewness, and this is reflected in the skewness index of $-.87$ that appears in Output 5.9. Once again, the Shapiro-Wilk test shows that the sample does not demonstrate a significant departure from normality.

Output 5.9 Tables and Stem-and-Leaf Plot from PROC UNIVARIATE for Sample F (Negative Skewness)

```
                       The UNIVARIATE Procedure
                          Variable:  AGE

                             Moments

N                          18     Sum Weights                18
Mean                72.2222222    Sum Observations         1300
Std Deviation       3.62273143    Variance            13.124183
Skewness            -0.8698258    Kurtosis           0.11009602
Uncorrected SS           94112    Corrected SS       223.111111
Coeff Variation     5.01608967    Std Error Mean     0.85388599


                    Basic Statistical Measures

          Location                        Variability

     Mean      72.22222      Std Deviation            3.62273
     Median    73.00000      Variance                13.12418
     Mode      75.00000      Range                   13.00000
                             Interquartile Range      5.00000


                      Tests for Normality

    Test                  --Statistic---       -----p Value------

    Shapiro-Wilk          W     0.929575    Pr < W        0.1909
    Kolmogorov-Smirnov    D     0.14221     Pr > D       >0.1500
    Cramer-von Mises      W-Sq  0.074395    Pr > W-Sq     0.2355
    Anderson-Darling      A-Sq  0.465209    Pr > A-Sq     0.2304


                    Quantiles (Definition 5)

                    Quantile        Estimate

                    100% Max           77
                    99%                77
                    95%                77
                    90%                76
                    75% Q3             75
                    50% Median         73
                    25% Q1             70


                   The UNIVARIATE Procedure
                      Variable:  AGE

                   Quantiles (Definition 5)

                    Quantile        Estimate

                    10%                66
                    5%                 64
                    1%                 64
                    0% Min             64
```

(continued on the next page)

Output 5.9 *(continued)*

```
                        Extreme Observations

        -----------Lowest-----------          -----------Highest----------

    Value    PARTICIPANT     Obs          Value    PARTICIPANT     Obs

      64           18        18            75            5          5
      66           17        17            75            6          6
      68           16        16            76            2          2
      69           15        15            76            3          3
      70           14        14            77            1          1


            Stem Leaf                      #          Boxplot
              76 000                       3             |
              74 00000                      5          +-----+
              72 0000                       4          *--+--*
              70 00                         2          +-----+
              68 00                         2             |
              66 0                          1             |
              64 0                          1             |
                 ----+----+----+----+
```

The stem-and-leaf plot in Output 5.9 reveals a long tail that points in the direction of lower values for AGE (such as 64 and 66). This, of course, is the type of plot that you would expect for a negatively skewed distribution.

Conclusion

Regardless of what other statistical procedures you use, always begin the data analysis process by performing the simple analyses described here. This will help ensure that the dataset and program do not contain any obvious errors that, if left unidentified, could lead to incorrect conclusions. Once the data have undergone this initial screening, you can move on to the more sophisticated procedures described in the remainder of this text.

References

Rummel, R. J. (1970). *Applied factor analysis.* Evanston, IL: Northwestern University Press.

SAS Institute Inc. (2004). *SAS/STAT 9.1 user's guide.* Cary, NC: Author.

Measures of Bivariate Association

6

> **Overview:** This chapter discusses procedures to test the significance of the relationship between two variables. Recommendations are made for choosing the correct statistic based on the level of measurement of variables. The chapter shows how to use PROC GPLOT to prepare bivariate scattergrams, how to use PROC CORR to compute Pearson correlations and Spearman correlations, and how to use PROC FREQ to perform the chi-square test of independence.

Introduction: Significance Tests versus Measures of Association

A **bivariate relationship** involves the relationship between two variables. For example, if you conduct an investigation in which you study the relationship between GRE verbal test scores and college grade point average (GPA), you are studying a bivariate relationship.

There are numerous statistical procedures that you can use to examine bivariate relationships. These procedures can provide a test of statistical significance, a measure of association, or both. A **test of statistical significance** allows you to test hypotheses about the relationship between variables in the population. For example, the Pearson product moment correlation coefficient allows you to test the null hypothesis that the correlation between two interval- or ratio-level variables is zero in the population. In a study, you might draw a sample of 200 participants and determine that the Pearson correlation coefficient between the GRE verbal test and GPA is .35 for this sample. You can then use this finding to test the null hypothesis that the correlation between GRE verbal and GPA is zero in the population. The resulting test might prove to be significant at $p < .01$ (depending on the sample size). This p value suggests that there is less than 1 chance in 100 of obtaining a sample correlation of $r = .35$ or larger from the overall population if the null hypothesis were true. You therefore reject the null hypothesis and conclude that the degree of relationship between variables does not occur due to chance alone.

In addition to serving as a test of statistical significance, a Pearson correlation coefficient can also serve as a measure of association. A **measure of association** reflects the strength of the relationship between variables (regardless of the statistical significance of the relationship). For example, the absolute value of a Pearson correlation coefficient reveals how strongly the two variables are related. A Pearson correlation can range from -1.00 through 0.00 through $+1.00$, with larger absolute values indicative of stronger relationships. For example, a correlation of 0.00 indicates no relationship between variables, a correlation of .20 (or $-.20$) indicates a weak relationship, and a correlation of .90 (or $-.90$) indicates a strong relationship. (A later section provides more detailed guidelines for interpreting Pearson correlations.)

Choosing the Correct Statistic

Levels of Measurement

When examining the relationship between variables, it is important to use the appropriate bivariate statistic, given the nature of the variables studied. In choosing this statistic, you must pay particular attention to the level of measurement used to assess both variables. This section briefly reviews the various levels of measurement that were discussed in Chapter 1 of this text. The following section will show how you can use the scale of measurement to identify the right statistic for a given research problem.

Chapter 1 described the four levels of measurement used with variables in social science research: nominal; ordinal; interval; and ratio. A variable measured on a **nominal scale** is a *classification* variable: it simply indicates the group to which a participant belongs. For example, "race" is a nominal-level variable: a participant may be classified as being African American, Asian American, Caucasian, and so forth.

A variable measured on an **ordinal scale** is a *ranking* variable. An ordinal variable indicates which participants have more of the construct being assessed and who have less. For example, if all students in a classroom were ranked with respect to verbal ability so that the best student was assigned the value "1," the next-best student was assigned the value "2," and so forth, this ranking variable would be assessed at the ordinal level. Ordinal scales, however, have a limitation in that equal differences in scale value do not necessarily have equal quantitative meaning. In other words, the difference in verbal ability between student #1 and student #2 is not necessarily the same as the difference in ability between student #2 and student #3.

With an **interval scale** of measurement, equal differences in scale values do have equal quantitative meaning. For example, imagine that you develop a test of verbal ability. Scores on this test can range from 0 through 100, with higher scores reflecting greater ability. If scores on this test are truly on an interval scale, then the difference between a score of 60 and 70 should be equal to the difference between a score of 70 and 80. In other words, the units of measurement should be the same across the full range of the scale. Nonetheless, an interval-level scale does not have a true zero-point. Among other things, this means that a score of zero on the test does not necessarily indicate a complete absence of verbal ability.

Finally, a **ratio scale** has all of the properties of an interval scale but also has a true zero-point. This makes it possible to make meaningful statements about the ratios between scale values. For example, body weight is assessed on a ratio scale: a score of "0 pounds" indicates no body weight at all. With this variable, it is possible to state that, "a person who weighs 200 pounds is twice as heavy as a person who weighs 100 pounds." Other examples of ratio-level variables include age, height, and income.

A Table of Appropriate Statistics

Once you identify the level of measurement at which your variables are assessed, it is relatively simple to determine the correct statistic for analyzing the relationship between variables. This section presents a figure to make this task easier. First, be warned that the actual situation is a bit more complex than the figure suggests. This is because when two variables are assessed at a given level of measurement, there may actually be more than one statistic that can be used to investigate the relationship between those variables. In keeping with the relatively narrow scope of this text, however, this chapter presents only a few of the available options. (In general, this chapter emphasizes what are probably the most commonly used statistics.) To learn about additional procedures and the special conditions under which they might be appropriate, consult a more comprehensive statistics textbook such as Hays (1988).

Figure 6.1 identifies some of the statistics that might be appropriate for pairs of variables assessed at given levels of measurement. The vertical columns of the figure indicate the scales of measurement used to assess the predictor variable of the pair, while the horizontal rows indicate the scale used to assess the criterion (or predicted) variable. The appropriate statistic for a given pair of variables is identified where a given row and column intersect.

Figure 6.1 Statistics for Pairs of Variables

		Predictor Variable			
		Nominal-Level	Ordinal-Level	Interval-Level	Ratio-Level
	Ratio-Level	ANOVA	Spearman Correlation	Pearson Correlation or Spearman Correlation	Pearson Correlation or Spearman Correlation
	Interval-Level	ANOVA	Spearman Correlation	Pearson Correlation or Spearman Correlation	
Criterion Variable	Ordinal-Level	Kruskal-Wallis Test	Spearman Correlation		
	Nominal-Level	Chi-Square Test			

The Chi-Square Test of Independence

For example, at the intersection of a nominal-level predictor variable and a nominal-level criterion variable, the figure indicates that an appropriate statistic might be the chi-square test (short for "chi-square test of independence"). To illustrate this situation, imagine that you are conducting research that investigates the relationship between place of residence and political party affiliation. Assume that you have hypothesized that people who live in Scotland are more likely to vote Labour, relative to those who live in England or Wales. To test this hypothesis, you gather data on two variables for each of 1,000 participants. The first variable is "place of residence," and each participant is coded as living either in Scotland, England, or Wales. It should be clear that place of residence is therefore a nominal-level variable that can assume one of three values. The second variable is "party affiliation," and each participant is coded as either being a Labour, Conservative, or Green Party supporter (among others). Obviously, party affiliation is also a nominal-scale variable that can assume one of three values in this instance.

If you analyzed your data with a chi-square test of independence and obtained a significantly large chi-square, this would tell you that there is a significant relationship between place of residence and party affiliation. Close inspection of the two-way classification table that is produced in the course of this analysis (to be discussed later) would then tell you whether those in Scotland were, in fact, more likely to vote Labour.

The Spearman Correlation Coefficient

Figure 6.1 shows that, where an ordinal-level predictor intersects with an ordinal-level criterion, the Spearman correlation coefficient is recommended. As an illustration, imagine that an author has written a book ranking the 100 largest universities in the European Union from best to worst, so that the best school was ranked #1, the second best was ranked #2, and so forth. In addition to providing an overall ranking for the institutions, the author also ranked them from best to worst with respect to a number of specific criteria such as "intellectual environment," "prestige," "quality of athletic programs," and so forth. Assume that you have formed the hypothesis that universities' overall rankings will demonstrate a strong positive correlation with the rankings of the quality of their library facilities.

To test this hypothesis, you compute the correlation between two variables: their overall rankings and the rankings of their library facilities. Both variables are clearly on an ordinal scale because both merely tell you about the ordinal rankings of the universities. For example, the library facilities variable might tell you that university #1 has a more extensive collection than university #2, but it does not tell you *how much* more extensive. Because both variables are on an ordinal scale, it is appropriate to analyze the relationship between them with the Spearman correlation coefficient.

The Pearson Correlation Coefficient

Finally, Figure 6.1 shows that, when both the predictor and criterion are assessed on either the interval- or ratio-level of measurement, the Pearson correlation coefficient is generally appropriate. For example, assume that you want to test the hypothesis that income is positively correlated with age. That is, you predict that older people will tend to have more income than younger people. To test this hypothesis, you obtain data from 200 participants on both variables. The first variable is AGE, which is simply the participants' age in years. The second variable is INCOME: annual income in dollars. It can assume values such as $0, $10,000, $1,200,000, and so forth.

You know that both variables in this study are assessed on a ratio scale, because, with both variables, equal intervals have equal quantitative meaning, and both variables have a true zero point (i.e., zero years and zero dollars indicate a complete absence of age and income). Because both are measured on a ratio scale, you can compute a Pearson correlation coefficient to assess the nature of the relationship between variables. (This assumes that certain other assumptions for the statistic have been met. These assumptions are discussed in a later section and are also summarized at the end of this chapter.)

Other Statistics

The statistics identified on the diagonal of the figure are statistics to be used when both predictor and criterion variables are assessed on the same level of measurement (e.g., you should use the chi-square test when both variables are nominal-level variables). The situation is more complex, however, when one variable is assessed at one level of measurement, and the second variable is assessed at another. For example, when the criterion variable is interval-level and the predictor variable is nominal-level, the appropriate procedure is ANOVA (for "analysis of variance"). This chapter discusses only three of the statistics identified in Figure 6.1: the Pearson correlation coefficient; the Spearman correlation coefficient; and the chi-square test of independence. (Analysis of variance is described in detail in Chapters 9 and 10.)

Exceptions to Figure 6.1

Remember that Figure 6.1 presents only *some* of the statistics that might be appropriate for a given combination of variables; there will be many exceptions to the general guidelines that it illustrates. For example, the figure assumes that all nominal-level variables assume only a relatively small number of values (perhaps two to six), and all ordinal-, interval-, and ratio-level variables are *continuous,* taking on a large number of values.

When these conditions do not hold, the correct statistic might differ from the one indicated by the table. For example, in some cases ANOVA might be the appropriate statistic when both criterion and predictor variables are assessed on a ratio level. This would be the case when conducting an experiment in which the criterion variable is "number of errors on a memory task" (a ratio-level variable), and the predictor is "amount of caffeine administered: 0 mg. versus 100 mg. versus 200 mg." In this case, the predictor variable (amount of caffeine administered) is clearly assessed on a ratio-level, but the correct statistic is ANOVA because this predictor takes on such a small number of values (just three). As you read the examples of research throughout this text, you will develop a better understanding of how to take these factors into account when choosing the correct statistic for a specific study.

Conclusion

The purpose of this section is to provide a simple strategy for choosing the correct measure of bivariate association under conditions that are frequently encountered in social science research. The following section reviews information about the Pearson correlation coefficient, and shows how to use PROC PLOT to view scattergrams to verify that the relation between two variables is linear. This and subsequent sections also provide a more detailed discussion of the three statistics emphasized in this chapter: the Pearson correlation; the Spearman correlation; and the chi-square test of independence. For each, the text provides an example in which the statistic would be appropriate, shows how to prepare the data and the necessary SAS program, and shows how to interpret the output.

Pearson Correlations

When to Use

You can use the Pearson product-moment correlation coefficient (symbolized by the letter r) to assess the nature of the relationship between two variables when both are measured on either an interval- or ratio-level of measurement. It is further assumed that both variables should include a relatively large number of values. For example, you would not use this statistic if one of the variables could assume only three values.

It would be appropriate to compute a Pearson correlation coefficient to investigate the nature of the relationship between GRE verbal test scores and grade point average (GPA). GRE verbal is assessed on an interval-level of measurement, and can assume a wide variety of values (i.e., possible scores range from 200 through 800). Grade point ratio is also assessed on an interval level and can also assume a wide variety of values from 0.00 through 4.00.

In addition to interval- or ratio-scale measurement, the Pearson statistic also assumes that the observed values are distributed normally. When one or both variables display a markedly non-normal distribution (e.g., when one or both variables are markedly skewed), it might be more appropriate to analyze the data with the Spearman correlation coefficient. A later section of this chapter discusses the Spearman coefficient. (Summarized assumptions of both the Pearson and Spearman correlations are at the end of this chapter.)

Interpreting the Coefficient

To more fully understand the nature of the relationship between the two variables studied, it is necessary to interpret two characteristics of a Pearson correlation coefficient. First, the **sign of the coefficient** tells you whether there is a positive or negative relationship between variables. A positive correlation indicates that as values for one variable increase, values for the second variable also increase. A **positive correlation** is illustrated in Figure 6.2 which shows the relationship between GRE verbal test scores and GPA in a fictitious sample of data.

Figure 6.2 A Positive Correlation

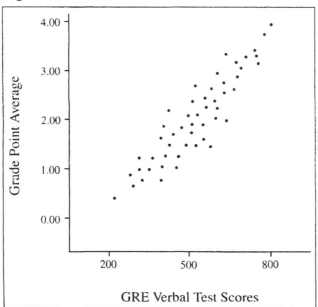

You can see that participants who received low scores on the predictor variable (GRE verbal) also received low scores on the criterion variable (GPA). At the same time, participants who received high scores on GRE verbal also received high scores on GPA. The two variables can therefore be said to be positively correlated.

With a **negative correlation**, as values for one variable increase, values for the second variable decrease. For example, you might expect to see a negative correlation between GRE verbal test scores and the number of errors that participants make on a vocabulary test (i.e., the students with high GRE verbal scores tend to make few mistakes, and the students with low GRE scores tend to make many mistakes). This relationship is illustrated with fictitious data in Figure 6.3.

Figure 6.3 A Negative Correlation

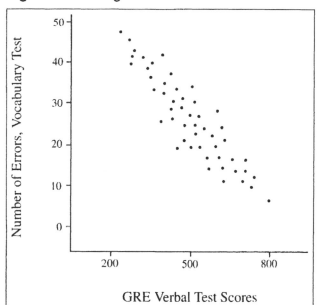

The second characteristic of a correlation coefficient is its **magnitude**: the greater the absolute value of a correlation coefficient, the stronger the relationship between the two variables. Pearson correlation coefficients can range in size from –1.00 through 0.00 through +1.00. Coefficients of 0.00 indicate no relationship between variables. For example, if there were a zero correlation between GRE scores and GPA, then knowing a person's GRE score would tell you nothing about his or her GPA. In contrast, correlations of –1.00 or +1.00 indicate perfect relationships. If the correlation between GRE scores and GPA were 1.00, it would mean that knowing someone's GRE score would allow you to predict his or her GPA with complete accuracy. In the real world, however, GRE scores are not that strongly related to GPA, so you would expect the correlation between them to be considerably less than 1.00.

The following is an approximate guide for interpreting the strength of the relationship between two variables, based on the absolute value of the coefficient:

```
±1.00 = Perfect correlation
 ±.80 = Strong correlation
 ±.50 = Moderate correlation
 ±.20 = Weak correlation
 ±.00 = No correlation
```

We recommend that you consider the magnitude of correlation coefficients as opposed to whether or not coefficients are statistically significant. This is because significance estimates are strongly influenced by sample sizes. For instance, an r value of .15 (weak correlation) would be statistically significant with samples in excess of 700 whereas a coefficient of .50 (moderate correlation) would not be statistically significant with a sample of only 15 participants.

Remember that one considers the *absolute value* of the coefficient when interpreting its size. This is to say that a correlation of –.50 is just as strong as a correlation of +.50, a correlation of –.75 is just as strong as a correlation of +.75, and so forth.

Linear versus Nonlinear Relationships

The Pearson correlation is appropriate only if there is a linear relationship between the two variables. There is a **linear relationship** between two variables when their scattergram follows the form of a straight line. For example, it is possible to draw a straight line through the center of the scattergram presented in Figure 6.4, and this straight line fits the pattern of the data fairly well. This means that there is a linear relationship between GRE verbal test scores and GPA.

Figure 6.4 A Linear Relationship

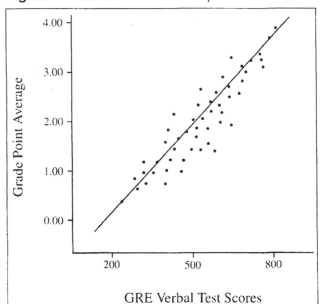

In contrast, there is a **nonlinear relationship** between two variables if their scattergram does not follow the form of a straight line. For example, imagine that you have constructed a test of creativity and have administered it to a large sample of college students. With this test, higher scores reflect higher levels of creativity. Imagine further that you obtain the GRE verbal test scores for these students, plot their GRE scores against their creativity scores, and obtain the scattergram presented in Figure 6.5.

Figure 6.5 A Nonlinear Relationship

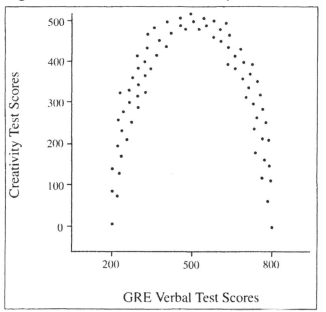

GRE Verbal Test Scores

The scattergram in Figure 6.5 reveals a nonlinear relationship between GRE scores and creativity. It shows that:

- students with low GRE scores tend to have low creativity scores;
- students with moderate GRE scores tend to have high creativity scores;
- students with high GRE scores tend to have low creativity scores.

It is not possible to draw a good-fitting straight line through the data points of Figure 6.5. This is why we say that there is a *nonlinear* (or perhaps a *curvilinear*) relationship between GRE scores and creativity scores.

When one uses the Pearson correlation to assess the relationship between variables reflecting a nonlinear relationship, the resulting correlation coefficient usually *underestimates* the actual strength of the relationship between variables. For example, computing the Pearson correlation between the GRE scores and creativity scores presented in Figure 6.5 might result in a coefficient of .10, which would indicate a very weak relationship between the two variables. From the diagram, however, there is clearly a fairly strong relationship between GRE scores and creativity. The figure shows that if you know someone's GRE score, you can accurately predict his or her creativity score.

The implication of this is that you should always verify that there is a linear relationship between two variables before computing a Pearson correlation for those variables. One of the easiest ways of verifying that the relationship is linear is to prepare a scattergram similar to those presented in the preceding figures. Fortunately, this is easily done by SAS with the GPLOT procedure.

Producing Scattergrams with PROC GPLOT

Here is the general form for requesting a scattergram with the PLOT procedure:

```
PROC GPLOT    DATA=dataset-name;
    PLOT    criterion-variable*predictor-variable ;
RUN;
```

The variable listed as the "criterion-variable" in the preceding program is plotted on the vertical axis, and the "predictor-variable" is plotted on the horizontal axis.

To illustrate this procedure, imagine that you have conducted a study dealing with the *investment model*, a theory of commitment in romantic associations (Rusbult, 1980). The investment model identifies a number of variables that are believed to influence a person's commitment to a romantic association. **Commitment** refers to the participant's intention to remain in the relationship. These are some of the variables that are predicted to influence participant commitment:

Satisfaction: The participant's affective response to the relationship

Investment size: The amount of time and personal resources that the participant has put into the relationship

Alternative value: The attractiveness of the participant's alternatives to the relationship (e.g., the attractiveness of alternate romantic partners)

Assume that you have developed a 16-item questionnaire to measure these four variables. The questionnaire is administered to 20 participants who are currently involved in a romantic relationship, and the participants are asked to complete the instrument while thinking about their relationship. When they have completed the questionnaire, it is possible to use their responses to compute four scores for each participant. First, each receives a score on the *commitment scale*. Higher values on the commitment scale reflect greater commitment to the relationship. Each participant also receives a score on the *satisfaction scale*, where higher scores reflect greater satisfaction with the relationship. Higher scores on the *investment scale* mean that the participant believes that he or she has invested a great deal of time and effort in the relationship. Finally, with the *alternative value scale*, higher scores mean that it would be attractive to the respondent to find a different romantic partner.

Once the data have been entered, you can use the PLOT procedure to prepare scattergrams for various combinations of variables. The following SAS program inputs some fictitious data and requests that a scattergram be prepared in which commitment scores are plotted against satisfaction scores:

```
1       DATA D1;
2          INPUT   #1   @1   COMMITMENT    2.
3                       @4   SATISFACTION  2.
4                       @7   INVESTMENT    2.
5                       @10  ALTERNATIVES  2.   ;
6       DATALINES;
7       20 20 28 21
8       10 12  5 31
9       30 33 24 11
10       8 10 15 36
11      22 18 33 16
12      31 29 33 12
13       6 10 12 29
14      11 12  6 30
15      25 23 34 12
16      10  7 14 32
17      31 36 25  5
18       5  4 18 30
19      31 28 23  6
20       4  6 14 29
21      36 33 29  6
22      22 21 14 17
23      15 17 10 25
24      19 16 16 22
25      12 14 18 27
26      24 21 33 16
27      ;
28      RUN;
29
30      PROC GPLOT   DATA=D1;
31         PLOT COMMITMENT*SATISFACTION;
32      RUN;
```

In the preceding program, scores on the commitment scale are entered in columns 1 to 2, and are given the SAS variable name COMMITMENT. Similarly, scores on the satisfaction scale are entered in columns 4 to 5, and are given the name SATISFACTION; scores on the investment scale appear in columns 7 to 8 and are given the name INVESTMENT; and scores on the alternative value scale appear as the last column of data and are given the name ALTERNATIVES.

The data for the 20 participants appear on lines 7 to 26 in the program. There is one line of data for each participant.

Line 30 of the program requests the GPLOT procedure, specifying that the dataset to be analyzed is dataset D1. The PLOT command on line 31 specifies COMMITMENT as the criterion variable and SATISFACTION as the predictor variable for this analysis. The results of this analysis appear in Output 6.1.

Output 6.1 Scattergram of Commitment Scores Plotted against Satisfaction Scores

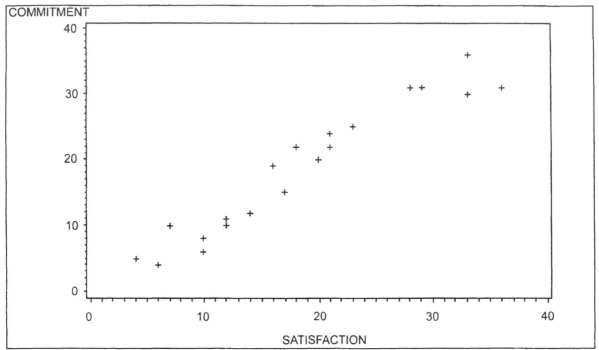

Notice that in this output, the criterion variable (COMMITMENT) is plotted on the vertical axis while the predictor variable (SATISFACTION) is plotted on the horizontal axis. The shape of the scattergram indicates that there is a linear relationship between SATISFACTION and COMMITMENT. This is evident from the fact that it would be possible to draw a relatively good-fitting straight line through the center of the scattergram. Given that the relationship is linear, it seems safe to proceed with the computation of a Pearson correlation coefficient for this pair of variables.

The general shape of the scattergram also suggests that there is a fairly strong relationship between the two variables: knowing where a participant stands on the SATISFACTION variable allows you to predict with some accuracy where that participant will stand on the COMMITMENT variable. Later, you will compute the correlation coefficient for these two variables to determine just how strong the relationship is.

Output 6.1 also indicates that the relationship between SATISFACTION and COMMITMENT is *positive* (i.e., large values on SATISFACTION are associated with large values on COMMITMENT and small values on SATISFACTION are associated with small values on COMMITMENT). This makes intuitive sense; you would expect that participants who are highly satisfied with their relationships would also be highly

committed to those relationships. To illustrate a negative relationship, you can plot
COMMITMENT against ALTERNATIVES. To do this, include the following
statements in the preceding program:

```
PROC GPLOT   DATA=D1;
   PLOT COMMITMENT*ALTERNATIVES;
RUN;
```

These statements are identical to the earlier statements except that ALTERNATIVES is
now specified as the predictor variable. These statements produce the scattergram
presented in Output 6.2.

Output 6.2 Scattergram of Commitment Scores Plotted against Alternative Value Scores

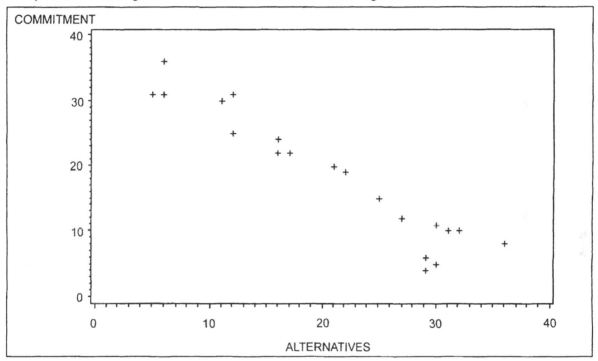

Notice that the relationship between these two variables is *negative*. This is what you
would expect as it makes intuitive sense that participants who indicate that alternatives to
their current romantic partner are attractive would not be overly committed to a current
partner. The relationship between ALTERNATIVES and COMMITMENT also appears
to be linear. It is therefore appropriate to assess the strength of the relationship between
these variables with the Pearson correlation coefficient.

Computing Pearson Correlations with PROC CORR

The CORR procedure offers a number of options regarding what *type* of coefficient will be computed as well as a number of options regarding the *way* they will appear on the printed page. Some of these options are discussed here.

Computing a Single Correlation Coefficient

In some instances, you might want to compute the correlation between just two variables. Here is the general form for the statements that will accomplish this:

```
PROC CORR    DATA=dataset-name    options;
   VAR    variable1    variable2;
RUN;
```

The choice of which variable is "variable1" and which is "variable2" is arbitrary. For a specific example, assume that you want to compute the correlation between commitment and satisfaction. These are the required statements:

```
PROC CORR    DATA=D1;
     VAR COMMITMENT SATISFACTION;
RUN;
```

This program command results in a single page of output, reproduced here as Output 6.3:

Output 6.3 Computing the Pearson Correlation between Commitment and Satisfaction

```
                            The CORR Procedure

                   2  Variables:    COMMITMENT    SATISFACTION


                              Simple Statist
Variable          N       Mean      Std Dev         Sum      Minimum      Maximum

COMMITMENT       20   18.60000     10.05459   372.00000      4.00000     36.00000
SATISFACTION     20   18.50000      9.51177   370.00000      4.00000     36.00000


                   Pearson Correlation Coefficients, N = 20
                       Prob > |r| under H0: Rho=0

                                 COMMITMENT          SATISFACTION

                  COMMITMENT        1.00000              0.96252
                                                         <.0001


                  SATISFACTION      0.96252              1.00000
                                    <.0001
```

The first part of Output 6.3 presents simple descriptive statistics for the variables being analyzed. This allows you to verify that everything looks appropriate (e.g., the correct number of cases were analyzed, no variables were out of range). The names of the variables appear below the "Variable" heading, and the statistics for the variables appear to the right of the variable names. These descriptive statistics show that 20 participants provided usable data for the COMMITMENT variable, that the mean for COMMITMENT is 18.6, and the standard deviation is 10.05. It is always important to review the "Minimum" and "Maximum" columns to verify that no impossible scores appear in the data. With COMMITMENT, the lowest possible score was 4 and the highest possible score was 36. The "Minimum" and "Maximum" columns of Output 6.3 show that no observed values were out of range, thus providing no evidence of misentered data. (Again, these proofing procedures do not *guarantee* that no errors were made in entering data but they are useful for identifying some types of errors.) Since the descriptive statistics provide no obvious evidence of entering or programming mistakes, you are now free to review the correlations themselves.

The bottom half of Output 6.3 provides the correlations requested in the VAR statement. There are actually four correlation coefficients in the output because your statement requested that the system compute every possible correlation between the variables COMMITMENT and SATISFACTION. This caused SAS to compute the correlation between COMMITMENT and SATISFACTION, between SATISFACTION and COMMITMENT, between COMMITMENT and COMMITMENT, and between SATISFACTION and SATISFACTION.

The correlation between COMMITMENT and COMMITMENT appears in the upper-left corner of the matrix of correlation coefficients in Output 6.3. You can see that the correlation between these variables is 1.00. This makes sense, because the correlation of any variable with itself is always equal to 1.00. Similarly, in the lower-right corner, you see that the correlation between SATISFACTION and SATISFACTION is also 1.00.

The coefficient you are actually interested in appears where the column headed COMMITMENT intersects with the row headed SATISFACTION. The top number in the "cell" where this column and row intersect is .96, which is the Pearson correlation between COMMITMENT and SATISFACTION (rounded to two decimal places).

Just below the correlation is the p value associated with the correlation. This is the significance estimate obtained from a test of the null hypothesis that the correlation between COMMITMENT and SATISFACTION is zero in the population. More technically, the p value gives us the probability that you would obtain a sample correlation this large (or larger) if the correlation between COMMITMENT and SATISFACTION were really zero in the population. For the present correlation coefficient of $r = .96$, the corresponding p value is less than .01. This means that, given your sample size, there is less than 1 chance in 100 of obtaining a correlation of .96 or larger from this population by chance alone. You might therefore reject the null hypothesis and tentatively conclude that COMMITMENT is related to SATISFACTION in the population. (The alternative hypothesis for this statistical test is that the correlation is not equal to zero in the population. This alternative hypothesis is two-sided which means that it does not predict whether the correlation coefficient is positive or negative, only that it is not equal to zero.)

Determining Sample Size

The size of the sample used in computing the correlation coefficient can appear in one of two places on the output page. First, if all correlations in the analysis were based on the same number of participants, the sample size appears only once on the page, in the line above the matrix of correlations. This line appears just below the descriptive statistics. In Output 6.3, the line says:

```
Pearson Correlation Coefficients, N = 20
```

The "N =" portion of this output indicates the sample size. In Output 6.3, the sample size is 20.

However, if one is requesting correlations between several different pairs of variables, it is possible that certain coefficients will be based on more participants than others (due to missing data). In this case, the sample size is printed for each correlation coefficient. Specifically, the sample size appears immediately below the correlation coefficient and its associated significance level (i.e., *p* value), following this format:

```
correlation
p value
N
```

Computing All Possible Correlations for a Set of Variables

Here is the general form for computing all possible Pearson correlation coefficients for a set of variables:

```
PROC CORR   DATA=dataset-name    options;
   VAR   variable-list;
RUN;
```

Each variable name in the preceding "variable-list" should be separated by at least one space. For example, assume that you now want to compute all possible correlations for the variables COMMITMENT, SATISFACTION, INVESTMENT, and ALTERNATIVES. The statements that request these correlations are as follows:

```
PROC CORR   DATA=D1;
   VAR COMMITMENT SATISFACTION INVESTMENT ALTERNATIVES;
RUN;
```

The preceding program produced the output reproduced here as Output 6.4:

Output 6.4 Computing All Possible Pearson Correlations

```
                         The CORR Procedure

      4  Variables:    COMMITMENT   SATISFACTION INVESTMENT   ALTERNATIVES


                         Simple Statistics

Variable       N      Mean     Std Dev          Sum      Minimum      Maximum

COMMITMENT    20   8.60000    10.05459    372.00000      4.00000     36.00000
SATISFACTION  20  18.50000     9.51177    370.00000      4.00000     36.00000
INVESTMENT    20  20.20000     9.28836    404.00000      5.00000     34.00000
ALTERNATIVES  20  20.65000     9.78869    413.00000      5.00000     36.00000


                 Pearson Correlation Coefficients, N = 20
                      Prob > |r| under H0: Rho=0

                    COMMITMENT      SATISFACTION     INVESTMENT    ALTERNATIVES

COMMITMENT            1.00000          0.96252         0.71043        -0.95604
                                       <.0001          0.0004          <.0001

SATISFACTION         0.96252          1.00000          0.61538        -0.93355
                      <.0001                            0.0039          <.0001

INVESTMENT           0.71043          0.61538          1.00000        -0.72394
                      0.0004           0.0039                           0.0003

ALTERNATIVES        -0.95604         -0.93355         -0.72394         1.00000
                      <.0001           <.0001           0.0003
```

You can interpret the correlations and significance values in this output in exactly the same way as with Output 6.3. For example, to find the correlation coefficient between INVESTMENT and COMMITMENT, you find the cell where the row for INVESTMENT intersects with the column for COMMITMENT. The top number in this cell is .71, which is the Pearson correlation coefficient between these two variables. Just below this correlation coefficient is the p value less than .01, meaning that there is less than 1 chance in 100 of observing a sample correlation this large if the population correlation is really zero. The observed correlation is statistically significant.

Notice that the pattern of the correlations supports some of the predictions of the investment model: commitment is positively related to satisfaction and investment size; and is negatively related to alternative value. With respect to magnitude, the correlations range from being moderately strong to very strong. (Remember that these data are fictitious.)

> **What happens if I omit the VAR statement?** It is possible to run PROC CORR without the VAR statement. This causes every possible correlation to be computed between all quantitative variables in the dataset. Use caution when doing this, however; with large datasets, leaving off the VAR statement can result in a very long printout.

Computing Correlations between Subsets of Variables

By using the WITH statement in the SAS program, it is possible to compute correlations between one subset of variables and a second subset of variables. The general form is as follows:

```
PROC CORR   DATA=dataset-name   options;
   VAR    variables-that-will-appear-as-columns;
   WITH   variables-that-will-appear-as-rows;
RUN;
```

Any number of variables can appear in the VAR statement and any number of variables can also appear in the WITH statement. To illustrate, assume that you want to prepare a matrix of correlation coefficients in which there is one column of coefficients representing the COMMIT variable, and there are three rows of coefficients representing the SATISFACTION, INVESTMENT, and ALTERNATIVES variables. The following statements would create this matrix:

```
PROC CORR   DATA=D1;
   VAR   COMMITMENT;
   WITH SATISFACTION INVESTMENT ALTERNATIVES;
RUN;
```

Output 6.5 presents the results generated by this program. Obviously, the correlations in this output are identical to those obtained in Output 6.4, though Output 6.5 is more compact. This is why it is often wise to use the WITH statement in conjunction with the VAR statement as this can produce smaller and more manageable printouts than obtained if you use only the VAR statement.

Output 6.5 Computing Pearson Correlations for Subsets of Variables

```
                              The CORR Procedure

           3 With Variables:    SATISFACTION INVESTMENT    ALTERNATIVES
           1      Variables:    COMMITMENT


                             Simple Statistics

   Variable        N      Mean    Std Dev        Sum    Minimum    Maximum

   SATISFACTION    20  18.50000    9.51177   370.00000   4.00000   36.00000
   INVESTMENT      20  20.20000    9.28836   404.00000   5.00000   34.00000
   ALTERNATIVES    20  20.65000    9.78869   413.00000   5.00000   36.00000
   COMMITMENT      20  18.60000   10.05459   372.00000   4.00000   36.00000

                 Pearson Correlation Coefficients, N = 20
                     Prob > |r| under H0: Rho=0

                                         COMMITMENT

              SATISFACTION               0.96252
                                          <.0001

              INVESTMENT                 0.71043
                                          0.0004

              ALTERNATIVES              -0.95604
                                          <.0001
```

Options Used with PROC CORR

The following items are some of the PROC CORR options that you might find especially useful when conducting social science research. Remember that the option names should appear *before* the semicolon that ends the PROC CORR statement:

ALPHA
:   prints coefficient alpha (a measure of scale reliability) for the variables listed in the VAR statement. (Chapter 7 deals with coefficient alpha in greater detail.)

COV
:   prints covariances between the variables. This is useful when you need a variance-covariance table, rather than a table of correlations.

KENDALL
:   prints Kendall's tau-b coefficient, a measure of bivariate association for variables assessed at the ordinal level.

NOMISS
:   drops from the analysis any observation (participant) with missing data on any of the variables listed in the VAR statement. Using this option ensures that all correlations are based on exactly the same observations (and, therefore, on the same *number* of observations).

NOPROB

prevents the printing of *p* values associated with the correlations.

RANK

for each variable, reorders the correlations from highest to lowest (in absolute value) and prints them in this order.

SPEARMAN

prints Spearman correlations, which are appropriate for variables measured on an ordinal level. Spearman correlations are discussed in the following section.

Spearman Correlations

When to Use

Spearman's rank-order correlation coefficient (symbolized as r_s) can be appropriate in a variety of circumstances. First, you can use Spearman correlations when both variables are assessed at an ordinal level of measurement. It is also correct when one variable is an ordinal-level variable, and the other is an interval- or ratio-level variable.

However, it can also be appropriate to use the Spearman correlation when both variables are on an interval or ratio scale. This is because the Spearman coefficient is a **distribution-free test**. Among other things, a distribution-free test is one that makes no assumptions concerning the shape of the distribution from which sample data were drawn. For this reason, researchers sometimes compute Spearman correlations when one or both of the variables are interval- or ratio-level but are markedly non-normal (i.e., skewed or kurtotic), such that a Pearson correlation would be inappropriate. The Spearman correlation is less useful than a Pearson correlation when both variables are normally distributed, but it is more useful when one or both variables are not normally distributed.

SAS computes a Spearman correlation by rank-ordering both variables and computing the correlation between the ranks. The resulting correlation coefficient can range from −1.00 through to +1.00, and is interpreted in the same way as a Pearson correlation coefficient.

Computing Spearman Correlations with PROC CORR

Here is the general form for computing a Spearman correlation between two variables. Notice that this is identical to the form used to compute Pearson correlations except that you specify the option SPEARMAN in the PROC CORR statement. (If you did not specify the SPEARMAN option, the program would have again produced Pearson correlations since Pearson correlations are the default output.)

```
PROC CORR   DATA=dataset-name   SPEARMAN   options;
   VAR   variable1   variable2;
  RUN;
```

To illustrate this statistic, assume that a teacher has administered a test of creativity to 10 students at Time 1. After reviewing the results, she ranks her students from 1 to 10, with "1" representing the "most creative student," and "10" representing the "least creative student." Two months later, at Time 2, she repeats the process, arriving at a slightly different set of rankings. She now wants to determine the correlation between her rankings made at Time 1 and Time 2. The data (rankings) are clearly ordinal level measures, so the correct statistic is the Spearman rank-order correlation coefficient.

This is the entire program that will input the fictitious data and compute the Spearman correlation:

```
1      DATA D1;
2         INPUT   #1   @1   TEST1   2.
3                      @4   TEST2   2.   ;
4      DATALINES;
5        1  2
6        2  3
7        3  1
8        4  5
9        5  4
10       6  6
11       7  7
12       8  9
13       9 10
14      10  8
15       ;
16     RUN;
17
18     PROC CORR   DATA=D1   SPEARMAN;
19        VAR TEST1 TEST2;
20     RUN;
```

This program provides the output presented as Output 6.6. Notice that the format is identical to that observed with the Pearson correlations in Output 6.4 except that the heading above the matrix of correlations indicates that Spearman correlations have been printed. Correlation coefficients and significance estimates (i.e., *p* values) are interpreted in the usual way.

Output 6.6 Computing the Spearman Correlation between Test 1 and Test 2

```
                        The CORR Procedure

              2  Variables:    TEST1    TEST2

                       Simple Statistics

Variable     N       Mean      Std Dev     Median    Minimum    Maximum

TEST1       10      5.50000    3.02765     5.50000   1.00000   10.00000
TEST2       10      5.50000    3.02765     5.50000   1.00000   10.00000


              Spearman Correlation Coefficients, N = 10
                    Prob > |r| under H0: Rho=0

                               TEST1          TEST2

              TEST1          1.00000        0.91515
                                            0.0002

              TEST2          0.91515        1.00000
                             0.0002
```

When requesting Spearman correlations, the VAR and WITH statements are used in the same way as when computing Pearson correlation coefficients. That is:

- using only the VAR statement results in the printing of all possible correlations for the listed variables;

- combining the VAR with the WITH statement results in the printing of correlations for subsets of variables;

- leaving off these statements results in the printing of all possible correlations for all numeric variables.

The Chi-Square Test of Independence

When to Use

The chi-square test of independence (sometimes called the chi-square test of association or homogeneity) is appropriate when both variables are assessed on a nominal level of measurement. That is, when both variables are *classification* variables. Theoretically, either of the variables can have any number of categories, but in practice the number of categories is usually relatively small, perhaps 2 to 10.

The Two-Way Classification Table

The nature of the relationship between two nominal-level variables is easiest to understand using a **two-way classification** table. This is a table in which the rows represent the categories of one variable, while the columns represent the categories of the second variable. This two-way classification table is important because, once it has been prepared, it is possible to review the number of observations that appear in the various cells of the table to determine if there is any pattern that would indicate some relationship between variables.

For example, assume that you want to prepare a table that plots one variable that contains two categories against a second variable with three categories. The general form for such a table appears in Figure 6.6:

Figure 6.6 General Form for a Two-Way Classification Table

		Column Variable		
		column 1	column 2	column 3
Row Variable	row 1	$cell_{11}$	$cell_{12}$	$cell_{13}$
	row 2	$cell_{21}$	$cell_{22}$	$cell_{23}$

The point at which a row and column intersects is called a **cell** and each cell is given a unique subscript. The first number in this subscript indicates *row* to which the cell belongs, and the second number indicates *column* to which the cell belongs. So the general form for cell subscripts is "$\text{cell}_{r\,by\,c}$," where r = row and c = column. This means that cell_{21} is at the intersection of row 2 and column 1, cell_{13} is at the intersection of row 1 and column 3, and so forth. One of the first steps in performing a chi-square test of independence is to determine exactly how many participants fall into each of the cells of the classification table (i.e., how many participants appear in each subgroup). The pattern shown by these subgroups will help you understand whether the two classification variables are related to one another.

An Example

To make this a bit more specific, assume that you are a university administrator preparing to purchase a large number of new personal computers for three of the schools that constitute your university: the School of Arts and Science; the School of Education; and the School of Business. For a given school, you can purchase either IBM compatible computers or Macintosh computers, and you need to know which type of computer the students within each school tend to prefer.

In general terms, your research question is "Is there a relationship between school of enrollment and computer preference?" The chi-square test of independence will help answer this question. If this test shows that there is a relationship between the two variables, you can review the two-way classification table to determine which type of computer most students in each of the three schools prefer.

To answer this question, you draw a representative sample of 370 students from the 8,000 students that constitute the three schools. Each student is given a short questionnaire that asks just two questions:

```
1.    In which school are you enrolled? (Circle one):

      a. School of Arts and Sciences.

      b. School of Business.

      c. School of Education.


2.    Which type of computer do you prefer that we purchase for your
      school?  (circle one):

      a.   IBM compatible.

      b.   Macintosh.
```

These two questions constitute the two nominal-level variables for your study. Question 1 allows you to create a "school of enrollment" variable that can take on one of three values (Arts & Science, Business, or Education), while question 2 allows you to create a "computer preference" variable that can take on one of two values (IBM compatible or Macintosh). Clearly, these are *nominal-level* variables as they indicate merely group membership and provide no quantitative information.

You can now prepare a two-way classification table that plots preference against school. This table (with fictitious data) appears as Figure 6.7. Notice that in the table, computer preference is the row variable, in that row 1 represents students who preferred IBM compatibles, while row 2 represents those who preferred Macintosh. In the same way, you can see that school of enrollment is the column variable:

Figure 6.7 Two-Way Classification Table Plotting Computer Preference against School of Enrollment

		School of Enrollment		
		Arts & Sciences	Business	Education
Computer Preference	IBM Compatible	$n=30$	$n=100$	$n=20$
	Macintosh	$n=60$	$n=40$	$n=120$

Figure 6.7 presents the number of students who appear in each cell. For example, the first row of the table shows that, among those students who preferred IBM compatibles, 30 were Arts and Science students, 100 were Business students, and 20 were Education majors.

Remember that the purpose of the study is to determine whether there is any relationship between the two variables (i.e., ascertain whether school of enrollment is related to computer preference). This is just another way of saying "if you know what school in which a student is enrolled, does that help you predict what type of computer that student is likely to prefer?" In the present case, this question is easiest to answer if you review the table just one column at a time. For example, review just the Arts and Sciences column of the table. Notice that most of the students ($n = 60$) preferred Macintosh computers, while fewer ($n = 30$) preferred IBM compatibles. The column for the Business students shows the opposite trend, however. Most business students ($n = 100$) preferred IBM compatibles. Finally, the pattern for the Education students was similar to that of the Arts and Sciences students as the majority ($n = 120$) preferred Macintosh.

In short, there appears to be a relationship between school of enrollment and computer preference, with Business students preferring IBM compatibles, and Arts and Sciences and Education students preferring Macintoshes. At this point, this is just a trend observed in the *sample*. Is this trend of sufficient magnitude to conclude that the degree of difference is sufficiently larger than might occur by chance alone? To determine this, you conduct the chi-square test of independence.

Tabular versus Raw Data

You can use SAS to compute the chi-square test of independence regardless of whether you are dealing with tabular data or raw data. With **raw data**, you are working with data that have not been summarized or tabulated in any way. For example, imagine that you administered your questionnaire to 370 students and have not yet tabulated their responses. You merely have 370 completed questionnaires. In this situation, you are working with raw data.

On the other hand, **tabular data** are summarized in a table. For example, imagine that it was actually another researcher who administered this questionnaire, summarized participant responses in a two-way classification table (similar to Figure 6.7), and provided this completed table in a published report. In this case, you are dealing with tabular data.

When computing the chi-square statistic, there is no real advantage to using one form of data or the other, though you generally have a lot less data to enter if your data are already in tabular form. The following section shows how to input the data and request the chi-square statistic for tabular data. A subsequent section deals with raw data.

Computing Chi-Square from Tabular Data

Inputting Tabular Data

Often, the data to be analyzed with a chi-square test of independence have already been summarized in a two-way classification table such as in Figure 6.7. In these situations, you must create a special type of INPUT statement to read the data. Here is the general form:

```
DATA    dataset-name;
   INPUT    row-variable-name        $
            column-variable-name     $
            number-variable-name   ;
DATALINES;
row-value   column-value   number-in-cell
row-value   column-value   number-in-cell
row-value   column-value   number-in-cell

...more data...
```

The INPUT statement in this program tells SAS that the dataset includes three variables. The names of these three variables are symbolized as "row-variable-name," "column-variable-name," and "number-variable-name." The first is a character variable that codes the rows of the classification table (in the present study, the row-variable was "computer preference"). The second variable is a character variable that codes the columns of the table (here, the column-variable was "school of enrollment"). Finally, the third variable (symbolized as "number-variable-name") is a quantitative variable that codes how many participants appear in a given cell. (You will give specific names to these variables in the program to be presented shortly.)

The preceding program is in *free format*, meaning that it did not specify in which column in the DATALINES section each variable is located. However, this should not cause problems so long as you remember to separate each value in the DATALINES section by at least one blank space and do not accidentally skip any values in the DATALINES section.

Each line of data in the DATALINES section corresponds to just one of the cells in the classification table. In the preceding general form, the "number-in-cell" in the DATALINES section represents the number of participants in that cell. Therefore, the number of lines in the DATALINES section will be equal to the number of cells in the two-way classification table. The present classification table included six cells, so there will be six data lines in the DATALINES statement.

This is the actual data input step for inputting the tabular data presented in the two-way classification table of Figure 6.7:

```
1       DATA D1;
2          INPUT      PREFERNECE $
3                     SCHOOL       $
4                     NUMBER    ;
5
6       DATALINES;
7       IBM    ARTS       30
8       IBM    BUSINESS  100
9       IBM    EDUCATION  20
10      MAC    ARTS       60
11      MAC    BUSINESS   40
12      MAC    EDUCATION 120
13      ;
14      RUN;
```

The preceding INPUT statement tells SAS that the dataset contains just three variables for each line of data. The first variable is a character variable named PREFERENCE (coding student preferences, the row-variable), the second is a character variable named SCHOOL (coding the student's school, the column-variable), and the third variable is a numeric variable called NUMBER (indicating how many students appear in a given cell). Compare the INPUT statement from the preceding program to the INPUT statement from the general form presented earlier to verify that you understand what each variable name represents.

The DATALINES portion of the program includes six lines of data, one for each cell. The first cell represents those students who preferred IBM compatibles and were in the School of Arts and Sciences. The value for NUMBER on this line shows that there were 30 participants in this cell. The next line shows that there were 100 participants who preferred IBM compatibles and were in the School of Business, and so forth. You should compare the six lines of data to the six cells of Figure 6.6 to verify how the data were coded.

Computing Chi-Square with PROC FREQ Using Tabular Data

By now, you might be wondering why there is so much emphasis on preparing a two-way classification table when you want to perform a chi-square test of independence. This is necessary because computing the chi-square statistic involves determining the **observed frequencies** in each cell of the table (the number of observations that actually appear in each cell), and comparing these to the **expected frequencies** in each cell (i.e., the number of observations that you would expect to appear in each cell if the row variable and the column variable were completely unrelated). Now that your two-way classification table has been completed and entered, you can request the chi-square statistic.

Here is the general form for a SAS program that creates a two-way classification table for two nominal-level variables when the data have been entered in tabular form. The options used with these statements (described after the program) allow you to request a chi-square test of independence along with additional information.

```
PROC FREQ    DATA=dataset-name;
   TABLES    row-variable-name*column-variable-name    /    options ;
   WEIGHT    number-variable-name;
RUN;
```

These are some of the options for the TABLES statement that can be especially useful in social science research:

ALL

Requests several significance tests (including the chi-square test of independence) and measures of bivariate association. Although several statistics are printed, not all will be appropriate for a given analysis. The choice of the correct statistic will depend on the level of measurement used with the variables along with other considerations.

CHISQ

Requests the chi-square test of independence, and prints a number of measures of bivariate association based on the chi-square statistic.

EXACT

Prints Fisher's exact test. This is printed automatically for 2 x 2 tables (provided that the CHISQ option is specified), but must be specifically requested for other tables.

EXPECTED

Prints the expected cell frequencies. That is, the cell frequencies that would be expected if the two variables were, in fact, independent or unrelated. This is a very useful option for determining the *nature* of the relationship between variables.

MEASURES

Requests several measures of bivariate association, along with their asymptotic standard errors. These include the Pearson and Spearman correlation coefficients, gamma, Kendall's tau-b, Stuart's tau-c, symmetric lambda, asymmetric lambda, uncertainty coefficients, as well as other measures. Again, some of these indices will not be appropriate for a given study. All of these measures are printed if you request the ALL option.

To illustrate, here is a complete SAS program that reads tabular data, creates a two-way classification table, and prints the statistics requested by the ALL option (including the chi-square statistic):

```
 1      DATA D1;
 2         INPUT    PREFERENCE    $
 3                  SCHOOL        $
 4                  NUMBER    ;
 5
 6      DATALINES;
 7      IBM    ARTS        30
 8      IBM    BUSINESS    100
 9      IBM    EDUCATION   20
10      MAC    ARTS        60
11      MAC    BUSINESS    40
12      MAC    EDUCATION   120
13      ;
14      RUN;
15
16      PROC FREQ    DATA=D1;
17         TABLES    PREFERENCE*SCHOOL    /    ALL;
18         WEIGHT    NUMBER;
19      RUN;
```

The preceding TABLES statement requests that PREFERENCE be the row-variable and SCHOOL be the column-variable in the printed table. This request is followed by a slash (/), the ALL option, and a semicolon.

In the WEIGHT statement, you provide the name of the variable that codes the number of participants in each cell. In this case, you specify the variable NUMBER.

The two-way classification table produced by this program appears here as Output 6.7:

Output 6.7 Two-Way Classification Table Requested by PROC FREQ

```
                    Table of PREFERENCE by SCHOOL

          PREFERENCE        SCHOOL

          Frequency|
          Percent  |
          Row Pct  |
          Col Pct  |ARTS     |BUSINESS|EDUCATIO|  Total
          ---------+--------+--------+--------+
          IBM      |     30 |    100 |     20 |    150
                   |   8.11 |  27.03 |   5.41 |  40.54
                   |  20.00 |  66.67 |  13.33 |
                   |  33.33 |  71.43 |  14.29 |
          ---------+--------+--------+--------+
          MAC      |     60 |     40 |    120 |    220
                   |  16.22 |  10.81 |  32.43 |  59.46
                   |  27.27 |  18.18 |  54.55 |
                   |  66.67 |  28.57 |  85.71 |
          ---------+--------+--------+--------+
          Total          90      140      140      370
                       24.32    37.84    37.84   100.00
```

In the 2 x 3 classification table reproduced in Output 6.7, the name of the row variable (PREFERENCE) appears in the upper-left corner. The label for each row appears on the far-left side of the appropriate row. The first row (labeled IBM) represents the participants who preferred IBM compatibles and the second row (labeled MAC) represents participants who preferred Macintoshes.

The name of the column-variable (SCHOOL) appears above the three columns and each, in turn, is headed with its label. Column 1 represents the Arts and Sciences students, column 2 represents the Business students, and column 3 represents the Education students.

Where a given row and column intersect, information regarding participants in that cell is provided. Within each cell, the following information is provided (in this sequence):

1. The "Frequency" or the raw number of participants in the cell.

2. The "Percent" or the percent of participants in that cell relative to the total number of participants (the number of participants in the cell divided by the total number of participants).

3. The "Row Pct" or the percent of participants in that cell, relative to the number of participants in that row. For example, there are 30 participants in the IBM ARTS cell, and 150 participants in the IBM row. Therefore, the row percent for this cell is 30 / 150 = 20%.

4. The "Col Pct" or the percent of participants in that cell, relative to the number of participants in that column. For example, there are 30 participants in the IBM ARTS cell, and 90 participants in the ARTS column. Therefore, the column percent for this cell is 30 / 90 = 33.33%.

In the present example, it is particularly revealing to review the classification table just one column at a time and to pay particular attention to the last entry in each cell: the "column percent." First, consider the ARTS column. The column percent entries show that only 33.33% of the Arts and Sciences students preferred IBM compatibles whereas 66.67% preferred Macintoshes. Next, consider the BUSINESS column that shows the reverse trend: 71.43% of the Business students preferred IBM compatibles while only 28.57% preferred Macintoshes. Finally, the trend of the Education students in the EDUCATION column is similar to that for the Arts and Sciences students: only 14.29% preferred IBM compatibles while 85.71% preferred Macintoshes.

These percentages reinforce the suspicion that there is a relationship between school of enrollment and computer preference. But is the relationship statistically significant (i.e., can these differences occur by chance alone)? To answer this, you must consult the chi-square test of independence, which (along with other information) is reproduced in Output 6.8.

Output 6.8 Chi-Square Test of Independence and Other Statistics Requested by the ALL Option

```
            Statistics for Table of PREFERENCE by SCHOOL

        Statistic                     DF      Value      Prob
        --------------------------------------------------------
        Chi-Square                    2       97.3853    <.0001
        Likelihood Ratio Chi-Square   2      102.6849    <.0001
        Mantel-Haenszel Chi-Square    1       16.9812    <.0001
        Phi Coefficient                       0.5130
        Contingency Coefficient               0.4565
        Cramer's V                            0.5130
```

The chi-square test of independence is the first statistic in the table. It tests the null hypothesis that, in the population, the two variables are independent, or unrelated. When the null hypothesis is true, expect the value of the chi-square statistic to be relatively small. The stronger the relationship between the two variables in the sample, the larger the chi-square statistic will be.

Output 6.8 shows that the obtained chi-square value was approximately 97.39, with 2 degrees of freedom. The degrees of freedom for the chi-square test are calculated as:

```
df = (r-1)(c-1)
```

where:

```
r =  number of categories for the row variable and;
c =  number of categories for the column variable.
```

For the current analysis, the row variable (PREFERENCE) had two categories and column variable (SCHOOL) had three categories, so the degrees of freedom are calculated as:

```
df = (2-1)(3-1)
   = (1)(2)
   = 2
```

At 97.39, the obtained value of the chi-square statistic is quite large, given the degrees of freedom. The probability, or p value, for this chi-square statistic is printed below the heading "Prob" in Output 6.8. This p value is less than .01, meaning that there is less than one chance in 100 of obtaining a chi-square value of this size (or larger) by chance alone. You can therefore reject the null hypothesis and tentatively conclude that school of enrollment is related to computer preferences.

Computing Chi-Square from Raw Data

Inputting Raw Data

If data to be analyzed are in raw form (i.e., if the data have not already been summarized in a two-way classification table), you can enter them following the procedures discussed in Chapter 3. For example, if the preceding questionnaire had been administered to the 370 participants, you could enter their data according to the following guide:

Column	Variable Name	Explanation
1-4	SCHOOL	School of enrollment, coded: ARTS BUSINESS EDUCATION
5	blank	
6-8	PREFERENCE	Computer preference, coded: IBM = "IBM compatible" MAC = "Macintosh"

The entire data input step, including a small portion of the sample data, is presented here:

```
1       DATA D1;
2          INPUT    #1   @1   SCHOOL        $4.
3                         @6   PREFERENCE    $3.    ;
4       DATALINES;
5       ARTS          IBM
6       BUSINESS      IBM
7       BUSINESS      MAC
8       EDUCATION     IBM
9       .
10      .
11      .
12      EDUCATION     MAC
13      ARTS          MAC
14      BUSINESS      IBM
15      ;
16      RUN;
```

The fictitious data for the participants begin on line 5, and there is one line of data for each. You can see that the first participant was an Arts and Sciences student and preferred an IBM compatible, the second student was a Business student and also preferred an IBM compatible, and so forth. For this program, there are 370 lines of data because there are 370 participants.

The preceding program specified SCHOOL and PREFERENCE as *character* variables with values such as ARTS and MAC, but it also would have been possible to code them as *numeric* variables. For example, SCHOOL could have been coded so that 1 = Arts and Sciences, 2 = Business, and 3 = Education. You could have proceeded with the analysis in the usual fashion though you would then have to make a record to remember exactly which group is represented by these numerical values, or use the VALUES statement of PROC FORMAT to attach meaningful value labels (e.g., "ARTS" and "BUSINESS") to the variable categories when they are printed. For the latter approach, see *SAS/STAT User's Guide*.

Computing Chi-Square with PROC FREQ Using Raw Data

Here is the general form of the statements that request a chi-square test of independence (along with other statistics) when the data are input in raw form:

```
PROC FREQ    DATA=dataset-name;
    TABLES    row-variable*column-variable    /    options ;
RUN;
```

This general form is identical to the general form used with tabular data except that the WEIGHT statement is deleted. The full program (including a portion of the data) to compute the chi-square test with raw data is presented here:

```
 1        DATA D1;
 2            INPUT   #1   @1   SCHOOL       $4.
 3                              @6   PREFERENCE   $3.    ;
 4        DATALINES;
 5        ARTS        IBM
 6        BUSINESS    IBM
 7        BUSINESS    MAC
 8        EDUCATION   IBM
 9        .
10        .
11        .
12        EDUCATION   MAC
13        ARTS        MAC
14        BUSINESS    IBM
15        ;
16        RUN;
17
18        PROC FREQ   DATA=D1;
19            TABLES   PREFERENCE*SCHOOL    /    ALL;
20        RUN;
```

From this point forward, the analysis proceeds in exactly the same manner as when the dataset was based on tabular data. You can request the same options, and you interpret the results in exactly the same way.

Special Notes Regarding the Chi-Square Test

Using Fisher's Exact Test for 2 x 2 Tables and Larger Tables

A 2 x 2 table contains just two rows and two columns. A two-way classification table for a chi-square study will be a 2 x 2 table if there are just two values for the row-variable and just two values for the column-variable. Imagine that you modified the preceding computer preference study so that there were just two values for the computer preference variable (i.e., IBM compatible and Macintosh) as before, but just two values for the school of enrollment variable (i.e., Arts and Sciences and Business). The two-way classification table that would result from this modified study would resemble the one portrayed in Figure 6.7 except that the column for the School of Education would be eliminated. The resulting table is called a 2 x 2 table because it consists of just two rows (IBM versus Macintosh) and two columns (Arts & Sciences versus Business).

When analyzing a 2 x 2 classification table, it is best to use Fisher's exact test rather than the standard chi-square test of independence. This test is printed automatically whenever a 2 x 2 table is analyzed and you request the FISHER option. In the SAS output, examine the probability value that appears to the right of the heading "Fisher's Exact Test." This estimates the probability of observing a table that gives at least as much evidence of association as the one actually observed, given that the null hypothesis is true. (See the *SAS/STAT User's Guide* for more information.) In other words, when the *significance level* for Fisher's exact test is less than .05, you can reject the null hypothesis that the two nominal-scale variables are independent in the population, and can conclude that they are, in fact, related.

In some situations, Fisher's exact test can also be appropriate for larger classification tables (i.e., for tables with more than two rows and/or columns). This is the case when the sample size is small and the sample size per degree of freedom is less than 5. With larger classification tables, you must specifically request the Fisher's exact test by specifying the EXACT option in the TABLES statement. See the *SAS/STAT User's Guide* for further details.

Minimum Cell Frequencies

The chi-square test might not be valid if the observed frequency in any of the cells is zero or if the expected frequency in any of the cells is less than five (use the EXPECTED option with the TABLES statement to compute expected cell frequencies). When these minimums are not met, consider gathering additional data or perhaps collapsing similar categories in order to increase cell frequencies.

Conclusion

Bivariate associations are the simplest types of associations studied and the statistics presented here (the Pearson correlation, the Spearman correlation, and the chi-square test of independence) are appropriate for examining most types of bivariate relationships encountered in the social sciences. With these relatively simple procedures behind you, you are now ready to proceed to other tests of group differences.

Assumptions Underlying the Tests

Assumptions Underlying the Pearson Correlation Coefficient

- **Interval-level measurement.** Both the predictor and criterion variables should be assessed on an interval- or ratio-level of measurement.
- **Random sampling.** Each participant in the file will contribute one score on the predictor variable, and one score on the criterion variable. These pairs of scores should represent a random sample drawn from the population of interest.
- **Linearity.** The relationship between the criterion and predictor variables should be linear. This means that the mean criterion scores at each value of the predictor variable should fall on a straight line. The Pearson correlation coefficient is not

appropriate for assessing the strength of the relationship between two variables with a curvilinear relationship.

- **Bivariate normal distribution.** The pairs of scores should follow a bivariate normal distribution. That is, scores on the criterion variable should form a normal distribution at each value of the predictor variable. Similarly, scores of the predictor variable should form a normal distribution at each value of the criterion variable. When scores represent a bivariate normal distribution, they form an *elliptical scattergram* when plotted (i.e., their scattergram is shaped like a rugby ball: fat in the middle and tapered on the ends).

Assumptions Underlying the Spearman Correlation Coefficient

- **Ordinal-level measurement.** Both the predictor and criterion variables should be assessed on an ordinal level of measurement. However, interval- or ratio-level variables are sometimes analyzed with the Spearman correlation coefficient when one or both variables are markedly skewed.

Assumptions Underlying the Chi-Square Test of Independence

- **Nominal-level measurement.** Both variables should be assessed on a nominal scale.

- **Random sampling.** Participants contributing data should represent a random sample drawn from the population of interest.

- **Independent cell entries.** Each participant should appear in one cell only. The fact that a given participant appears in one cell should not affect the probability of another appearing in any other cell (i.e., independence of observations).

- **Expected frequencies of five or more.** When analyzing a 2 x 2 classification table, no cell should display an expected frequency of less than 5. With larger tables (e.g., 3 x 4 tables), no more than 20% of the cells should have expected frequencies less than 5.

References

Hays, W. L. (1988). Statistics (4th ed.). New York: Holt, Rinehart, and Winston.

SAS Institute Inc. (2004). SAS/STAT 9.1 user's guide. Cary, NC: Author.

Assessing Scale Reliability with Coefficient Alpha

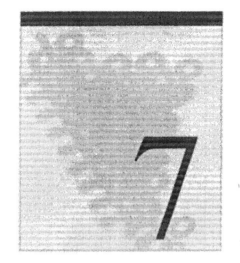

Overview. This chapter shows how to use PROC CORR to compute the coefficient alpha reliability index for a multiple-item scale. It reviews basic issues regarding the assessment of reliability, and describes the circumstances under which a measure of internal consistency is likely to be high. Fictitious questionnaire data are analyzed to demonstrate how you can use the results of PROC CORR to perform an item analysis, thereby improving the reliability of scale responses.

Introduction: The Basics of Scale Reliability

You can compute coefficient alpha when you have administered a multiple-item rating scale to a group of participants and want to determine the internal consistency of responses to the scale. The items constituting the scale can be scored dichotomously (scored as "right" or "wrong") or the items can have a multiple-point rating format (e.g., participants can respond to each item using a 7-point "agree-disagree" rating scale).

This chapter shows how to use the SAS PROC CORR procedure to compute the coefficient alpha for the types of scales that are often used in social science research. However, this chapter does not show how to actually *develop* a multiple-item scale for use in research. To learn about recommended approaches for creating summated rating scales, see Spector (1992).

Example of a Summated Rating Scale

A **summated rating scale** usually consists of a short list of statements, questions, or other items to which participants respond. Very often, the items that constitute the scale are statements, and participants indicate the extent to which they agree or disagree with each statement by selecting some response on a rating scale (e.g., a 7-point rating scale in which 1 = "Strongly Disagree" and 7 = "Strongly Agree"). The scale is called a *summated* scale because the researcher typically sums responses to all selected responses to create an overall score on the scale. These scales are often referred to as **Likert-type** scales.

Imagine that you are interested in measuring job satisfaction in a sample of employees. To do this, you might develop a 10-item scale that includes items such as "in general, I am satisfied with my job." Employees respond to these items using a 7-point response format in which 1 = "Strongly Disagree" and 7 = "Strongly Agree."

You administer this scale to 200 employees and compute a job satisfaction score for each by summing his or her responses to the 10 items. Scores can range from a low of 10 (if the employee circled "Strongly Disagree" for each item) to a high of 70 (if the employee circled "Strongly Agree" for each item). Given the way these scores were created, higher scores indicate higher levels of job satisfaction. With the job satisfaction scale now developed and administered to a sample, you hope to use it as a predictor or criterion variable in research. However, the people who later read about your research are going to have questions about the psychometric properties of responses to your scale. At the very least, they will want to see empirical evidence that responses to the scale are reliable. This chapter discusses the meaning of scale reliability and shows how SAS can be used to obtain an index of internal consistency for summated rating scales.

True Scores and Measurement Error

Most observed variables measured in the social sciences (e.g., scores on your job satisfaction scale) actually consist of two components: a **true score** that indicates where the participant actually stands on the variable of interest along with **measurement error**. Almost all observed variables in the social sciences contain at least some measurement error, even variables that seem to be objectively measured.

Imagine that you assess the observed variable "age" in a group of participants by asking them to indicate their age in years. To a large extent, this observed variable (what the participants wrote down) is influenced by the true score component. To a large extent, what they write will be influenced by how old they actually are. Unfortunately, however, this observed variable will also be influenced by measurement error. Some will write down the wrong age because they don't know how old they are, some will write the wrong age because they don't want the researcher to know how old they are, and other participants will write the wrong age because they didn't understand the question. In short, it is likely that there will not be a perfect correlation between the observed variable (what the participants write down) and their true scores on the underlying construct (i.e., their actual age).

This can occur even though the "age" variable is relatively objective and straightforward. If a question such as this is going to be influenced by measurement error, imagine how much more error results when more subjective constructs are measured (e.g., items that constitute your job satisfaction scale).

Underlying Constructs versus Observed Variables

In applied research, it is useful to draw a distinction between underlying constructs versus observed variables. An **underlying construct** is the hypothetical variable that you want to measure. In the job satisfaction study, for example, you wanted to measure the underlying construct of job satisfaction within a group of employees. The **observed variable**, on the other hand, consists of the responses that you actually obtained. In that example, the observed variable consisted of scores on the 10-item measure of job satisfaction. These scores may or may not be a good measure of the underlying construct.

Reliability Defined

With this understanding, it is now possible to provide some definitions. A **reliability coefficient** can be defined as the percent of variance in an observed variable that is accounted for by true scores on the underlying construct. For example, imagine that in the study just described, you were able to obtain two scores for the 200 employees in the sample: their observed scores on the job satisfaction questionnaire; and their true scores on the underlying construct of job satisfaction. Assume that you compute the correlation between these two variables. The square of this correlation coefficient represents the reliability of responses to your job satisfaction scale; it is the percent of variance in observed job satisfaction scores that is accounted for by true scores on the underlying construct of job satisfaction.

The preceding was a technical definition for reliability but this definition is of little use in practice because it is generally not possible to obtain true scores for a variable. For this reason, reliability is usually defined in terms of the **consistency** of the scores that are obtained on the observed variable. An instrument is said to be reliable if it is shown to provide consistent scores upon repeated administration, upon administration by alternate forms, and so forth. A variety of methods of estimating scale reliability are used in practice.

Test-Retest Reliability

Assume that you administer your measure of job satisfaction to a group of 200 employees at two points in time: once in January and again in March. If responses to the instrument were indeed reliable, you would expect that the participants who provided high scores in January will tend to provide high scores again in March, and that those who provided low scores in January will also provide low scores in March. These results would support the test-retest reliability of responses to the scale. Test-retest reliability is assessed by administering the same instrument to the same sample of participants at two points in time, and then computing the correlation between the sets of scores.

But what is an appropriate interval over which questionnaires should be administered? Unfortunately, there is no hard-and-fast rule here; the interval depends on what is being measured. For enduring constructs such as personality variables, test-retest reliability has been assessed over several decades. For other constructs such as depressive symptomatology, the interval tends to be much shorter (e.g., weeks) due to the fluctuating course of depression and its symptoms. Generally speaking, the test-retest interval should not be too short so that respondents recall their responses to specific items (e.g., less than a week) but not so long as to measure natural variability in the construct (e.g., bona fide change in depressive symptoms). The former leads to an overstatement of test-retest reliability whereas the latter leads to understatement of test-retest reliability.

Internal Consistency

A further problem with the test-retest reliability procedure is the time that it requires. What if you do not have time to perform two administrations of the scale? In such situations, you are likely to turn to reliability indices that may be obtained with only one administration. In research that involves the use of questionnaire data, the most popular of these are the internal consistency indices of reliability. Briefly, **internal consistency** is the extent to which the individual items that constitute a test correlate with one another or with the test total. In the social sciences, the most widely used index of internal consistency is the coefficient alpha symbolized by the Greek letter α (Cronbach, 1951)[1].

---

[1] Usage of the Greek letter alpha (α) to represent an index of internal consistency should not be confused with the alpha used to specify significance levels for other statistical analyses described in this text.

Coefficient Alpha

Formula

Coefficient alpha is a general formula for scale reliability based on internal consistency. It provides the lowest estimate of reliability that can be expected for an instrument.

The formula for coefficient alpha is as follows:

$$\alpha_{xx} = \left(\frac{N}{N-1} \right) \left(\frac{S^2 - \Sigma S_i^2}{S^2} \right)$$

where

α_{xx} = Coefficient alpha

N = Number of items constituting the instrument

S^2 = Variance of the summated scale scores (e.g., assume that you compute a total score for each participant by summing responses to the items that constitute the scale; the variance of this total score variable would be S^2)

ΣS_i^2 = The sum of the variances of the individual items that constitute this scale

When Will Coefficient Alpha Be High?

Other factors held constant, coefficient alpha will be high to the extent that many items are included in the scale, and the items that constitute the scale are highly correlated with one another.

To understand why a coefficient alpha is high when the items are highly correlated with one another, consider the second term in the preceding formula:

$$\left(\frac{S^2 - \Sigma S_i^2}{S^2} \right)$$

This term shows that the variance of the summated scales scores is (essentially) divided by itself to compute coefficient alpha. However, the combined variance of the individual items is first subtracted from this variance before division is performed. This part of the equation shows that, if combined variance of the individual items is a small value, then coefficient alpha will be a relatively larger value.

This is important because (with other factors held constant), the stronger the correlations between the individual items, the smaller the $\sum S^2_i$ term. This is why coefficient alpha for responses to a given scale is likely to be large to the extent that the variables constituting that scale are strongly correlated.

Assessing Coefficient Alpha with PROC CORR

Imagine that you have conducted research in the area of prosocial behavior and have developed an instrument designed to measure two separate underlying constructs: helping others and financial giving. **Helping others** refers to prosocial activities performed to help coworkers, relatives, and friends whereas **financial giving** refers to giving money to charities or the homeless. (See Chapter 15, "Principal Component Analysis," for a more detailed description of these constructs.) In the following questionnaire, items 1 to 3 are designed to assess helping others and items 4 to 6 are designed to assess financial giving.

```
Instructions:  Below are a number of activities in which people
sometimes engage.  For each item, please indicate how frequently
you have engaged in this activity over the past six months.
Provide your response by circling the appropriate number to the
left of the item, and use the following response key:

    7 = Very Frequently
    6 = Frequently
    5 = Somewhat Frequently
    4 = Occasionally
    3 = Seldom
    2 = Almost Never
    1 = Never


    1 2 3 4 5 6 7    1.  Went out of my way to do a favor for a
                         coworker.

    1 2 3 4 5 6 7    2.  Went out of my way to do a favor for a
                         relative.

    1 2 3 4 5 6 7    3.  Went out of my way to do a favor for a
                         friend.

    1 2 3 4 5 6 7    4.  Gave money to a religious charity.

    1 2 3 4 5 6 7    5.  Gave money to a charity not associated with
                         a religion.

    1 2 3 4 5 6 7    6.  Gave money to a panhandler.
```

Assume that you have administered this 6-item questionnaire to 50 participants. For the moment, we are concerned only with the reliability of the scale that includes items 1 through 3 (i.e., the items that assess helping others).

Let us further assume that you have made a mistake in assessing the reliability of this scale. Assume that you erroneously believed that the helping others construct was assessed by items 1 through 4 (whereas, in reality, the construct was assessed by items 1 through 3). It will be instructive to see what you learn when you mistakenly include item 4 in the analysis.

General Form

Here is the general form for the SAS statements that estimate the coefficient alpha (internal consistency) for a summated rating scale:

```
PROC CORR    DATA=dataset-name    ALPHA    NOMISS;
    VAR  list-of-variables;
RUN;
```

In the preceding program, the ALPHA option requests that the coefficient alpha be computed for the group of variables included in the VAR statement. The NOMISS option is required to compute the coefficient alpha. The VAR statement should list only the variables (items) that constitute the scale in question. You must perform a separate CORR procedure for each scale whose reliability you want to assess.

A 4-Item Scale

Here is an actual program, including the DATA step to analyze fictitious data from your study. Only a few sample lines of data appear here. The complete dataset appears in Appendix B. Ordinarily, one would not compute Cronbach's alpha in this case as internal consistency is often underestimated with so few items. Cronbach's alpha also tends to overestimate the internal consistency of responses to scales with 40 or more items (Cortina, 1993). The following examples are provided simply to illustrate the computation and meaning of Cronbach's alpha.

```
1     DATA D1;
2        INPUT    #1    @1    (V1-V6)    (1.)   ;
3
4     DATALINES;
5     556754
6     567343
7     777222
8     .
9     .
10    .
11    767151
12    455323
13    455544
14    ;
15    RUN;
16
17    PROC CORR    DATA=D1    ALPHA    NOMISS;
18        VAR V1 V2 V3 V4;
19    RUN;
```

The results of this analysis appear as Output 7.1. Page 1 of these results provides the means, standard deviations, and other descriptive statistics that you should review to verify that the analysis proceeded as expected. Page 2 provides the results pertaining to the reliability of responses to the scale.

Output 7.1 Simple Statistics and Coefficient Alpha Results for Analysis of Scale That Includes Items 1 through 4, Prosocial Behavior Study

```
                              The SAS System                              1

                            The CORR Procedure

              4  Variables:     V1       V2       V3       V4


                            Simple Statistics

Variable     N       Mean    Std Dev         Sum     Minimum     Maximum
V1          50    5.18000    1.39518   259.00000     1.00000     7.00000
V2          50    5.40000    1.10657   270.00000     3.00000     7.00000
V3          50    5.52000    1.21622   276.00000     2.00000     7.00000
V4          50    3.64000    1.79296   182.00000     1.00000     7.00000
```

```
                                                                          2

                     Cronbach Coefficient Alpha

               Variables                  Alpha
               ---------------------------
               Raw                      0.490448
               Standardized             0.575912


          Cronbach Coefficient Alpha with Deleted Variable

                  Raw Variables            Standardized Variables

Deleted       Correlation                Correlation
Variable      with Total       Alpha     with Total        Alpha
-------------------------------------------------------------------------
V1             0.461961      0.243936      0.563691       0.326279
V2             0.433130      0.318862      0.458438       0.420678
V3             0.500697      0.240271      0.546203       0.342459
V4            -.037388       0.776635     -.030269        0.773264


               Pearson Correlation Coefficients, N = 50
                     Prob > |r| under H0: Rho=0

                    V1            V2            V3            V4

     V1        1.00000       0.49439       0.71345      -0.10410
                             0.0003        <.0001        0.4719

     V2        0.49439       1.00000       0.38820       0.05349
               0.0003                      0.0053        0.7122

     V3        0.71345       0.38820       1.00000      -0.02471
               <.0001        0.0053                      0.8648

     V4       -0.10410       0.05349      -0.02471       1.00000
               0.4719        0.7122        0.8648
```

On page 2 of Output 7.1, to the right of the heading "Cronbach Coefficient Alpha (Raw)," you see that the reliability coefficient for the scale that includes items 1 through 4 is only .49 (rounded to two decimal places). Reliability estimates for raw variables are normally reported in published reports as opposed to the standardized alphas.

How Large Must a Reliability Coefficient Be to Be Considered Acceptable?
A widely used rule of thumb of .70 has been suggested by Nunnally (1978). In contrast, reliability coefficients less than .70 are generally seen as inadequate. However, remember that this is only a rule of thumb, and social scientists sometimes report coefficient alphas under .70 (and sometimes even under .60)!

Is a larger alpha coefficient always better than a smaller one? Not necessarily. An ideal estimate of internal consistency is believed to be between .80 and .90 (i.e., $.90 \geq \alpha \geq .80$; Clark & Watson, 1995; DeVellis, 1991). This is because estimates in excess of .90 are suggestive of item redundancy or inordinate scale length.

Back to our example, the coefficient alpha of .49 reported in Output 7.1 is not acceptable; obviously, it should be possible to significantly improve this coefficient. But how?

In some situations, the reliability of responses to a multiple-item scale is improved by deleting those items with poor item-total correlations. An **item-total correlation** is the correlation between an individual item and the sum of the remaining items that constitute the scale. If an item-total correlation is small, this can be seen as evidence that the item is not measuring the same construct measured by the other scale items. You might therefore choose to discard items exhibiting small item-total correlations (assuming that data have been entered correctly).

Consider Output 7.1. Under the "Correlation with Total" (Raw Variables) heading, you can see that items 1 through 3 each demonstrate reasonably strong correlations with the sum of the remaining items on the scale. However, item V4 demonstrates an item-total correlation of approximately –.04. This suggests that item V4 is not measuring the same construct as items V1 to V3.

In Output 7.1 under the "Alpha" heading, you find an estimate of what alpha would be if a given variable (item) was deleted from the scale. To the right of "V4", PROC CORR estimates that alpha would be approximately .78 if V4 were deleted. (This value appears where the row headed "V4" intersects with the column headed "Alpha" in the "Raw Variables" section.) This makes sense because variable V4 demonstrates a correlation with the remaining scale items of only –.04. You could substantially improve this scale by removing the item that is not measuring the same construct assessed by the other items.

A 3-Item Scale
Output 7.2 reveals the results of PROC CORR when coefficient alpha is requested for just variables V1 to V3. This is done by specifying only V1 to V3 in the VAR statement.

Output 7.2 Simple Statistics and Coefficient Alpha Results for Analysis of Scale That Includes Items 1 through 3, Prosocial Behavior Study

```
                          The SAS System                          3

                         The CORR Procedure

           3  Variables:    V1       V2       V3


                         Simple Statistics

  Variable      N        Mean      Std Dev         Sum      Minimum      Maximum

  V1           50     5.18000      1.39518   259.00000      1.00000      7.00000
  V2           50     5.40000      1.10657   270.00000      3.00000      7.00000
  V3           50     5.52000      1.21622   276.00000      2.00000      7.00000
```

```
                                                                  4
                     Cronbach Coefficient Alpha

                 Variables              Alpha
                 ---------------------------
                 Raw                 0.776635
                 Standardized        0.773264


          Cronbach Coefficient Alpha with Deleted Variable

               Raw Variables                Standardized Variables

  Deleted    Correlation                  Correlation
  Variable   with Total       Alpha       with Total       Alpha
  --------------------------------------------------------------------
  V1          0.730730      0.557491      0.724882      0.559285
  V2          0.480510      0.828202      0.476768      0.832764
  V3          0.657457      0.649926      0.637231      0.661659


             Pearson Correlation Coefficients, N = 50
                  Prob > |r| under H0: Rho=0

                         V1            V2            V3

            V1      1.00000       0.49439       0.71345
                                  0.0003        <.0001

            V2      0.49439       1.00000       0.38820
                    0.0003                      0.0053

            V3      0.71345       0.38820       1.00000
                    <.0001        0.0053
```

Page 4 of Output 7.2 provides a raw-variable coefficient alpha of .78 for the three variables included in this analysis. This value appears to the right of the heading "Cronbach Coefficient Alpha" to the right of the heading "Raw." This coefficient exceeds the recommended minimum value of .70 (Nunnally, 1978) and approaches the ideal range of $.90 \geq \alpha \geq .80$ (Clark & Watson, 1995; DeVellis, 1991). Clearly, responses to the helping others subscale demonstrate a much higher level of reliability with item V4 deleted.

Summarizing the Results

Summarizing the Results in a Table

Researchers typically report the reliability of a scale in a table that reports simple descriptive statistics for the study's variables such as means, standard deviations, and intercorrelations. In these tables, coefficient alpha estimates are usually reported on the diagonal of the correlation matrix, within parentheses. Such an approach appears in Table 7.1.

Table 7.1

Means, Standard Deviations, Intercorrelations, and Coefficient Alpha Estimates for Study Variables

Variables	Mean	SD	1	2	3
1. Authoritarianism	13.56	2.54	(.90)		
2. Helping others	15.60	3.22	.37	(.78)	
3. Financial giving	12.55	1.32	.25	.53	(.77)

Note. N = 200. Reliability estimates appear on the diagonal.

In the preceding table, information for the authoritarianism variable is presented in both the row and the column that is headed "1." Where the row headed "1" intersects with the column headed "1," you find the coefficient alpha for the authoritarianism scale; you can see that this index is .90. In the same way, you can find the coefficient alpha for helping others where row 2 intersects with column 2 ($\alpha = .78$) and you can find the coefficient alpha for financial giving where row 3 intersects with column 3 ($\alpha = .77$).

Preparing a Formal Description of the Results for a Paper

When reliability estimates are computed for a relatively large number of scales, it is common to report them in a table (such as Table 7.1) and make only passing reference to them within the text of the paper when within acceptable parameters. For example, within the section on instrumentation, you might indicate:

Estimates of internal consistency as measured by Cronbach's alpha all exceeded .70 and are reported on the diagonal of Table 7.1.

When reliability estimates are computed for only a small number of scales, it is possible to instead report these estimates within the body of the text itself. Here is an example of how this might be done:

```
Internal consistency of scale responses was assessed by
Cronbach's alpha.  Reliability estimates were .90, .78, and .77
for responses to the authoritarianism, helping others, and
financial giving subscales, respectively.
```

Conclusion

Assessing scale reliability with the coefficient alpha (or some other reliability index) should be one of the first tasks you undertake when conducting questionnaire research. If responses to selected scales are not reliable, there is no point performing additional analyses. You can often improve suboptimal reliability estimates by deleting items with poor item-total correlations in keeping with the procedures discussed in this chapter. When several subscales on a questionnaire display poor reliability, it might be advisable to perform a principal component analysis or an exploratory factor analysis on responses to all questionnaire items to determine which tend to group together empirically. If many items load on each retained factor and if the factor pattern obtained from such an analysis displays a simple structure, chances are good that responses to the resulting scales will demonstrate adequate internal consistency.

References

Clark, L. A., & Watson, D. (1995). Constructing validity: Basic issues in objective scale development. *Psychological Assessment, 7,* 309-319.

Cortina, J. M. (1993). What is coefficient alpha? An examination of theory and applications. *Journal of Applied Psychology, 78,* 98-104.

Cronbach, L. J. (1951). Coefficient alpha and the internal structure of tests. *Psychometrika, 16,* 297-334.

DeVellis, R. F. (1991). *Scale development: Theory and applications.* Newbury Park, CA: Sage.

Nunnally, J. (1978). *Psychometric theory.* New York: McGraw-Hill.

Spector, P. E. (1992). *Summated rating scale construction: An introduction.* Newbury Park, CA: Sage.

t Tests: Independent Samples and Paired Samples

> **Overview.** This chapter describes the differences between the independent-samples *t* test and the paired-samples *t* test, and shows how to perform both types of analyses. It develops an example of a research design that would provide data appropriate for a *t* test. With respect to the independent-samples test, the chapter shows how to write the necessary program using PROC TTEST, determine whether the equal-variances or unequal variances *t* test is appropriate, and interpret the results. With respect to the paired-samples test, it provides examples of paired-samples research designs, discusses problems with these designs, and shows how to perform the analysis using PROC MEANS and PROC TTEST.

Introduction: Two Types of *t* Tests

A *t* test is appropriate when your analysis involves a single predictor variable that is measured on a nominal scale and assumes only one of two values (e.g., sex), and a single criterion variable that is measured on an interval or ratio scale (e.g., GRE scores). This statistical procedure is usually viewed as a test of group differences. For example, when experimental condition is the predictor variable and scores on an attitude scale are the criterion variable, you might want to know whether there is a significant difference between the experimental group and the control group with respect to their mean attitude scores. A *t* test can determine this.

There are actually two types of *t* tests that are appropriate for different experimental designs. First, the **independent-samples *t* test** is appropriate if the observations obtained under one treatment condition are independent of (i.e., unrelated to) the observations obtained under the other treatment condition. For example, imagine that you draw a random sample of participants and randomly assign each participant to either Condition 1 or Condition 2 in your experiment. After manipulating the independent variable, you determine scores on the attitude scale for participants in both conditions, and use an independent-samples *t* test to determine whether the mean attitude score is significantly different for the participants in Condition 1 as compared to those in Condition 2. The independent-samples *t* test is appropriate here because the observations (attitude scores) in Condition 1 are completely unrelated to the observations in Condition 2 (i.e., Condition 1 consists of one group of people, and Condition 2 consists of a different group of people who are not related to, or affected by, the people in Condition 1).

The second type of test is the **paired-samples *t* test**. This statistic is appropriate if each observation in Condition 1 is paired in a meaningful way with a corresponding observation in Condition 2. There are several ways that this pairing can be achieved. For example, imagine that you draw a random sample of participants and decide that each participant will provide two attitude scores: one score after being exposed to Condition 1; and a second score after being exposed to Condition 2. In a sense, you still have two samples of observations (the sample from Condition 1 versus that from Condition 2), but the observations from the two samples are now *related* (i.e., from the same set of participants at two points in time). For example, this means that if a given participant scored relatively high on the attitude scale under Condition 1, it is likely that participant will also score relatively high under Condition 2. When analyzing the data, it therefore makes sense to pair each participant's score from Condition 1 with his or her score from Condition 2. Because of this pairing, a paired-samples *t* statistic is calculated differently than an independent-samples *t* statistic.

This chapter has two major sections. The first deals with the independent-samples *t* test, and the second deals with the paired-samples *t* test. These sections provide additional examples of situations in which the two procedures might be appropriate.

Earlier, you read that a *t* test is appropriate when the analysis involves a nominal-scale predictor variable and an interval/ratio-scale criterion. A number of additional assumptions should also be met for the test to be valid (e.g., responses to the dependent variable are normally distributed). These assumptions are summarized in an appendix at the end of this chapter. When these assumptions are violated, consider using a nonparametric statistic instead. For help, see the *SAS/STAT User's Guide*.

The Independent-Samples *t* Test

Example: A Test of the Investment Model

The use of the independent-samples *t* test is illustrated by testing a hypothesis that could be derived from the investment model (Rusbult, 1980). As discussed in earlier chapters, the investment model identifies a number of variables that are predicted to affect commitment to romantic relationships (as well as to other types of relationships). *Commitment* can be defined as the participant's intention to remain in the relationship and to maintain the relationship. One version of the investment model predicts that commitment will be affected by four variables: rewards; costs; investment size; and alternative value. These variables are briefly defined below:

Rewards:	the number of "good things" that the participant associates with the relationship (i.e., the positive aspects of the relationship);
Costs:	the number of "bad things" or hardships associated with the relationship;
Investment Size:	the amount of time and personal resources that the participant puts into the relationship;
Alternative value:	the attractiveness of alternatives to the relationship (e.g., the attractiveness of other potential romantic partners).

At least four testable hypotheses could be derived from the investment model as described here: (a) rewards have a causal effect on commitment; (b) costs have a causal effect on commitment; (c) investment size has a causal effect on commitment; (d) alternative value has a causal effect on commitment. If you were ambitious, you might design a single study that tests all four of these hypotheses simultaneously. Such a study would be quite complex, however. To keep things simple, this chapter focuses on testing only the first hypothesis (i.e., the prediction that the level of rewards affects commitment).

Rewards refer to the positive aspects of the relationship. Your relationship would score high on rewards if your partner was physically attractive, intelligent, kind, fun, rich, and so forth. Your relationship would score low on rewards if your partner was unattractive, unintelligent, unfeeling, dull, and so forth. It can be seen that the hypothesized relationship between rewards and commitment makes good intuitive sense (i.e., an increase in rewards *should* result in an increase in commitment). The predicted relationship between these variables is illustrated with Figure 8.1:

Figure 8.1 Hypothesized Causal Relationship between Rewards and Commitment

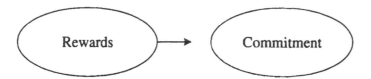

There are several ways that you might test the hypothesis that rewards have a causal effect on commitment. One approach would involve an experimental procedure in which participants are given written descriptions of different fictitious romantic partners and asked to rate their likely commitment to these partners. The descriptions could be manipulated so that a given fictitious partner could be described as a "high-reward" partner to one group of participants, and as a "low-reward" partner to a second group of participants. If the hypothesis concerning the relationship between rewards and commitment was correct, you would expect to see higher commitment scores for the high-reward partner. This part of the chapter describes a fictitious study that utilizes such a procedure and then tests the relevant null hypothesis using an independent-samples *t* test.

The Study

Assume that you have drawn a sample of 20 participants and have randomly assigned 10 to the high-reward condition and 10 to the low-reward condition. All participants were given a packet of materials, and the following instructions appeared on the first page:

```
In this study, you are asked to imagine that you are single and not
involved in any romantic relationships.  You will read descriptions
of 10 different "partners" (i.e., people with whom you might be
involved in a romantic relationship).  For each description, imagine
that you are involved in a romantic relationship with that person.
Think about what it would be like to date that person, given his/her
positive features, negative features, and other considerations.
After you have thought about it, rate how committed you would be to
maintaining your romantic relationship with that person.  Each
"partner" is described on a separate sheet of paper, and at the
bottom of each sheet there are four items with which you can rate
your commitment to that particular relationship.
```

On the subsequent 10 pages were descriptions of the 10 partners, with each partner described on a separate page. The paragraph describing a given partner provides information about the extent to which the relationship with that person was rewarding and costly. It also provides information relevant to the investment size and alternative value associated with the relationship.

The Dependent Variable

The dependent variable in this study is "participant commitment to a specific romantic partner." It would be ideal if you could arrive at a single score that indicates how committed a given participant is to a given partner. High scores on this variable would reveal high commitment to the partner whereas low scores would suggest the opposite. This section describes one way that you could use rating scales to arrive at such a score.

At the bottom of the sheet that described a given partner, the participant was provided with four items that used a 9-point Likert-type rating format. Participants are asked to respond to these items to indicate the strength of their commitment to the partner described on that page. The items used in making these ratings are presented here:

```
PLEASE RATE YOUR COMMITMENT TO THIS PARTNER BY CIRCLING YOUR
RESPONSE TO EACH OF THE FOLLOWING ITEMS:

How committed would you be to remaining in this relationship?

   Not at all      1  2  3  4  5  6  7  8  9      Extremely
   Committed                                      Committed

How likely is it that you would maintain this relationship?

   Definitely Plan   1  2  3  4  5  6  7  8  9    Definitely
   Not to Maintain                                Plan to
                                                  Maintain

How likely is it that you would break up with this partner in the near future?

   Extremely       1  2  3  4  5  6  7  8  9      Extremely
   Likely                                         Unlikely

"I would feel totally committed to this partner."

   Disagree        1  2  3  4  5  6  7  8  9      Agree
   Strongly                                       Strongly
```

Notice that, with each of the preceding items, circling a higher response number (i.e., closer to "9") reveals a higher level of commitment to the relationship. For a given partner, responses to these four items were summed to arrive at a final commitment score for that partner. This score could range from a low of 4 (if the participant had circled the "1" on each item) to a high of 36 (if the participant had circled the "9" on each item). These scores will serve as the dependent variable in your study.

Manipulating the Independent Variable

The independent variable in this study was "level of rewards associated with a specific romantic partner." This independent variable was manipulated by varying the descriptions of partners shown to the two treatment groups.

The partner descriptions given to the high-reward group were identical to those given to the low-reward group, but this was true only for the first nine descriptions. For partner

10, there was an important difference between the descriptions provided to the two groups. The sheet given to the high-reward group described a relationship with a relatively high level of rewards, whereas the one given to the low-reward group described a relationship with a relatively low level of rewards. Below is the description seen by participants in the high-reward condition:

```
PARTNER 10:  Imagine that you have been dating partner 10 for about
a year, and you have put a great deal of time and effort into this
relationship.  There are not very many attractive potential partners
where you live, so it would be difficult to replace this person with
someone else.  Partner 10 lives in the same neighborhood as you, so
it is easy to see him/her as often as you like.  This person enjoys
the same recreational activities that you enjoy and is also very
good-looking.
```

Notice how the preceding description provides information relevant to the four investment model variables previously discussed. The first sentence provides information dealing with investment size ("...you have put a great deal of time and effort into this relationship"), and the second sentence deals with alternative value ("There are not very many attractive potential partners where you live..."). The third sentence indicates that this is a low-cost relationship because "...it is easy to see him/her as often as you like." In other words, there are no hardships associated with seeing this partner. (If the descriptions said that the partner lives in a distant city, this would have been a high-cost relationship.)

However, it is the last sentence in this description that you are most interested in, because this last sentence describes the level of rewards associated with the relationship. The relevant sentence is "This person enjoys the same recreational activities that you enjoy and is also very good-looking." It is this statement that establishes partner 10 as a high-reward partner for participants in the high-reward group.

In contrast, consider the following description of partner 10 given to participants assigned to the low-reward group. Notice that it is identical to the description given to the high-reward group with regard to the first three sentences. The last sentence, however, deals with rewards, so this last sentence is different for the low-reward group. It describes a low-reward relationship:

```
PARTNER 10:  Imagine that you have been dating partner 10 for about
a year, and you have put a great deal of time and effort into this
relationship.  There are not very many attractive potential partners
where you live, so it would be difficult to replace this person with
someone else.  Partner 10 lives in the same neighborhood as you, so
it is easy to see him/her as often as you like.  This person does
not enjoy the same recreational activities that you enjoy and is not
very good-looking.
```

For this study, note that the vignette for partner 10 is the *only scenario in which you are interested.* You will analyze ratings of participants' commitment to partner 10 and disregard responses to the previous nine. (These were only included to give participants practice at the task before coming to partner 10.)

Also notice the logic behind these experimental procedures; both groups of participants were treated in exactly the same way with respect to everything except the independent variable. For the first nine partners, the descriptions were identical in the two groups. Even the description of partner 10 was identical with respect to everything except the level of rewards associated with the relationship. Therefore, if the participants in the high-reward group are significantly more committed to partner 10 than those in the low-reward group, you can be reasonably confident that it was the level of reward manipulation that affected their commitment ratings. It would be difficult to explain the results in any other way.

In summary, you began your investigation with a group of 20 participants. You randomly assigned 10 to the high-reward condition, and 10 to the low-reward condition. After they completed their task, you disregarded their responses to the first nine scenarios but recorded their responses to partner 10. These responses are the subject of the following analysis.

Writing the SAS Program

Remember that an independent-samples *t* test is appropriate for comparing two samples of observations. It allows you to determine whether there is a significant difference between the two with respect to their mean scores on this criterion. More technically, it allows you to test the null hypothesis that, in the population, there is no difference between the two groups with respect to their scores on the criterion variable. This section shows how to write a SAS program with PROC TTEST to test this null hypothesis for the current fictitious study.

There is one predictor variable in this study (i.e., level of reward). This variable could assume one of two values: participants were either in the high-reward group or in the low-reward group. Since this variable simply codes group membership, you know that it is measured on a nominal scale (i.e., experimental versus control conditions). In coding the data, you will assign participants a score of "2" if they were in the high-reward condition, and a score of "1" if they were in the low-reward condition. You will need a short SAS variable name for this variable, so call it REWGRP ("reward group").

There is one criterion variable in this study: "commitment" (i.e., participants' ratings of how committed they would be to a relationship with partner 10). When entering the data, you enter the sum of the rating numbers that have been circled by each participant in response to partner 10. This variable could assume values from 4 through 36 as measured on an interval scale; give it the SAS variable name of COMMIT.

The general form for the SAS program to perform an independent-samples *t* test is as follows:

```
PROC TTEST   DATA=dataset-name   ALPHA=confidence-level-alpha;
   CLASS   predictor-variable;
   VAR     criterion-variables;
RUN;
```

Here is the entire program—including the DATA step—to analyze fictitious data from the preceding study.

```
 1          DATA D1;
 2             INPUT  #1  @1  REWGRP  1.
 3                        @3  COMMIT  2.  ;
 4          DATALINES;
 5          1 12
 6          1 10
 7          1 15
 8          1 13
 9          1 16
10          1  9
11          1 13
12          1 14
13          1 15
14          1 13
15          2 25
16          2 22
17          2 27
18          2 24
19          2 22
20          2 20
21          2 24
22          2 23
23          2 22
24          2 24
25          ;
26          RUN;
27
28          PROC TTEST   DATA=D1   ALPHA=.05;
29             CLASS REWGRP;
30             VAR COMMIT;
31          RUN;
```

Notes Regarding the SAS Program

Each line of data contains responses for one participant. Assume that you enter your data using the following format:

Line	Column	Variable Name	Explanation
1	1	REWGRP	Codes group membership, so that 1 = low-reward condition, and 2 = high-reward condition
	2	blank	
	3-4	COMMIT	Commitment ratings obtained when participants rated partner 10

The classification variable, REWGRP, was coded using 1s and 2s but the choice of these values was arbitrary (i.e., you could code using different numbers). If you prefer, you could even use character values such as "L" to code the low-reward condition and "H" to code the high-reward condition.

The criterion variable, COMMIT, is a two-digit variable because it could take on a value as high as 36, a two-digit number. Data from the 10 low-reward participants were first entered (in lines 5 to 14), followed by data from the 10 high-reward participants (in lines 15 to 24). It is not really necessary to sort the data from the two groups in this way, however. Data from low-reward and high-reward participants could have been entered in a random sequence.

Remember to always follow the CLASS statement with the name of the study's predictor variable. In other words, always follow it with the name of the nominal-scale variable that indicates group membership (REWGRP, in this case). It might be helpful to remember that CLASS stands for "classification variable" which is another way of saying "nominal-scale variable."

In contrast, always follow the VAR statement with the name of the interval- or ratio-level variable that serves as the study's criterion (COMMIT, in this case). You can list more than one criterion variable, and a separate *t* test will be performed for each.

Results from the SAS Output

Output 8.1 presents the results obtained from the preceding program. The name of the criterion variable (COMMIT, in this case) appears below "Variable" on the left side of the output. The name of the predictor variable (REWGRP) appears beside "Variable."

Output 8.1 Results of PROC TTEST: Significant Differences Observed

```
                               The TTEST Procedure

                                   Statistics

                      Lower CL              Upper CL  Lower CL            Upper CL
  Variable  REWGRP   N    Mean     Mean       Mean    Std Dev  Std Dev   Std Dev   Std Err

  COMMIT             10   11.418     13      14.582   1.5209   2.2111    4.0366    0.6992
            1
  COMMIT             10   21.908   23.3      24.692   1.3389   1.9465    3.5536    0.6155
            2
  COMMIT   Diff (1-2)    -12.26   -10.3      -8.343   1.5739   2.083     3.0804    0.9315


                                    T-Tests

             Variable    Method          Variances      DF    t Value    Pr > |t|

             COMMIT      Pooled          Equal          18    -11.06      <.0001
             COMMIT      Satterthwaite   Unequal       17.7   -11.06      <.0001


                              Equality of Variances

             Variable    Method      Num DF    Den DF   F Value    Pr > F

             COMMIT      Folded F        9         9     1.29      0.7103
```

The output is divided into three sections. The first contains a table labeled "Statistics" which provides simple univariate statistics for the commitment variable. This table provides means, standard deviations, the standard error of the difference between means, and confidence limits (CL). The next section contains a table headed "T-Tests," which provides the results of the independent-samples *t* test. In fact, the results of two *t* tests are presented here; one test assumes equal variances, and a second assumes unequal variances. (More on this later.) The last section labeled "Equality of Variances" presents the results of an *F* statistic that tests the null hypothesis that the two samples have equal variances. (Again, more on this later.)

Steps in Interpreting the Output

1. Make Sure That Everything Looks Right

Before reviewing the results of the *t* tests, you should always first examine the results of the two tables headed "Statistics" to verify that there were no obvious errors in preparing your SAS program. For example, you should verify that the number of participants in each condition (as reported in the output) is correct. Reviewing these tables will also provide you with an understanding of the general trend in your results.

Below the heading "Variable," you find the name of the criterion variable in your analysis. In this case, the criterion variable is COMMIT.

Below the heading "REWGRP," you see the names of the values that were used to identify the two treatment conditions. In the present analysis, you can see that these two values were REWGRP 1 (used to identify participants in the low-reward condition) and REWGRP 2 (used to identify participants in the high-reward condition). To the right of REWGRP 1, you will see statistics relevant to the low-reward condition (e.g., this sample's mean and standard deviation on the criterion variable). To the right of the entry REWGRP 2, you find statistics relevant to the high-reward condition.

The third entry down in the REWGRP column is "Diff (1 – 2)." This row provides information about the difference between REWGRP 1 (the low-reward condition) and REWGRP 2 (the high-reward condition). Among other things, this row reports the difference between the means of these two samples on the criterion variable COMMIT.

The column labeled "N" indicates the number of participants in both groups. You can see that there were 10 participants in the low-reward condition (REWGRP 1) and 10 participants in the high-reward condition (REWGRP 2).

The column headed "Mean" provides the average score for both groups on the criterion variable. Output 8.1 shows that the mean score for participants in the low-reward condition was 13 and that the mean score for those in the high-reward condition was 23.3. This means that, on the average, participants in the high-reward condition expressed greater perceived commitment to partner 10. At this point, however, you do not know whether this difference is statistically significant. You will learn that later when you review the results of the *t* test.

The third entry in the "Mean" column [to the right of "Diff (1 – 2)"] is –10.3. This indicates that the difference between the means of the low- versus the high-reward conditions is equal to –10.3 (i.e., REWGRP 1 – REWGRP 2 or 13 – 23.3).

The column headed "Std Dev" provides the estimated standard deviations for the two samples (computed separately). You can see that the estimated standard deviation for the low-reward condition is 2.21 and the corresponding statistic for the high-reward condition is 1.95 (rounded to two decimal places).

In the third entry "Std Dev" column, to the right of "Diff (1 – 2)", is the pooled estimate standard deviation (s_p). You will need this statistic when you later compute the index of effect size, d. For the current analysis, Output 8.1 shows that the pooled estimate standard deviation is 2.08.

Output 8.1 shows another the column headed "Std Err" which signifies standard errors. Of greatest interest is the third entry down which appears to the right of "Diff (1 – 2)". This value is the **standard error of the difference between the means**. For the current analysis, you can see that the standard error of the difference is 0.93.

---

2. Determine Which *t* Statistic Is Appropriate

Output 8.1 shows that the TTEST procedure actually provides two *t* statistics but only one of these will be relevant for a given analysis. The first *t* statistic is the standard statistic based on the assumption that the two samples have equal variances. In other words, the distribution of scores around the means for both samples is comparatively similar. The second *t* statistic is based on the assumption that the two samples were drawn from populations with unequal variances (i.e., distinct distribution of scores around their respective means).

Fortunately, the TTEST procedure automatically performs a folded form of the F statistic that tests the similarity of variances in the two samples. The results of this test are reported in the bottom section of the output under the heading "Equality of Variances" which is reproduced here:

Equality of Variances					
Variable	Method	Num DF	Den DF	F Value	Pr > F
COMMIT	Folded F	9	9	1.29	0.7103

This analysis begins with the null hypothesis that, in the population, there is no difference between the low-reward participants and the high-reward participants with respect to their variances on the commitment variable. It computes a special F statistic to test this hypothesis. If the significance or p value for the resulting F test is less than .05, you will reject the null hypothesis of no differences and conclude that the variances are unequal. If the p value is greater than .05, you will tentatively conclude that the variances are equal.

For the present analysis, the F value was only 1.29 with a corresponding p value of .71 (this p value appears below "Pr > F"). This means that the probability of obtaining an F value this large or larger when the population variances are equal was quite large, that is .71. You therefore fail to reject the null hypothesis of equal variances and tentatively conclude that the variances are statistically equivalent for the two conditions (REWGRP 1 and 2). This means that you will interpret the equal variances *t* statistic that appears to the right of the word "Equal" in Output 8.1 (if the PR > F had been less than .05, you would have interpreted the *t* statistic to the right of the word "Unequal").

In summary, when the PR > F is nonsignificant (greater than .05), report the *t* test based on equal variances. When the PR > F is significant (less than .05), report the *t* test based on unequal variances. In this example, the two values are identical. With other datasets, however, the values might have differed considerably. Thus, it is important to first examine the equality of variance *F* statistic before interpreting the results of the TTEST procedure.

---

3. Review the Appropriate *t* Statistic and Its Associated Probability Value

The middle section of the output provides the information that is of primary interest (reproduced below)—that is, the obtained *t* statistic along with its corresponding degrees of freedom and probability value. As determined in the preceding section, you will review the equal-variances *t* statistic for the current analysis.

```
                              The TTEST Procedure

                                Statistics

                      Lower CL          Upper CL  Lower CL            Upper CL
Variable  REWGRP    N    Mean    Mean      Mean   Std Dev  Std Dev   Std Dev  Std Err

COMMIT            10    11.418     13    14.582   1.5209   2.2111    4.0366   0.6992
              1

COMMIT            10    21.908   23.3    24.692   1.3389   1.9465    3.5536   0.6155
              2

COMMIT   Diff (1-2)    -12.26   -10.3    -8.343   1.5739   2.083     3.0804   0.9315


                                 T-Tests

          Variable   Method          Variances     DF    t Value    Pr > |t|

          COMMIT     Pooled          Equal         18     -11.06     <.0001
          COMMIT     Satterthwaite   Unequal       17.7   -11.06     <.0001
```

In the column headed with a "t Value" and to the right of the word "Equal," you can see that the obtained *t* statistic is –11.06 which is quite large. It is associated with 18 degrees of freedom; under "Pr > | t |," you can see that the *p* value associated with this *t* is less than .01.

But what does this *p* value really mean? This *p* value is the probability that you would obtain a *t* statistic this large or larger (in absolute magnitude) if the null hypothesis were true.

You can state the null hypothesis in this study as follows: "In the population, there is no difference between the low- and high-reward groups with respect to their mean scores on the commitment variable." Symbolically, the null hypothesis can be represented in this way:

$$H_0: M_1 = M_2$$

where M_1 = mean commitment score for the population of people in the low-reward condition, and M_2 = mean commitment score for the population of people in the high-reward condition.

If this null hypothesis were true, you should have obtained a *t* statistic closer to zero. In contrast, however, your obtained *t* statistic was –11.06; this is much larger (in its absolute magnitude) than you would expect if the null hypothesis were true. In fact, given the degrees of freedom for this test, the TTEST procedure indicates that the probability of getting a *t* value this large was less than 1 in 10,000! Remember that, anytime you obtain a *p* value less than .05, you reject the null hypothesis. Because your *p* value is so small in this case, you are able to reject the null hypothesis of no commitment differences between groups. You can therefore conclude that there is likely a difference in mean commitment in the population between people in the high-reward condition as compared to those in the low-reward condition.

4. Review the Sample Means

The significant *t* statistic indicates that the two samples are significantly different from each other. But which is *higher* on the commitment variable? To determine this, you review the output under the column headed "Mean." To the right of the value "1," and under the heading "Mean," you see that the low-reward group had a mean score of 13 on commitment (remember that the value "1" coded the low-reward participants). To the right of the value "2," and under the heading "Mean," you see that the high-reward group had a mean score of 23.3. It is therefore evident that, as predicted, the high-reward group exhibits a higher level of commitment compared to the low-reward group.

5. Review the Confidence Interval for the Difference between the Means

With PROC TTEST, SAS computes a confidence interval for the difference between the means. As you will recall, the PROC TTEST statement (line 28) for the current analysis contained the following ALPHA option:

```
ALPHA=0.05
```

This option causes SAS to compute the 95% confidence interval. If you had instead wanted the 99% confidence interval, you would have instead used this option:

```
ALPHA=0.01
```

The 95% confidence interval appears in the "Statistics" table in the PROC TTEST output. The information that you need at this point appears in Output 8.1 in the row headed "COMMIT Diff (1-2)." This row provides information about the difference between REWGRP 1 (low-reward condition) versus REWGRP 2 (high-rewarded condition). Where the row headed "COMMIT Diff (1-2)" intersects with the column headed "Mean," you can see that the observed difference between the samples means is –10.3. This difference is computed by starting with the sample mean for REWGRP 1 (13) and subtracting from it the sample mean for REWGRP 2 (23.3).

A confidence interval extends from a lower confidence limit to an upper confidence limit. To find the lower confidence limit for the difference, find the location where the row headed "COMMIT Diff (1-2)" intersects with the column headed "Lower CL Mean." There, you can see that the lower confidence limit for the difference is –12.26. To find the upper confidence limit for the difference, find the location where the row headed "COMMIT Diff (1-2)" intersects with the column headed "Upper CL Mean." There, you can see that the upper confidence limit for the difference is –8.34.

Combined these findings indicate that the 95% confidence interval for the difference between means extends from –12.26 to –8.34. This means that you can estimate with a 95% probability that in the population, the actual difference between the mean of the low-reward condition and the mean of the high-reward condition is somewhere between –12.26 and –8.34. Notice that this interval does not contain the value of zero. This is consistent with rejection of the null hypothesis in the previous section (i.e., you rejected the null hypothesis which stated "In the population, there is no difference between the low- and high-reward groups with respect to their mean scores on the commitment variable.")

6. Compute the Index of Effect Size

In this example, you obtained a *p* value less than the standard criterion of .05. You therefore rejected the null hypothesis. You know that there is a statistically significant difference between the observed commitment levels for the high- and low-reward conditions. But is it a relatively large difference? The null hypothesis test alone does not tell you whether the difference is large or small. In fact, if your sample is very large, you can obtain statistically significant results even if the difference is relatively trivial.

Because of this limitation of null hypothesis testing, many researchers now supplement statistics such as *t* tests with measures of effect size. The exact definition of effect size will vary depending upon the type of analysis that you are performing. For an independent samples *t* test, we can define **effect size** as the degree to which one sample mean differs from a second sample mean, stated in terms of standard deviation units.

The figure below illustrates the meaning of an effect size. (For descriptive purposes, the shape of both distributions is identical). The mean for sample 2 (M_2), to the right side of the figure falls two standard deviations above the mean for sample 1 (M_1). As well, the mean for sample 1 falls two standard deviations below the mean for sample 2. This between group difference therefore reflects an effect size of 2. You can see that the greater the distance between sample means, the greater the likelihood that samples were drawn from different populations. In contrast, if the distribution of sample scores overlaps and they share the same mean value, it is likely participants were drawn from the same population. In this instance, the effect size would be zero.

Figure 8.2 Difference between Mean Scores Expressed as Standard Deviation Units

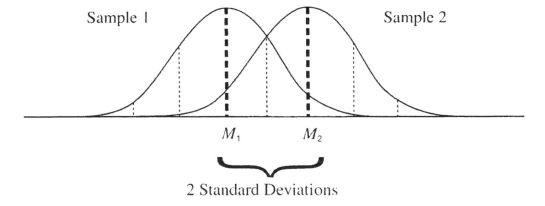

The symbol for effect size is *d* and the formula is as follows:

$$d = \frac{|M_1 - M_2|}{s_p}$$

where:

M_1= the observed mean of sample 1 (i.e., participants in treatment condition 1);

M_2 = the observed mean of sample 2 (i.e., participants in treatment condition 2);

s_p = the pooled estimate of the population standard deviation.

Although SAS does not automatically compute effect size, you can do so easily using the information that appears in the "Statistics" table from the PROC TTEST output. In the preceding formula, M_1 is the observed mean of sample 1 (which in the present study is the average level of commitment among low-reward participants). From Output 8.1, in the columns headed "Mean," you can see that the mean commitment level for this group is 13. In the formula, M_2 is the observed mean of sample 2 (i.e., high-reward participants). From Output 8.1, in the columns headed "Mean," you can see that the mean commitment level for this group is 23.3.

In the preceding formula, s_p represents the pooled estimate of the population standard deviation. This statistic appears in Output 8.1 where the row headed "COMMIT Diff (1–2)" intersects with the column headed "Std Dev." For the present analysis, you can see that the pooled estimate of the population standard deviation is 2.08.

You can now insert these statistics into the formula and compute the index of effect size in this way:

$$d = \frac{|M_1 - M_2|}{s_p}$$

$$d = \frac{|13 - 23.3|}{2.08}$$

$$d = \frac{|-10.3|}{2.08}$$

$$d = \frac{10.3}{2.08}$$

$d = 4.9519$

$d \approx 4.94$

And so the obtained index of effect size for the current analysis is 4.95. This means that the sample mean for the low-control condition differs from the sample mean of the high-reward condition by 4.95 standard deviations. To determine whether this is a relatively large or small difference, you can consult the guidelines provided by Cohen (1992). Cohen's guidelines for *t* tests are reproduced below in Table 8.1:

Table 8.1

Guidelines for Interpreting t Test Effect Sizes

Effect size	Obtained *d* statistic
Small effect	$d = .20$
Medium effect	$d = .50$
Large effect	$d = .80$

Your obtained *d* statistic of 4.95 is larger than the "large effect" value of .80 that appears in Table 8.1. This means that the difference between low- and high-reward participants in commitment levels for partner 10 produced both a statistically significant and very large effect.

Summarizing the Analysis Results

In performing an independent-samples *t* test, the following format can be used to summarize the research problem and results:

A) Statement of the problem
B) Nature of the variables
C) Statistical test
D) Null hypothesis (H_0)
E) Alternative hypothesis (H_1)
F) Obtained statistic
G) Obtained significance or probability (*p*) value
H) Conclusion regarding the null hypothesis
I) Confidence interval
J) Effect size
K) Figure representing the results
L) Formal description of results for a paper

As an illustration, a summary of the preceding analysis, according to this format, follows:

A) Statement of the problem: The purpose of this study was to determine whether there was a difference between people in a high-reward relationship and those in a low-reward relationship with respect to their mean commitment to the relationship.

B) Nature of the variables: This analysis involved two variables. The predictor variable was level of rewards, which was measured on a nominal scale and could assume two values: a low-reward condition (coded as 1); and a high-reward condition (coded as 2). The criterion variable was commitment which was measured on an interval scale.

C) Statistical test: Independent-samples *t* test.

D) Null hypothesis (H_0): $M_1 = M_2$. In the population, there is no difference between participants in a low-reward condition and those in a high-reward condition with respect to their mean commitment levels.

E) Alternative hypothesis (H_1): $M_1 \neq M_2$. In the population, there is a difference between participants in a low-reward condition and those in a high-reward condition with respect to their mean commitment levels.

F) Obtained statistic: $t = -11.06$.

G) Obtained probability (*p*) value: $p < .01$.

H) Conclusion regarding the null hypothesis: Reject the null hypothesis.

I) Confidence interval: Subtracting the mean of the high-reward condition from the mean of the low-reward condition resulted in an observed difference of -10.3. The 95% confidence interval for this difference extends from -12.26 to -8.343.

J) Effect size: $d = 4.95$ (large effect size).

K) Figure representing the results:

Figure 8.3 Mean Levels of Commitment Observed for Participants in High-Reward versus Low-Reward Conditions (Significant Differences Observed)

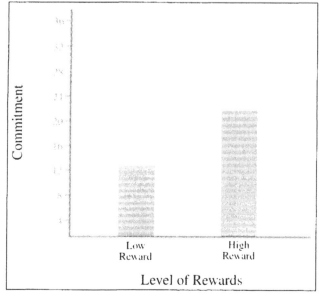

L) Formal description of results for a paper: Most chapters of this text show you how to summarize the results of an analysis in a way that would be appropriate if you were preparing a paper to submit for publication in a scholarly research journal. These summaries follow the format recommended in the publication manual of the American Psychological Association (2001), a format required by many journals in the social sciences. Here is an example of how the current results could be summarized according to this format:

> Results were analyzed using an independent-samples *t* test. This analysis revealed a significant difference between the two groups, *t*(18) = -11.06; *p* < . 01. The sample means are displayed in Figure 8.3, which shows that participants in the high-reward condition scored significantly higher on commitment than did participants in the low-reward condition (for high-reward group, *M* = 23.30, *SD* = 1.95; for low-reward group, *M* = 13.00, *SD* = 2.21). The observed difference between means was -10.30 and the 95% confidence interval for the difference between means extended from -12.26 to -8.34. The effect size was computed as *d* = 4.94. According to Cohen's (1992) guidelines for *t* tests, this represents a large effect.

An Example with Nonsignificant Differences

Obviously, researchers do not always obtain significant results when performing investigations such as the one described here. This section repeats the analyses reported previously, this time using fictitious data that will result in a nonsignificant *t* test. The conventions for summarizing nonsignificant results are then presented.

Program and Output

Below is the SAS program presented earlier in the chapter. In the following program, however, the data have been modified so that the two groups will not differ significantly on mean levels of commitment.

```
1          DATA D1;
2             INPUT  #1  @1  REWGRP  1.
3                        @3  COMMIT  2.  ;
4          DATALINES;
5          1 23
6          1 22
7          1 25
8          1 19
9          1 24
10         1 20
11         1 22
12         1 22
13         1 23
14         1 27
15         2 25
16         2 22
17         2 27
18         2 24
19         2 22
20         2 20
21         2 24
```

```
22          2 23
23          2 22
24          2 24
25          ;
26          RUN;
27
28          PROC TTEST    DATA=D1    ALPHA=.05;
29             CLASS REWGRP;
30             VAR COMMIT;
31          RUN;
```

Simply "eyeballing" the data reveals that very similar commitment scores seem to be displayed by participants in the two conditions. Nonetheless, a formal statistical test is required to determine whether a significant difference exists. The results of the program analyzing the preceding data appear as Output 8.2.

Output 8.2 Results of PROC TTEST: Nonsignificant Differences Observed

```
                          The TTEST Procedure

                             Statistics

                     Lower CL           Upper CL  Lower CL              Upper CL
Variable  REWGRP   N    Mean    Mean      Mean    Std Dev  Std Dev    Std Dev  Std Err

COMMIT           10    21.046   22.7    24.354    1.5901   2.3118     4.2205   0.7311
          1
COMMIT           10    21.908   23.3    24.692    1.3389   1.9465     3.5536   0.6155
          2
COMMIT    Diff (1-2)   -2.608   -0.6     1.4078   1.6147    2.137     3.1602   0.9557


                               T-Tests

        Variable    Method           Variances    DF    t Value   Pr > |t|

        COMMIT      Pooled           Equal        18     -0.63    0.5380
        COMMIT      Satterthwaite    Unequal    17.5     -0.63    0.5382


                         Equality of Variances

          Variable    Method    Num DF    Den DF   F Value   Pr > F

          COMMIT      Folded F      9         9      1.41    0.6166
```

Remember that the first step in this analysis is to determine whether the equal-variances *t* test or the unequal variances *t* test is appropriate. The *F* test reported in the "Equality of Variances" section at the bottom of the output is 1.41, and its corresponding *p* value is .62. Because this *p* value is greater than .05, you cannot reject the null hypothesis of equal variances in the population. This means that it is appropriate to refer to the equal-variances *t* test.

The equal-variances *t* statistic appears in the "T-Tests" section of Output 8.2 under the heading "t Value" and to the right of "Equal." This equal-variances *t* statistic is quite small at −0.63. The *p* value for this *t* statistic is quite large at 0.54. Obviously, this *p* value is greater than the standard cutoff of .05, meaning that the *t* statistic is nonsignificant. These results mean that you cannot reject the null hypothesis of equal population means on commitment. In other words, you conclude that there is not a significant difference between mean levels of commitment in the two samples.

The "Statistics" section of Output 8.2 provides mean commitment scores for both conditions. Below the heading "Mean" and to the right of "1" and "2" respectively, you can see that the mean commitment scores were 22.7 for the low-reward condition and 23.3 for the high-reward condition. Below the heading "Mean" and to the right of the heading COMMIT Dif (1-2), you can see that the "observed difference between the low- and high-reward conditions is –.6. The "Lower CL Mean" column shows that the lower confidence limit for this difference is –2.61 and the "Upper CL Mean" column shows that the upper confidence limit for this difference is 1.41. Combined, this indicates that the 95% confidence interval for the difference between means extends from –2.61 to 1.41. Notice that this interval includes the value of zero, consistent with your finding that the difference between means is nonsignificant.

You have already seen that the mean for the low-reward condition is 22.7, and the mean for the high-reward condition is 23.3. The only other piece of information that you need in order to compute the effect size is s_p (i.e., the pooled estimate standard deviation). This appears in Output 8.2 as the third entry in the column headed "Std Dev." There, you can see that the pooled estimate is 2.14. You can now insert these statistics into the formula for effect size:

$$d = \frac{|M_1 - M_2|}{s_p}$$

$$d = \frac{|22.7 - 23.3|}{2.14}$$

$$d = \frac{|-.6|}{2.14}$$

$$d = \frac{.6}{2.14}$$

$$d = .2804$$

$$d \approx .28$$

Thus, the index of effect size for the current analysis is .28. According to Cohen's guidelines appearing in Table 8.1, this falls somewhere between a small and medium effect.

Summarizing the Results of the Analysis

For this analysis, the statistical interpretation format would appear as follows. Since this is the same study, you would complete items A through E in exactly the same way as previously indicated:

F) Obtained statistic: $t = -0.63$.

G) Obtained probability (p) value: $p = .54$.

H) Conclusion regarding the null hypothesis: Fail to reject the null hypothesis.

I) Confidence interval: Subtracting the mean of the high-reward condition from the mean of the low-reward condition resulted in an observed difference of $-.6$. The 95% confidence interval for this difference extended from -2.61 to 1.41.

J) Effect size: $d = .28$ (small to medium effect size).

K) Figure representing the results:

Figure 8.4 Mean Levels of Commitment Observed for Participants in High-Reward versus Low-Reward Conditions (Nonsignificant Differences Observed)

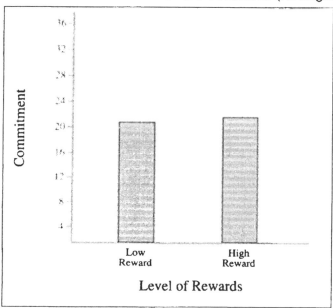

J) Formal description of results for a paper:

```
     Results were analyzed using an independent-samples t
test.  This analysis failed to reveal a significant
difference between the two groups, t(18) = -0.63; p = .54.
The sample means are displayed in Figure 8.4, which shows
```

```
that participants in the high-reward condition demonstrated
scores on commitment that were quite similar to those shown
by participants in the low-reward condition (for high-reward
group, M = 23.30, SD = 1.95; for low-reward group, M = 22.70,
SD = 2.31).  The observed difference between means was -.6
and the 95% confidence interval for the difference between
means extended from -2.61 to 1.41.  The effect size was
computed as d = .28.  According to Cohen's (1992) guidelines
for t tests, this represents a small to medium effect.
```

The Paired-Samples *t* Test

Paired Samples versus Independent Samples

The paired-samples *t* test (sometimes referred to as the *correlated-samples t test* or *matched-samples t test*) is similar to the independent-samples test in that both procedures involve comparing two samples of observations, and determining whether or not the mean of one sample significantly differs from the mean of another. With the independent-samples procedure, the two groups of scores are completely independent (i.e., an observation in one sample is not related to any observation in the other). In experimental research, this is normally achieved by drawing a sample of participants and randomly assigning each to either a treatment or control condition. Because each participant contributes data under only one condition, the two samples are empirically independent.

With the paired-samples procedure, in contrast, each score in one sample is *paired* in some meaningful way with a score in the other sample. There are several ways that this can be achieved. The following examples illustrate some of the most commonly used approaches. One word of warning: the following fictitious studies merely illustrate paired sample designs, and do not necessarily represent sound research methodology from the perspective of internal or external validity. Problems with some of these designs are reviewed in a later section.

Examples of Paired-Samples Research Designs

Each Participant Is Exposed to Both Treatment Conditions

Earlier sections of this chapter described an analogue experiment in which *level of reward* was manipulated to see if it affected participants' *level of commitment* to a romantic relationship. The procedure used in that study required participants to review 10 vignettes and then rate their commitment to each fictitious romantic partner. The dependent variable in the investigation was the rated amount of commitment that participants displayed toward partner 10. The independent variable was manipulated by varying the description of partner 10 that was provided to the participants; those in the "high-reward" condition read that partner 10 had several positive attributes while participants in the "low-reward" condition read that partner 10 did not have these attributes. This study was as an independent-samples study because each participant was assigned to either a high-reward condition or a low-reward condition, and no one was assigned to both.

This example could easily be modified to instead follow a paired-samples research design. You could do this by conducting the study with only one group of participants (rather than two), and having each participant rate partner 10 twice: once after reading the low-reward version of partner 10 and a second time after reading the high-reward version of partner 10.

It would be appropriate to analyze data derived from such a study using the paired-samples *t* test because it would be possible to meaningfully pair observations obtained under the two conditions. For example, participant 1's rating of partner 10 under the low-reward condition could be paired with his or her rating of partner 10 under the high-reward condition, participant 2's rating of partner 10 under the low-reward condition could be paired with his or her rating of partner 10 under the high-reward condition, and so forth. Table 8.1 shows how the resulting data could be arranged in tabular form:

Table 8.1

Fictitious Data from a Study Using a Paired Samples Procedure

	Commitment Ratings of Partner 10	
Participant	Low-Reward Condition	High-Reward Condition
Paul	9	20
Vilem	9	22
Pavel	10	23
Sunil	11	23
Maria	12	24
Fred	12	25
Jirka	14	26
Eduardo	15	28
Asher	17	29
Shirley	19	31

Remember that your dependent variable is still the commitment ratings for partner 10. For participant 1 (Paul), you have obtained two scores on this dependent variable: a score of 9 obtained in the low-reward condition, and a score of 20 obtained in the high-reward condition. This is what it means to have the scores *paired* in a meaningful way: Paul's score in the low-reward condition is paired with his score from the high-reward condition. The same is true for the remaining participants as well.

Participants Are Matched

The preceding study used a type of **repeated measures** approach: only one sample of participants participated; and repeated measurements on the dependent variable (commitment) were taken from each. That is, each participant contributed one score under the low-reward condition and a second score under the high-reward condition.

A different approach could have used a type of matching procedure. With a **matching procedure,** a given participant provides data under only one experimental condition; however, each is matched with a different participant who provides data under the other experimental condition. The participants are matched on some variable that is expected to be related to the dependent variable, and matching is done prior to manipulation of the independent variable.

For example, imagine that it is possible to administer an "emotionality scale" to participants and that prior research has shown that scores on this scale are strongly correlated with scores on romantic commitment (i.e., the dependent variable in your study). You could administer this emotionality scale to 20 participants, and use their scores on the scale to match them; that is, to place them in pairs according to their similarity on the emotionality scale.

For example, scores on the emotionality scale might range from a low of 100 to a high of 500. Assume that John scores 111 on this scale, and Petr scores 112. Because their scores are very similar, you pair them together, and they become participant pair 1. Dov scores 150 on this scale and Lukas scores 149. Because their scores are very similar, you pair them together as participant pair 2. Table 8.2 shows how you could arrange these fictitious pairs of participants:

Table 8.2

Fictitious Data from a Study Using a Matching Procedure

		Commitment Ratings of Partner 10	
Participant Pair		Low-Reward Condition	High-Reward Condition
Participant pair 1	(John and Petr)	8	19
Participant pair 2	(Dov and Lukas)	9	21
Participant pair 3	(Luis and Marco)	10	21
Participant pair 4	(Bjorn and Jorge)	10	23
Participant pair 5	(Ion and André)	11	24
Participant pair 6	(Martita and Kate)	13	26
Participant pair 7	(Blanche and Jane)	14	27
Participant pair 8	(Reuben and Joe)	14	28
Participant pair 9	(Mike and Otto)	16	30
Participant pair 10	(Sean and Seamus)	18	32

Within each pair, one participant is randomly assigned to the low-reward condition and the other is assigned to the high-reward condition. Assume that, for each of the participant pairs in Table 8.2, the person listed first had been randomly assigned to the low condition and the person listed second had been assigned to the high condition. The study then proceeds in the usual way, with participants rating the various hypothetical partners.

Table 8.2 shows that John saw partner 10 in the *low-reward* condition and provided a commitment rating of 8. Petr saw partner 10 in the *high-reward* condition, and provided a commitment score of 19. When analyzing the data, you pair John's score on the commitment variable with Petr's score on commitment. The same will be true for the remaining participant pairs. A later section shows how to write a SAS program that does this.

> **When should the matching take place?** Remember that participants are placed together in pairs on the basis of some matching variable *before the independent variable is manipulated*. They are *not* placed together in pairs on the basis of their scores on the dependent variable. In the present case, participants were paired based on the similarity of their scores on the emotionality scale. Later, the independent variable was manipulated and their commitment scores were recorded. Although they are not paired on the basis of their scores on the dependent variable, the researcher normally assumes that their scores on the dependent variable will be correlated. More on this in a later section.

Pretest and Posttest Measures Are Taken

Consider now a different type of research problem. Assume that an educator believes that taking a foreign language course improves critical thinking among college students. To test this hypothesis, she administers a test of critical thinking to a single group of college students at two points in time. A pretest is administered at the beginning of the semester (prior to taking the language course), and a posttest is administered at the end of the semester (after completing the course). The data obtained from the two administrations appear in Table 8.3:

Table 8.3

Fictitious Data from Study Using a Pretest-Posttest Procedure

	Scores on Test of Critical Thinking Skills	
Participant	Pretest	Posttest
Paul	34	55
Vilem	35	49
Pavel	39	59
Sunil	41	63
Maria	43	62
Fred	44	68
Jirka	44	69
Eduardo	52	72
Asher	55	75
Shirley	57	78

You can analyze these data using the paired-samples *t* test because you can pair together the various scores in a meaningful way. That is, you can pair each participant's score on the pretest with his or her score on the posttest. When the data are analyzed, the results will indicate whether or not there was a significant change in critical thinking scores over the course of the semester.

Problems with the Paired-Samples Approach

Some of the studies described in the preceding section utilize fairly weak experimental designs. This means that, even if you had conducted the studies, you might not have been able to draw firm conclusions from the results because alternate explanations could be offered for those results.

For example, consider the first study in which each participant was exposed to both the low-reward version of partner 10 as well as the high-reward version of partner 10. If you design this study poorly, it might suffer any of a number of confounds. For example, what if you designed the study so that each participant rated the low-reward version first and the high-reward version second? If you then analyzed the data and found that higher commitment ratings were observed for the high-reward condition, you would not know whether to attribute this finding to the manipulation of the independent variable (level of rewards) or to **order effects** (i.e., the possibility that the order in which the treatments were presented influenced scores on the dependent variable). For example, it is possible that participants tend to give higher ratings to partners that are rated later in serial order. If this is the case, the higher ratings observed for the high-reward partner might simply reflect such an order effect.

The third study described previously (which investigated the effects of a language course on critical thinking skills) also displays a weak experimental design: the single-group, pretest-posttest design. Assume that you administer the test of critical thinking to the students at the beginning and again at the end of the semester. Assume further that you observe a significant increase in their skill levels over this period. This would be consistent with your hypothesis that the foreign language course helps develop critical thinking.

Unfortunately, this would not be the *only* reasonable explanation for the findings. Perhaps the improvement was simply due to the process of **maturation** (i.e., changes that naturally take place as people age). Perhaps the change is simply due to the general effects of being in college, independent of the effects of the foreign language course. Because of the weak design used in this study, you will probably never be able to draw firm conclusions about what was really responsible for the students' improvement.

This is not to argue that researchers should never obtain the type of data that can be analyzed using the paired-samples *t* test. For example, the second study described previously (the one using the matching procedure) was reasonably sound and might have provided interpretable results. The point here is that research involving paired-samples must be designed very carefully in order to avoid the problems discussed here. You can deal with most of these difficulties through the appropriate use of counterbalancing, control groups, and other strategies. Problems inherent in repeated measures and matching designs, along with the procedures that can be used to handle these problems, are discussed in Chapter 12, "One-Way ANOVA with One Repeated-Measures Factor," and Chapter 13, "Factorial ANOVA with Repeated-Measures Factors and Between-Subjects Factors."

When to Use the Paired-Samples Approach

When conducting a study that involves two treatment conditions, you will often have the choice of using either the independent-samples approach or the paired-samples approach. A number of considerations will influence your decision to use one design in place of the other. One of the most important considerations is the extent to which the paired-samples procedure results in a more sensitive test; that is, the extent to which the paired-samples approach makes it more likely to detect significant differences when they actually exist.

It is important to understand that the paired-samples *t* test has one important weakness when it comes to test sensitivity: the paired-samples test has only *half* the degrees of freedom as the equivalent independent-samples test. (A later section shows how to compute these degrees of freedom.) Because the paired-samples approach has fewer degrees of freedom, it must display a larger *t* value to attain statistical significance (compared to the independent-samples *t* test).

Then why use this approach? Because, under the right circumstances, the paired-samples approach results in a smaller standard error of the mean (the denominator in the formula used to compute the *t* statistic). Other factors held equal, a smaller standard error results in a more sensitive test.

However, there is a catch: the paired-samples approach will result in a smaller standard error only if scores on the two sets of observations are positively correlated. This concept is easiest to understand with reference to the pretest-posttest study described previously. Table 8.4 again reproduces the fictitious data obtained in this study:

Table 8.4

Fictitious Data from Study Using a Pretest-Posttest Procedure

| Participant | Scores on Test of Critical Thinking Skills | |
	Pretest	Posttest
Paul	34	55
Vilem	35	49
Pavel	39	59
Sunil	41	63
Maria	43	62
Fred	44	68
Jirka	44	69
Eduardo	52	72
Asher	55	75
Shirley	57	78

Notice that, in Table 8.4, scores on the pretest appear to be positively correlated with scores on the posttest. That is, participants who obtained relatively low scores on the pretest (such as Paul) also tended to obtain relatively low scores on the posttest. Similarly, participants who obtained relatively high scores on the pretest (such as Shirley) also tended to obtain relatively high scores on the posttest. Although the participants might have displayed a general improvement in critical thinking skills over the course of the semester, their ranking relative to one another remained relatively constant. Participants with the lowest scores at the beginning of the term still tended to have the lowest scores at the end of the term.

The situation described here is the type of situation that makes the paired-samples *t* test the optimal procedure. Because pretest scores are correlated with posttest scores, the paired-samples approach should yield a fairly sensitive test.

The same logic applies to the other studies described previously. For example, Table 8.5 again reproduces the data obtained from the study in which participants were assigned to pairs based on matching criteria:

Table 8.5

Fictitious Data from a Study Using a Matching Procedure

Participant Pair		Commitment Ratings of Partner 10	
		Low-Reward Condition	High-Reward Condition
Participant pair 1	(John and Petr)	8	19
Participant pair 2	(Dov and Lukas)	9	21
Participant pair 3	(Luis and Marco)	10	21
Participant pair 4	(Bjorn and Jorge)	10	23
Participant pair 5	(Ion and André)	11	24
Participant pair 6	(Martita and Kate)	13	26
Participant pair 7	(Blanche and Jane)	14	27
Participant pair 8	(Reuben and Joe)	14	28
Participant pair 9	(Mike and Otto)	16	30
Participant pair 10	(Sean and Seamus)	18	32

Again, there appears to be a correlation between scores obtained in the low-reward condition and those obtained in the high-reward condition. This is apparently because participants were first placed into pairs based on the similarity of their scores on the emotionality scale, and the emotionality scale is predictive of how participants respond to the commitment scale. For example, both John and Petr (pair 1) display relatively low scores on commitment, presumably because they both scored low on the emotionality scale that was initially used to match them. Similarly, both Sean and Seamus (participant pair 10) scored relatively high on commitment, presumably because they both scored high on emotionality.

This illustrates why it is so important to select *relevant* matching variables when using a matching procedure. There is a correlation between the two commitment variables above because (presumably) emotionality is related to commitment. If, instead, you had assigned participants to pairs based on some variable that is not related to commitment (e.g., participant shoe size), the two commitment variables would not be correlated and the paired-samples *t* test would not provide a more sensitive test. Under those circumstances, you would achieve more statistical power by, instead, using the independent-samples *t* test and capitalizing on the greater degrees of freedom.

Example: An Alternative Test of the Investment Model

The remainder of this chapter shows how to write SAS programs that perform paired-samples *t* tests and how to interpret the results. The first example is based on the first fictitious study described earlier, which examined the effect of levels of reward on commitment to a romantic relationship. The study included 10 participants who each rated partner 10 twice: once after reviewing the low-reward version of partner 10 and once after reviewing the high-reward version. Table 8.6 reproduces the data obtained from the participants:

Table 8.6

Fictitious Data from the Investment Model Study

	Commitment Ratings of Partner 10	
Participant	Low-Reward Condition	High-Reward Condition
Paul	9	20
Vilem	9	22
Pavel	10	23
Sunil	11	23
Maria	12	24
Fred	12	25
Jirka	14	26
Eduardo	15	28
Asher	17	29
Shirley	19	31

These data were keyed according to the following format:

Line	Column	Variable Name	Explanation
1	1–2	LOW	Commitment ratings obtained when participants rated the low-reward version of partner 10
	3	blank	
	4–5	HIGH	Commitment ratings obtained when participants rated the high-reward version of partner 10

Notice from the preceding format that no variable codes "group membership" or "treatment condition." Instead, two variables include commitment ratings: one variable includes commitment ratings obtained when participants reviewed the low-reward version of partner 10; and the second includes commitment ratings obtained when participants reviewed the high-reward version.

The Difference Score Variable

A paired-samples *t* test is performed by creating a **difference score** variable, and determining whether the average difference score is significantly different from zero. If the average difference score is significantly different from zero, you can conclude that your independent variable has a significant effect on the dependent variable (assuming that the study was well-designed and that certain other conditions hold).

In the present study, this difference score variable is created by starting with participants' commitment scores obtained in the high-reward condition and subtracting from the high-reward commitment score their commitment score obtained in the low-reward condition.

If a participant's commitment score from the high-reward condition is approximately equal to his or her commitment score from the low-reward condition, the resulting difference score is approximately equal to zero. This suggests that the participant's level of commitment was not affected by the level of reward manipulation. If none of the participants are affected by the manipulation (on average), then the average difference across participants should be approximately zero. Therefore, if the average difference score is not significantly different from zero, you will fail to reject the null hypothesis and will, instead, conclude that the manipulation had no effect on mean level of commitment.

On the other hand, assume that your manipulation *does* have the expected effect on commitment. This would mean that, for most participants, commitment ratings obtained under the high-reward condition would tend to be greater than commitment ratings under the low-reward condition. Subtracting low-reward commitment scores from high-reward commitment scores under these circumstances would tend to produce positive difference scores (rather than zero or negative difference scores). Therefore, if your manipulation has the predicted effect, the average difference score should be both positive and significantly different from zero. This is important to remember when you later review the results of your analyses.

The DATA Step

The following SAS program statements input the data from Table 8.6:

```
 1          DATA D1;
 2               INPUT   #1    @1    LOW    2.
 3                             @4    HIGH   2.    ;

 5          DATALINES;
 6            9 20
 7            9 22
 8           10 23
 9           11 23
10           12 24
11           12 25
12           14 26
13           15 28
14           17 29
15           19 31
16            ;
17          RUN;
```

Here is the general form for the SAS statements to perform a paired-samples *t* test:

```
PROC MEANS;
    VAR criterion-variable1 criterion-variable2;
RUN;

PROC TTEST    DATA=dataset name
              HO=comparison number
              ALPHA=alpha level;
    PAIRED criterion-variable1*criterion-variable2;
RUN;
```

In the preceding, `criterion-variable1` and `criterion-variable2` are the variables that include scores on the dependent variable under the two treatment conditions. In the present study, they correspond to the variables LOW and HIGH respectively.

PROC MEANS computes descriptive statistics for these variables. This allows you to check the data for possible errors in entering the data or writing the input statement. It also provides the means and standard deviations for both variables.

In the preceding general form, the PROC TTEST statement contains the following option:

```
HO=comparison-number
```

The "comparison-number" that appears in this option should be the mean difference score expected under the null hypothesis. In most cases when you perform a paired-sample *t* test, the mean difference score expected under the null hypothesis is zero. Therefore, you should generally use the following option when performing a paired-samples *t* test:

```
HO=0
```

Note that the "0" that appears in the preceding option "H0" is a zero (0) and is not an uppercase letter "O." In addition, the "0" that appears to the right of the equals sign is also a zero and is not the uppercase letter "O."

If you omit the H0 option from the PROC TTEST statement, the default comparison number is zero. This means that, in most cases, there is no harm in omitting this option.

The general form of the PROC TTEST statement also contains the following option:

```
ALPHA=confidence alpha-level
```

This ALPHA option allows you to specify the size of the confidence limits that you will estimate around the difference between means. Specifying ALPHA=0.01 produces 99% confidence limits, specifying ALPHA=0.05 produces 95% confidence limits, and specifying ALPHA=0.1 produces 90% confidence limits. Assume that in this analysis, you wish to create 95% confidence limits. This means that you will include the following option in the PROC TTEST statement:

```
ALPHA=0.05
```

The preceding general form also includes the following PAIRED statement:

```
PAIRED criterion-variable1*criterion-variable2;
```

In the PAIRED statement, you should list the names of the two SAS variables that contain the scores on the criterion variable obtained under the two treatment conditions. Notice that there is an asterisk (*) that separates the two variable names.

When SAS performs a paired-samples *t* test, it subtracts scores obtained under one condition from scores obtained under the other to create a new variable consisting of the difference scores. (This is done for you automatically.) The order in which you type your criterion variable names in the PAIRED statement determines how these difference scores are created. Specifically, SAS subtracts scores on `criterion-variable2` from scores on `criterion-variable1` in the example above. In other words, it subtracts scores on the variable on the right side of the asterisk from scores on the variable on the left side of the asterisk.

Actual SAS Statements for PROC TTEST

Below are the actual statements that request that SAS perform a paired-samples *t* test on the present data set.

```
PROC TTEST    DATA=D1   H0=0    ALPHA=.05;
   PAIRED    HIGH*LOW;
RUN;
```

In this PROC TTEST statement, you can see that you requested the option H0=0. This requests that SAS test the null hypothesis that the difference between the low-and high-reward conditions equals zero.

The PROC TTEST statement also includes the option ALPHA=0.05. This requests that SAS compute the 95% confidence interval for the difference between the means. (This is also the default alpha level, meaning that the SAS program computes the 95% confidence interval automatically even if not specifically requested.)

The PAIRED statement lists the SAS variable HIGH on the left side of the asterisk and LOW on the right. This means that scores on low-reward condition will be subtracted from scores on high-reward condition when computing difference scores. Given this format, if the mean difference score is a positive number, you will know that the participants displayed higher commitment scores under the high-reward condition than under the low-reward condition on average. Conversely, if the mean difference score is a negative number, you will know that participants displayed higher commitment scores under the low-reward condition than the high-reward condition.

The following is the entire program, including the DATA step, to analyze the fictitious data from Table 8.6. Notice how the actual variable names LOW and HIGH appear in the appropriate locations in lines 17 to 24.

```
1          DATA D1;
2              INPUT   #1   @1   LOW   2.
3                           @4   HIGH  2.   ;
4
5          DATALINES;
6           9 20
7           9 22
8          10 23
9          11 23
10         12 24
11         12 25
12         14 26
13         15 28
14         17 29
15         19 31
16          ;
17         RUN;
18         PROC MEANS    DATA=D1;
19            VAR LOW HIGH;
20         RUN;
21
22         PROC TTEST   DATA=D1   HO=0   ALPHA=.05;
23             PAIRED HIGH*LOW;
24         RUN;
```

Interpreting the SAS Output

Output 8.3 presents the results obtained from the preceding program, both the PROC MEANS and PROC TTEST procedures. Review the results to first verify that everything ran as expected. Under the column headed "N," you see that there were 10 observations for both variables. This is as expected, since there were 10 participants providing data under both conditions. Under the column headed "Mean," you can see that the average commitment score in the low-reward condition was 12.8 while the average in the high-reward condition was 25.1. Participants therefore displayed higher levels of commitment for the high-reward version of partner 10, consistent with your hypothesis. (Later, you will determine whether these differences are statistically significant.)

Output 8.3 Results of the Paired-Samples *t* Test, Investment Model Study

```
                          The MEANS Procedure

    Variable    N           Mean        Std Dev        Minimum          Maximum
    ----------------------------------------------------------------------------
    LOW        10      12.8000000      3.3928028      9.0000000      19.0000000
    HIGH       10      25.1000000      3.4140234     20.0000000      31.0000000
    ----------------------------------------------------------------------------


                          The TTEST Procedure

                             Statistics

                  Lower CL          Upper CL  Lower CL           Upper CL
    Difference    N    Mean   Mean     Mean   Std Dev  Std Dev   Std Dev   Std Err

    HIGH - LOW   10   11.817   12.3   12.783    0.4643   0.6749    1.2322    0.2134


                              T-Tests

                  Difference        DF    t Value    Pr > |t|

                  HIGH - LOW         9      57.63      <.0001
```

Earlier in the chapter, you learned that possible scores on the commitment scale can range from 4 to 36. You can now review the values in the "Minimum" and "Maximum" columns to verify that no observed values fell outside of this range. (Values exceeding these limits could indicate an error in entering the data or writing the input statement.) The output shows that observed scores on LOW range from 9 to 19 and that observed scores on HIGH range from 20 to 31. These values fall within your expected range, so there is no obvious evidence of errors. With this done, you are now free to review the output results relevant to your null hypothesis.

The bottom half of Output 8.3 provides the results produced by PROC TTEST. This section also provides results pertaining to the difference score between low-and high-reward conditions. In the middle section of the table, you can see that the average difference score was 12.3. This value is found in the column labeled "Mean" to the right along the row labeled "HIGH – LOW." To determine whether this is in error, you can manually compute the difference between means and compare it against the mean difference core of 12.3. From the first section of the output, you can see that the mean score for the low-reward condition was 12.8 and the mean score for the high-reward condition was 25.1. Subtracting the former mean from the latter results in the following:

$$25.1 - 12.8 = 12.3$$

Therefore, you can see that using the means from PROC MEANS to compute the mean difference by hand results in the same difference as was reported in the output of PROC TTEST. Again, this suggests that there were no errors made when entering the data or when writing the SAS program itself.

This positive value indicates that, on the average, scores on HIGH tended to be higher than scores on LOW. The direction of this difference is consistent with your prediction that higher rewards are associated with greater levels of commitment.

Also presented in the middle of this table are the confidence limits for this mean difference. As you remember from earlier in this chapter, a confidence interval extends from a lower confidence limit to an upper confidence limit. To find the lower confidence limit for the current difference between means, look below the label "Lower CL Mean." There, you can see that the lower confidence limit for the difference is 11.82. To find the upper confidence limit, look below the label "Upper CL Mean." There, you can see that the upper confidence limit for the difference is 12.78. This means that you can estimate with a 95% probability that the actual difference between the mean of the low-reward condition and the mean of the high-reward condition in the population is somewhere between 11.82 and 12.78.

Notice that this interval does not contain the value of zero. This indicates that you can reject the null hypothesis (i.e., there is no difference between the low- and high-reward conditions)." If the null hypothesis were true, you would have expected the confidence interval to include a value of zero (i.e., a difference score of zero). The fact that your confidence interval does not contain a value of zero indicates rejection of the null hypothesis. To determine the level of statistical significance, review the *t* test results below in Output 8.3.

The *t* statistic in a paired-samples *t* test is computed using the following formula:

$$t = \frac{M_d}{SE_d}$$

where:

M_d =   the mean difference score and;

SE_d=   the standard error of the mean for the difference scores (the standard deviation of the sampling distribution of means of difference scores).

The final section in Output 8.3 contains the relevant *t* statistic, under the heading "t Value". This *t* value of 57.63 was obtained by dividing the mean difference score of 12.3 (under the heading "Mean") by the standard error of the mean or .21 (under the heading "Std Err"). The *t* statistic had an associated *p* value less than . 01 (under the heading "Pr > |t|"). This *p* value is much lower than the standard cutoff of .05 indicating that the mean difference score of 12.3 was significantly different from zero. You can therefore reject the null hypothesis that the difference score is zero and can conclude that the mean commitment score of 25.1 observed with the high-reward version of partner 10 was significantly greater than the mean score of 12.8 observed with low-reward version of partner 10. In other words, you can tentatively conclude that the level of reward manipulation had an effect on rated commitment.

The degrees of freedom associated with this *t* test are equal to $N-1$, where $N =$ the number of pairs of observations in the study. This is analogous to stating that N is equal to the number of difference scores that are analyzed. If the study involves taking repeated measures from a single sample, N equals the number of participants. However, if the study involves two sets of participants who are matched to form participant pairs, N will equal the number of participant pairs (i.e., half the total number of participants).

The present study involved taking repeated measures from a single sample of 10 participants. Therefore, $N = 10$ in this study, and the degrees of freedom are equal to $10 -1$, or 9. This value appears below the heading "DF" in output 8.3.

Earlier in this chapter, you learned that an **effect size** can be defined as the degree to which a mean score obtained under one condition differs from the mean score obtained under a second. The symbol for effect size is *d*. When performing a paired-samples *t* test, the formula for effect size is as follows:

$$d = \frac{|M_1 - M_2|}{s_p}$$

where:

$M_1 =$ the observed mean of the sample of scores obtained under Condition 1;

$M_2 =$ the observed mean of the sample of scores obtained under Condition 2 and;

$s_p =$ the estimated standard deviation of the population of difference scores.

Although SAS does not automatically compute effect sizes, you can easily do so yourself using the information that appears in the output of PROC MEANS and PROC TTEST. First, you need the mean commitment level scores for both treatment conditions. These means appear in the upper section of Output 8.3.

In the preceding formula, M_1 represents the observed sample mean for scores obtained under Condition 1 (low-reward condition). In Output 8.3, you can see that the mean commitment score for these participants was 12.8. In the preceding formula, M_2 represents the observed mean obtained under Condition 2 (high-reward condition). You can see that the mean commitment score obtained under this condition was 25.1. Substituting these two means in the formula results in the following:

$$d = \frac{|12.8 - 25.1|}{s_p}$$

In the formula for d, S_p represents the estimated standard deviation of difference scores. This statistic appears in the "Statistics" table from the results of PROC TTEST. The estimated standard deviation of difference scores appears below the heading "Std Dev." For the current study, you can see that this standard deviation is .67. Substituting this value in the formula results in the following:

$$d = \frac{\mid 12.8 - 25.1 \mid}{.67}$$

$$d = \frac{12.3}{.67}$$

$$d = 18.3582$$

$$d \approx 18.36$$

Thus, the obtained index of effect size for the current study is 18.36. This means that the mean commitment score obtained under the low-reward condition differs from the mean commitment score obtained under the high-reward condition by 18.36 standard deviations. To determine whether this is a large or small difference, refer back to the guidelines provided by Cohen (1992) in Table 8.1. Your obtained d statistic of 18.36 is larger than the "large effect" value of .80. This means that the manipulation in your study produced a very large effect.

Summarizing the Results of the Analysis

You could summarize the results of the present analysis following the same format used with the independent groups t test as presented earlier in this chapter (e.g., statement of the problem, nature of the variables). Figure 8.5 illustrates the mean commitment scores obtained under the two conditions manipulated in the present study:

Figure 8.5 Mean Levels of Commitment Observed for Participants in High-Reward versus Low-Reward Conditions, Paired-Samples Design

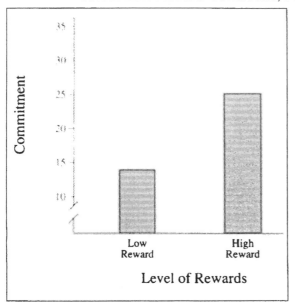

You could describe the results of the analysis in a paper in the following way:

> Results were analyzed using a paired-samples *t* test. This analysis revealed a significant difference between mean levels of commitment observed in the two conditions, *t*(9) = 57.63; *p* < .01. The sample means are displayed in Figure 8.4, which shows that mean commitment scores were significantly higher in the high-reward condition (*M* = 25.1, *SD* = 3.41) than in the low-reward condition (*M* = 12.8, *SD* = 3.39). The observed difference between these mean scores was 12.3 and the 95% confidence interval for the difference between means extended from 11.82 to 12.78. The effect size was computed as *d* = 18.36. According to Cohen's (1992) guidelines for *t* tests, this represents a very large effect.

Example: A Pretest-Posttest Study

An earlier section presented the hypothesis that taking a foreign language course will lead to an improvement in critical thinking among college students. To test this hypothesis, assume that you conducted a study in which a single group of college students took a test of critical thinking skills both before and after completing a semester-long foreign language course. The first administration of the test constituted the study's pretest, and the second administration constituted the posttest. Table 8.7 again reproduces the data obtained in the study:

Table 8.7

Fictitious Data from Study Using a Pretest-Posttest Procedure

	Scores on Test of Critical Thinking Skills	
Participant	Pretest	Posttest
Paul	34	55
Vilem	35	49
Pavel	39	59
Sunil	41	63
Maria	43	62
Fred	44	68
Jirka	44	69
Eduardo	52	72
Asher	55	75
Shirley	57	78

You could enter the data for this study according to the following format:

Line	Column	Variable Name	Explanation
1	1-2	PRETEST	Scores on the test of critical thinking obtained at the first administration
	3	blank	
	4-5	POSTTEST	Scores on the test of critical thinking obtained at the second administration

Here is the general form for the PROC MEANS and PROC TTEST statements to perform a paired-samples *t* test using data obtained from a study using a pretest-posttest design:

```
PROC MEANS;
   VAR pretest posttest;
RUN;

PROC TTEST    DATA=dataset name
              H0=comparison number
              ALPHA=alpha level;
   PAIRED posttest*pretest;
RUN;
```

Notice that these statements are identical to the general form statements presented earlier in this chapter, except that the "pretest" and "posttest" variables have been substituted for "variable1" and "variable2," respectively.

The following is the entire SAS program to input the data from Table 8.7 and perform a paired-samples *t* test.

```
1            DATA D1;
2               INPUT   #1   @1   PRETEST   2.
3                            @4   POSTTEST  2.   ;
4
5            DATALINES;
6            34 55
7            35 49
8            39 59
9            41 63
10           43 62
11           44 68
12           44 69
13           52 72
14           55 75
15           57 78
16           ;
17           RUN;
18
19           PROC MEANS    DATA=D1;
20              VAR PRETEST POSTTEST;
21           RUN;
22
23           PROC TTEST    DATA=D1
24                 H0=0
25                 ALPHA=.05;
26                 PAIRED POSTTEST*PRETEST;
27           RUN;
```

The preceding program results in the analysis of two variables: PRETEST (each participant's score on the pretest) and POSTTEST (each participant's score on the posttest). Once again, a difference variable was created by subtracting each participant's score on PRETEST from his or her POSTTEST score. Given the way that difference scores were created in the preceding program, a positive mean difference score would indicate that the average posttest score was higher than the average pretest score. Such a finding would be consistent with your hypothesis that the foreign language course would cause an improvement in critical thinking. If the average difference score is not significantly different from zero, however, your hypothesis would receive no support. (Again, remember that any results obtained from the present study would be difficult to interpret given the lack of an appropriate control group.)

You can interpret the results from the preceding program in the same manner as with Example 1, earlier. In the interest of space, those results do not appear again here.

Conclusion

The *t* test is one of the most commonly used statistics in the social sciences, in part, because some of the simplest investigations involve the comparison of just two treatment conditions. When an investigation involves more than two conditions however, the *t* test is no longer appropriate, and you should generally substitute it with the *F* test obtained from an analysis of variance (ANOVA). The simplest ANOVA procedure, the one-way ANOVA with one between-subjects factor, is the topic of the following chapter.

Assumptions Underlying the *t* Test

Assumptions Underlying the Independent-Samples *t* Test

- **Level of measurement.** The criterion variable should be assessed on an interval- or ratio-level of measurement. The predictor variable should be a nominal-level variable that includes just two categories (i.e., two groups).

- **Independent observations.** A given observation should not be dependent on any other observation in either group. In an experiment, you normally achieve this by drawing a sample and randomly assigning each participant to only one of the two treatment conditions. This assumption would be violated if a given participant contributed scores on the criterion variable under both treatment conditions.

  The independence assumption is also violated when one participant's behavior influences another's behavior within the same condition. For example, if participants are given experimental instructions in groups of five and are allowed to interact in the course of providing scores on the criterion variable, it is likely that their scores will not be independent. That is, each participant score is likely to be affected by the others in that group. In these situations, scores from participants constituting a given group of five should be averaged and these average scores should constitute the unit of analysis. None of the tests discussed in this text are robust against violations of the independence assumption.

- **Random sampling.** Scores on the criterion variable should represent a random sample drawn from the populations of interest.

- **Normal distributions.** The distribution of observed values for all continuous variables should approximate normal distributions.

- **Homogeneity of variance.** To use the equal-variances *t* test, you should draw the samples from populations with equal variances on the criterion. If the null hypothesis of equal population variances is rejected, you should use the unequal-variances *t* test.

Assumptions Underlying the Paired-Samples *t* Test

- **Level of measurement.** The criterion variable should be assessed on an interval- or ratio-level of measurement. The predictor variable should be a nominal-level variable that includes just two categories.

- **Paired observations.** A given observation appearing in one condition must be paired in some meaningful way with a corresponding observation appearing in the other condition. You can accomplish this by having each participant contribute one score under Condition 1 and a separate score under Condition 2. Observations could also be paired by using a matching procedure.

- **Independent observations.** A given participant's score in one condition should not be affected by any other participant's score in either condition. It is, of course, acceptable for a given participant's score in one condition to be dependent upon his or her own score in the other. This is another way of stating that it is acceptable for participants' scores in Condition 1 to be correlated with their scores in Condition 2.

- **Random sampling.** Participants contributing data should represent a random sample drawn from the populations of interest.

- **Normal distribution for difference scores.** Differences in paired scores should be normally distributed. These difference scores are normally created by beginning with a given participant's score on the dependent variable obtained under one treatment condition and subtracting from it that participant's score on the dependent variable obtained under the other treatment condition. It is not necessary that the individual dependent variables be normally distributed so long as difference scores are normally distributed.

- **Homogeneity of variance.** The populations represented by the two conditions should have equal variances on the criterion.

References

American Psychological Association (2001). *Publication manual of the American Psychological Association* (5th ed.). Washington, DC: Author.

Cohen, J. (1992). A power primer. *Psychological Bulletin, 112,* 155-159.

Rusbult, C. E. (1980). Commitment and satisfaction in romantic associations: A test of the investment model. *Journal of Experimental Social Psychology, 16,* 172-186.

SAS Institute Inc. (2004). *SAS/STAT 9.1 user's guide.* Cary, NC: Author.

One-Way ANOVA with One Between-Subjects Factor

> **Overview.** In this chapter, you learn how to enter data and prepare SAS programs that perform a one-way analysis of variance (ANOVA) using the GLM procedure. This chapter focuses on the between-subjects design, in which each participant is exposed to only one condition under the independent variable. The use of a multiple comparison procedure (Tukey's HSD test) is described, and guidelines for summarizing the results of the analysis in tables and text are provided.

Introduction: The Basics of One-Way ANOVA, Between-Subjects Design

One-way analysis of variance (ANOVA) is appropriate when an analysis involves:

- a single predictor variable that is measured on a nominal scale and can assume two or more values;

- a single criterion variable that is measured on an interval or ratio scale.

In Chapter 8, "*t* Tests: Independent Samples and Paired Samples," you learned about the independent samples *t* test, which you can use to determine whether there is a significant difference between two groups with regard to their respective scores on an interval- or ratio-scale criterion variable. But what if you are conducting a study in which you must compare more than just two groups? In those situations, it is often appropriate to analyze your data using a one-way ANOVA.

The analysis that this chapter describes is called *one-way* ANOVA because you use it to analyze data from studies in which there is only one predictor variable (or independent variable). In contrast, Chapter 10, "Factorial ANOVA with Two Between-Subjects Factors," presents a statistical procedure that is appropriate for studies with two predictor variables.

The Aggression Study

To illustrate a situation for which a one-way ANOVA might be appropriate, imagine that you are conducting research on aggression in children. Assume that a review of prior research has led you to believe that consuming sugar causes children to behave more aggressively. You therefore want to conduct a study to test the following hypothesis:

> The amount of sugar consumed by eight-year-old children has a positive effect on the levels of aggression that they subsequently display.

To test your hypothesis, you conduct an investigation in which each child in a group of 60 children is assigned to one of three experimental conditions. Assignments are made in the following way:

- 20 children are assigned to the "0 grams of sugar" control condition;

- 20 children are assigned to the "20 grams of sugar" treatment condition;

- 20 children are assigned to the "40 grams of sugar" treatment condition.

The independent variable in the study is "the amount of sugar consumed." You manipulate this variable by controlling the amount of sugar that is contained in the lunch that each child receives. In this way, you ensure that the children in the "0 grams of sugar" group are actually consuming 0 grams of sugar, that the children in the "20 grams of sugar" group are actually consuming 20 grams, and so forth.

The dependent variable in the study is "level of aggression" displayed by each child. To measure this variable, a pair of observers watches each child for a set period of time each day after lunch. These observers tabulate the number of aggressive acts performed by each child during this time. The total number of aggressive acts performed over a two-week period serves as each child's score on the dependent variable.

You can see that the data from this investigation are appropriate for a one-way ANOVA because:

- the study involves a single predictor variable that is measured on a nominal scale (i.e., "amount of sugar consumed");

- the predictor variable assumes more than two values (i.e., the "0-gram," the "20-gram," and the "40-gram" groups);

- the study involves a single criterion variable (number of aggressive acts) that is measured on an interval or ratio scale.

Between-Subjects Designs versus Repeated-Measures Designs

The research design that this chapter discusses is referred to as a **between-subjects** design because each participant appears in only one group, and comparisons are made *between* different groups of participants. For example, in the experiment just described, a given participant is assigned to just one treatment condition (e.g., the 20-gram group), and provides data on the dependent variable for only that specific experimental condition.

A distinction, therefore, is made between a between-subjects design and a repeated-measures design. With a **repeated-measures** design, a given participant provides data under each treatment condition in the study. (It is called a "repeated-measures" design because each participant provides repeated measurements on the dependent variable.)

In short, a one-way ANOVA with one between-subjects factor is directly comparable to the independent-samples *t* test from Chapter 8. The main difference is that you can use a *t* test to compare just two groups, while you can use a one-way ANOVA to compare two or more groups. In the same way, a one-way ANOVA with one repeated-measures factor is very similar to the paired-samples *t* test from Chapter 8. Again, the main difference is that you can use a *t* test to analyze data from just two treatment conditions, but you can use a repeated-measures ANOVA with data from two or more treatment conditions. (The repeated-measures ANOVA is covered in Chapter 12, "One-Way ANOVA with One Repeated-Measures Factor.")

Multiple Comparison Procedures

When you analyze data from an experiment with a between-subjects ANOVA, you can state the null hypothesis as follows:

> In the population, there is no difference between the various conditions with respect to their mean scores on the dependent variable.

For example, with the preceding study on aggression, you might state a null hypothesis that, in the population, there is no difference between the 0-gram, the 20-gram, and the 40-gram groups with respect to the mean number of aggressive acts performed. This null hypothesis could be represented symbolically in this way:

$$H_0: M_1 = M_2 = M_3$$

where M_1 represents the mean level of aggression shown by the 0-gram group, M_2 represents mean aggression shown by the 20-gram group, and M_3 represents mean aggression shown by the 40-gram group.

When you analyze your data, SAS's PROC GLM tests this null hypothesis by computing an F statistic. If the F statistic is sufficiently large (and the p value associated with the F statistic is sufficiently small), you can reject the null hypothesis. In rejecting the null, you tentatively conclude that, in the population, at least one of the three conditions differs from at least one other condition on the measure of aggression.

However, this leads to a problem: which pairs of treatment groups are significantly different from one another? There are various possibilities. Perhaps the 0-gram group is different from the 40-gram group but is not different from the 20-gram group. Perhaps the 20-gram group is different from the 40-gram group but is not different from the 0-gram group. Perhaps all three groups are significantly different from each other.

Faced with this problem, researchers routinely rely on **multiple comparison procedures**. These are statistical tests used in studies with more than two groups to help determine which pairs of groups are significantly different from one another. Several different multiple comparison procedures are available with SAS's PROC GLM, including Duncan's multiple-range test, the Scheffe test, and the Student-Newman-Keuls test. This chapter shows how to request and interpret Tukey's studentized range test, sometimes called Tukey's HSD test ("honestly significant difference"). The Tukey test is especially useful when the various groups in the study have unequal numbers of participants (which is often the case). For a description of the various multiple comparison tests that are available with PROC GLM, see the *SAS/STAT User's Guide*.

Statistical Significance versus the Magnitude of the Treatment Effect

This chapter also shows how to calculate the R^2 statistic from the output provided in an analysis of variance. In an ANOVA, R^2 represents the percent of variance in the criterion that is accounted for by the predictor variable. In a true experiment, you can view R^2 as an index of the magnitude of the treatment effect. It is a measure of the strength of the relationship between the predictor and the criterion. Values of R^2 can range from 0 through 1, with values closer to 0 indicating a weak relationship between the predictor and criterion, and values closer to 1 indicating a stronger relationship.

For example, assume that you conduct the preceding study on aggression in children. If your independent variable (amount of sugar consumed by the children) has a very weak effect on the level of aggression displayed by the children, R^2 is a small value, perhaps .02 or .04. On the other hand, if your independent variable has a very strong effect on their level of aggression, R^2 will be a larger value, perhaps .20 or .40 (exactly how large R^2 must be to be considered "large" depends on a number of factors that are beyond the scope of this chapter).

It is good practice to report R^2 or some other measure of the magnitude of effect in research papers because researchers like to draw a distinction between results that are merely statistically significant versus those that are truly meaningful. The problem is that researchers frequently obtain results that are statistically significant but not meaningful in terms of the magnitude of the treatment effect. This is especially likely to happen when conducting research with very large samples. When the sample is very large (say, several hundred participants), you can obtain results that are statistically significant even though your independent variable has a very weak effect on the dependent variable. This occurs because many statistical tests become very sensitive to minor group differences when samples are large.

For example, imagine that you conduct the preceding aggression study with 500 children in the 0-gram group, 500 children in the 20-gram group, and 500 children in the 40-gram group. It is possible that you would analyze your data with a one-way ANOVA, and obtain an F value that is significant at $p < .05$. Normally, this might lead you to rejoice. But imagine that you then calculate R^2 for this effect, and learn that R^2 is only .03. This means that only 3% of the variance in aggression is accounted for by the amount of sugar consumed. Obviously, your manipulation has had a very weak effect. Even though your independent variable is statistically significant, most researchers would argue that it does not account for a meaningful amount of variance in children's aggression.

This is why it is helpful to always provide a measure of the magnitude of the treatment effect (such as R^2) along with your test of statistical significance. In this way, your readers can assess whether your results are truly meaningful in terms of the strength of the relationship between the predictor and criterion variables.

Example with Significant Differences between Experimental Conditions

To illustrate one-way ANOVA, imagine that you replicate the study that examined the effect of rewards on commitment in romantic relationships. (This study was described in Chapter 8.) To give the study an added twist, however, in this investigation you use three experimental conditions instead of the two conditions described in the last chapter.

Recall that in the preceding chapter it was hypothesized that the rewards people experience in a romantic relationship will have a causal effect on their commitment to those relationships. You tested this prediction by conducting an experiment with 20 participants. All 20 participants were asked to read the descriptions of 10 potential romantic partners. For each partner, participants imagined what it would be like to date this person and rated how committed they would be to a relationship with that person. For the first nine partners, every participant saw exactly the same description. There were some important differences with respect to partner 10, however:

- Half of the participants had been assigned to a "high-reward" condition and these participants were told that partner 10 "...enjoys the same recreational activities that you enjoy, and is very good looking."

- The other half were assigned to the "low-reward" condition and were told that partner 10 "...does not enjoy the same recreational activities that you enjoy, and is not very good looking."

You are now going to repeat this experiment using essentially the same procedure as before, but this time you will add a third experimental condition called the "mixed-reward" condition. Here is the description of partner 10 that will be read by those assigned to this group:

```
PARTNER 10:   Imagine that you have been dating partner 10 for about
1 year, and you have put a great deal of time and effort into this
relationship.   There are not very many potential partners where you
live, so it would be difficult to replace this person with someone
else.   Partner 10 lives in the same neighborhood as you, so it is
easy to see him/her as often as you like.   Sometimes this person
seems to enjoy the same recreational activities that you enjoy and
sometimes s/he does not.   Sometimes partner 10 seems to be very good
looking and sometimes s/he does not.
```

Notice that the first three sentences of the preceding description are identical to the descriptions read by participants in the low-reward and high-reward conditions in the experiment described in the preceding chapter. These three sentences deal with investment size, alternative value, and costs, respectively. However, the last two sentences deal with rewards, and they are different from the sentences dealing with rewards in the other two conditions. In this mixed-reward condition, partner 10 is portrayed as somewhat of a chameleon: sometimes s/he is rewarding to be around and sometimes s/he is not.

The purpose of the present study is to determine how this new mixed-reward version of partner 10 will be rated. Will this version be rated more positively than the version provided in the low-reward condition? Will it be rated more negatively than the version provided in the high-reward condition?

To conduct this study, you begin with a total of 18 participants. You randomly assign six to the high-reward condition, and they read the high-reward version of partner 10 as it was described in Chapter 8. You randomly assign six participants to the low-reward condition, and they read the low-reward version of partner 10. Finally, you randomly assign six to the mixed-reward condition, and they read the mixed-reward version just presented. As was the case in the previous investigation, you analyze participants' ratings of how committed they would be to a relationship with partner 10 and ignore their ratings of the other nine partners.

You can see that this study is appropriate for analysis with a one-way ANOVA, between groups design because:

- it involves a single predictor variable assessed on a nominal scale (type of rewards);
- it involves a single criterion variable assessed on an interval scale (rated commitment);
- it involves three treatment conditions (low-reward, mixed-reward, and high-reward), making it inappropriate for analysis with an independent-groups *t* test.

The following section shows how to write a program in which PROC GLM analyzes some fictitious data for this study using a one-way ANOVA with one between-subjects factor.

Writing the SAS Program

There was one predictor variable in this study: "type of rewards." For purposes of writing the program, you will assign participants a value of "3" if they were in the high-reward condition, a value of "2" if they were in the mixed-reward condition, and a value of "1" if they were in the low-reward condition. You need a SAS variable name for this variable, so you call it REWARDGROUP.

There is one criterion variable in this study, "commitment": the participants' rating of how committed they would be to a relationship with partner 10. This variable is actually based on the sum of responses to four questionnaire items described in the last chapter. This variable could assume values from 4 through 36 and was measured on an interval scale. You give it a SAS variable name of COMMITMENT.

There are a number of ways to perform a one-way ANOVA with SAS. In fact, there is even a procedure called PROC ANOVA that is designed specifically for analysis of variance. However, this manual presents the use of PROC GLM instead. PROC GLM is more appropriate when some of your experimental groups have more participants than other experimental groups. Statistics textbooks sometimes refer to these situations as studies with **unequal cells** or studies with an **unbalanced design**. PROC GLM is appropriate for either balanced or unbalanced designs. At the end of this chapter, PROC GLM with unequal cells is discussed further.

Earlier, you learned that you would use a multiple comparison procedure when your ANOVA reveals significant results and you want to determine which pairs of groups are significantly different. The program provided here includes a statement to request Tukey's HSD test so that, if the overall ANOVA is significant, the multiple comparison tests are included in the output, ready for interpretation. However, remember that you only interpret a multiple comparison test if the *F* for the overall ANOVA is significant.

This is the syntax for the SAS program that performs a one-way ANOVA followed by Tukey's HSD Test:

```
PROC GLM DATA=filename;
   CLASS  predictor-variable;
   MODEL  criterion-variable = predictor-variable;
   MEANS  predictor-variable / TUKEY;
   MEANS  predictor-variable;
RUN;
```

Here is the entire program—including the DATA step—to analyze fictitious data from the preceding study:

```
 1          DATA D1;
 2             INPUT  #1  @1   REWARDGROUP  1.
 3                        @3   COMMITMENT   2.  ;
 4          DATALINES;
 5          1 13
 6          1 06
 7          1 12
 8          1 04
 9          1 09
10          1 08
11          2 33
12          2 29
13          2 35
14          2 34
15          2 28
16          2 27
17          3 36
18          3 32
19          3 31
20          3 32
21          3 35
22          3 34
23          ;
24          RUN;
25
26          PROC GLM DATA=D1;
27             CLASS REWARDGROUP;
28             MODEL COMMITMENT = REWARDGROUP;
29             MEANS REWARDGROUP / TUKEY;
30             MEANS REWARDGROUP;
31          RUN;
```

Notes Regarding the SAS Program

The data in the preceding program were entered so that there is one line of data for each participant. They were entered according to the following format:

Line	Column	Variable Name	Explanation
1	1	REWARDGROUP	Codes group membership, so that 1 = low-reward condition 2 = mixed-reward condition 3 = high-reward condition
	2	blank	
	3-4	COMMITMENT	Commitment ratings obtained when participants rated partner 10

You can see that it was necessary to create just two columns of variables to perform the ANOVA. The first column of data (in column 1) included each participant's score on REWARDGROUP, the variable that coded group membership. REWARDGROUP was coded using 1s, 2s and 3s, but you could have used any numbers (or even letters such as Ls, Ms, and Hs).

The second column of data (in columns 3 to 4) included participant scores on the commitment variable, COMMITMENT. COMMITMENT is a two-digit variable because it could take on a value as high as 36 (i.e., a two-digit number). All values on COMMITMENT were entered as two-digit numbers, putting a zero in front of single-digit numbers where appropriate (e.g., 06, 09). Using zeros in this fashion can make it easier to keep your place when entering multiple-digit variables.

In the program, data for the six low-reward participants were entered in lines 5 to 10, data for the six mixed-reward participants were entered on lines 11 through 16, and data for the six high-reward participants were entered on lines 17 through 22. It is not necessary to separate data for the three groups in this way, however; you could have entered the data in a random sequence if desired.

The PROC GLM step that requests the analysis of variance appears on lines 26 through 31 of the program. The CLASS statement on line 27 specifies the name of the nominal-scale predictor variable for the analysis. In this analysis, REWARDGROUP is the predictor.

The MODEL statement appears on line 28 of the program. With the MODEL statement, the name of the interval-level or ratio-level criterion variable should appear to the left of the "=" sign, and the name of the nominal-level predictor should appear to the right. If you want, you can list more than one criterion variable to the left of the "=" sign, and separate ANOVAs will be performed for each. Do not list more than one predictor variable to the right of the "=" sign, however, or the analysis performed will not be a one-way ANOVA.

The MEANS statement on line 29 requests that the Tukey HSD test be performed. The second MEANS statement (on line 30) merely requests that the mean commitment scores for the three treatment groups be printed.

Results from the SAS Output

With LINESIZE=80 and PAGESIZE=60 in the OPTIONS statement (first line), the preceding program would produce the following output. The contents of each page are summarized here:

- Page 1 provides class level information and the number of observations in the dataset.

- Page 2 provides the analysis of variance table.

- Page 3 provides the results of the Tukey HSD test.

- Page 4 provides the means and standard deviations requested by the second MEANS statement.

The output created by the program is reproduced here as Output 9.1.

Output 9.1 Results of One-Way ANOVA: Significant Differences Observed

```
                          The SAS System                              1


                          The GLM Procedure

                      Class Level Information

                  Class          Levels    Values

                  REWARDGROUP        3     1 2 3


                  Number of observations    18
```

(continued on the next page)

Output 9.1 *(continued)*

```
                                                                           2
                        The GLM Procedure

Dependent Variable: COMMITMENT

                                Sum of
Source                  DF      Squares    Mean Square   F Value   Pr > F

Model                    2   2225.333333   1112.666667    122.12   <.0001

Error                   15    136.666667      9.111111

Corrected Total         17   2362.000000

          R-Square    Coeff Var    Root MSE    COMMITMENT Mean

          0.942139    12.40464     3.018462       24.33333

Source                  DF     Type I SS    Mean Square   F Value   Pr > F

REWARDGROUP              2   2225.333333   1112.666667    122.12   <.0001

Source                  DF    Type III SS   Mean Square   F Value   Pr > F

REWARDGROUP              2   2225.333333   1112.666667    122.12   <.0001
```

```
                                                                           3
                        The GLM Procedure

         Tukey's Studentized Range (HSD) Test for COMMITMENT

NOTE: This test controls the Type I experimentwise error rate, but it generally
      has a higher Type II error rate than REGWQ.


              Alpha                                  0.05
              Error Degrees of Freedom                 15
              Error Mean Square                   9.111111
              Critical Value of Studentized Range  3.67338
              Minimum Significant Difference        4.5266


      Means with the same letter are not significantly different.


      Tukey Grouping         Mean       N     REWARDGROUP

                    A       33.333      6     3
                    A
                    A       31.000      6     2

                    B        8.667      6     1
```

(continued on the next page)

Output 9.1 *(continued)*

```
                                                                    4
                        The GLM Procedure

        Level of              ------------COMMITMENT-----------
        REWARDGROUP      N           Mean            Std Dev

        1                6        8.6666667         3.44480285
        2                6       31.0000000         3.40587727
        3                6       33.3333333         1.96638416
```

Steps in Interpreting the Output

1. Make Sure That Everything Looks Right

Page 1 of the PROC GLM output provides class level information for the analysis. Verify that the name of your predictor variable appears under the heading "Class." Under the heading "Levels," the output should indicate how many groups were included in your study (in this case, 3). Under the heading "Values," the output should indicate the specific numbers or letters that you used to code this predictor variable (in this case, the values were 1, 2 and 3). Finally, the last line indicates the number of observations in the dataset. The present example used three groups with six participants each for a total of $N = 18$. If any of this does not look right, you probably made an error in either entering your data or writing your INPUT statement.

Page 2 of the output provides the analysis of variance table. Near the top of page 2 on the left side, the name of the criterion variable should appear to the right of the heading "Dependent Variable." In this case, the dependent variable is COMMITMENT. Below this heading is the ANOVA summary table.

The first column of the ANOVA Summary table is headed "Source" and below the heading "Source" are three subheadings: "Model," "Error," and "Corrected Total." Look to the right of the heading "Corrected Total" and under the column headed "DF." In this case, you see the number "17." This number represents the corrected total degrees of freedom. This number should always be equal to $N - 1$, where N represents the total number of participants for whom you have a complete set of data. In this study, N is 18 and $18 - 1 = 17$, so it looks like everything is correct so far. (A **complete set of data** means the number of participants for whom you have scores on both the predictor and criterion variables.)

The means for the three experimental groups are found on page 3 of the output. Toward the bottom of the page, the third column from the right is headed "Mean." Below this heading are the group means. Notice that, in this case, they range from 8.67 through 33.33. Remember that your participants used a scale that could range from 4 through 36, so these group means appear reasonable. If these means had taken on values such as 0.21 or 88.33, it would probably mean that you had made an error in either entering the data, or writing the INPUT statements.

2. Review the Appropriate *F* Statistic and Its Associated *p* Value

Once you verify that there are no obvious errors in the output, you can review your results. First you should review the *F* statistic from the ANOVA to see if you can reject the null hypothesis. You can state the null hypothesis for the present study as follows: In the population, there is no difference between the high-reward, the mixed-reward, and the low-reward groups with respect to mean commitment scores. Symbolically, you can represent the null hypothesis this way:

$$H_0: \ M_1 = M_2 = M_3$$

where M_1 is the mean commitment score for the low-reward condition, M_2 is the mean commitment score for the mixed-reward condition, and M_3 is the mean commitment score the high-reward condition.

You will generally use the degrees of freedom, sum of squares, *F* value, and *p* value associated with the Type III sum of squares. This is an important point because when you actually run a program, the output provides two ANOVA summary tables at the bottom of page 2. For this example, the Type I and Type III sums of squares are identical because this is a balanced design (i.e., equal numbers of participants in each of the three conditions). Use of the Type III sum of squares is particularly important with unbalanced designs.

The first summary table provides the results using something called the "Type I sum of squares," and has the following headings:

```
DF        Type I SS        Mean Square       F Value        Pr > F
```

Below this, you see results for a second ANOVA table that uses the "Type III sum of squares" and has the following headings:

```
DF        Type III SS      Mean Square       F Value        Pr > F
```

The *F* statistic appropriate to test your null hypothesis appears toward the bottom of page 2 in the section that provides the Type III sum of squares. Notice that there is a heading called "Source" that refers to "source of variation." Below this heading you will find the name of your predictor variable, which in this case is REWARDGROUP. To the right, you see an ANOVA summary table using the Type III sum of squares.

In this ANOVA summary table at the bottom of the page, look to the right of the word REWARDGROUP. Under the heading "DF," you see that there are 2 degrees of freedom associated with the predictor variable. The degrees of freedom associated with a predictor are equal to $k - 1$, where k represents the number of groups. Under the heading "Type III SS," you find the sum of squares associated with your predictor variable, REWARDGROUP, which in this case is 2225.33. Under the heading "F Value," you find the *F* statistic appropriate for the test of your null hypothesis. In this case, $F = 122.12$, which is very large. The last heading is "Pr > F," which gives you the probability of obtaining an *F* this large or larger if the null hypothesis were true. In the present case, this *p* value is very small, less than .01. When a *p* value is less than .05, you can reject the null hypothesis. In this case, the null hypothesis of no population differences is rejected. In other words, you conclude that there is a significant effect for the type of rewards independent variable.

3. Prepare Your Own Version of the ANOVA Summary Table

Before moving on to interpret the sample means for the three experimental groups, you should prepare your own version of the ANOVA summary table. All of the information that you need either is presented on page 2 of the output or you can easily calculate it by hand.

Table 9.1 provides the completed ANOVA summary table for the current analysis:

Table 9.1

ANOVA Summary Table for Study Investigating the Relationship between Type of Rewards and Commitment

Source	df	SS	MS	F	R^2
Type of rewards	2	2225.33	1112.67	122.12 *	.94
Within groups	15	136.67	9.11		
Total	17	2362.00			

Note. $N = 18$.

* $p < .01$

In completing this table, you should first summarize the information associated with the predictor variable, type of rewards. On the bottom of page 2 of Output 9.1, look to the right of the name of the predictor variable (REWARDGROUP, in this case). Here you find the results associated with the Type III sum of squares. Remember that all of the information related to "type of rewards" in Table 9.1 must come from this section of the output that deals with the Type III sum of squares. Under "DF," find the between-subjects degrees of freedom, two in this case. Note where these degrees of freedom are placed in Table 9.1: they appear in the column headed *df*, to the right of "Type of rewards."

On the output, next find the sum of squares associated with REWARDGROUP under the heading "Type III SS." This value (2225.33) goes under SS in Table 9.1.

Next in Table 9.1, under the heading *MS*, you find the mean square associated with REWARDGROUP. This mean square is also known as the "mean square between groups." It is calculated by PROC GLM using this simple formula:

$$MS_{\text{between groups}} = \frac{\text{Type III SS}}{\text{DF}} = \frac{2225.33}{2} = 1112.67$$

In the above formula, Type III SS represents the Type III sum of squares associated with REWARDGROUP and DF represents the degrees of freedom associated with REWARDGROUP. This mean square has already been calculated by PROC GLM and appears in Output 9.1 under the heading "Mean Square." It should appear in Table 9.1 under *MS*.

Next, find the *F* statistic appropriate for the test of your null hypothesis. This is also on page 2 of Output 9.1 under the heading "F Value" and to the right of "REWARDGROUP." This statistic is 122.12 and is placed under the heading *F* in Table 9.1.

Under the heading "Pr > F" from Output 9.1, find the *p* value associated with this *F*. However, you do not have a separate column in Table 9.1 for this *p* value. Instead, you can mark the *F* value with an asterisk ($*$) to indicate that the *F* is statistically significant. At the bottom of the table, you place a note that indicates the level of significance attained by this *F* value. In this case, the output indicated that $p < 0.01$ so add a note at the bottom of the table that indicates that the effect for type of reward was significant at the $p < .01$ level:

$*\ p < .01$

If the *F* value had been significant at the .05 level, your note would have looked like this:

$*\ p < .05$

If the *F* value had not been statistically significant, you would not have put an asterisk next to it or placed a note at the bottom of the table.

Finally, you need to calculate the R^2 for this predictor variable. As mentioned before, the R^2 statistic indicates the percent of variance in the criterion (COMMITMENT) that is accounted for by the predictor (type of reward). The formula for R^2 is also straightforward:

$$R^2 = \frac{\text{Type III SS}}{\text{Corrected Total Sum of Squares}}$$

In the above formula, Type III SS represents the Type III sum of squares associated with REWARDGROUP and Corrected Total Sum of Squares represents the total sum of squares from the analysis. This total sum of squares appears on page 2 of Output 9.1 to the right of the heading, "Corrected Total," and under the heading, "Sum of Squares." Here, the appropriate figures from page 2 of Output 9.1 are inserted in the formula:

$$R^2 = \frac{\text{Type III SS}}{\text{Corrected Total Sum of Squares}} = \frac{2225.33}{2362.00} = .94$$

With $R^2 = .94$, this indicates that 94% of the variance in commitment is accounted for by the "type of rewards" independent variable. This is a very high value of R^2, much higher than you are likely to obtain in an actual investigation. (Remember that the present data are fictitious.)

If you look closely at Output 9.1, you will notice that the R^2 value is already presented on page 2, just below the ANOVA summary table. Nevertheless, this chapter still advises that you calculate R^2 values by hand because, depending on the type of analysis that you do, the output of PROC GLM will not always provide the R^2 value in which you are interested. This is particularly true when you have more than one predictor variable. Therefore, it is good practice to ignore the R^2 provided in the output and, instead, calculate it manually.

Now that you have filled in the line headed "Type of rewards" in Table 9.1, you fill in the line headed "Within groups." This line deals with the error variance in your sample and everything you need appears on page 2 of Output 9.1 to the right of the heading "Error." Under "DF" on the output, you can see that there are 15 degrees of freedom associated with this estimate of error variance. Note where this goes in Table 9.1: where the column headed *df* intersects with the row headed "Within groups."

Under the heading "Sum of Squares" on page 2 of Output 9.1, you can see that the sum of squares associated with the error term is 136.67. This term is transferred to its corresponding spot in Table 9.1 in the column headed *SS*.

Finally, under the heading "Mean Square" on the output, you can see that the mean square error (also called the mean square within groups) is 9.11. This term is transferred to Table 9.1 under the column headed *MS*.

The last line in Table 9.1 deals with the total variability in the dataset. You will find all of the information that you need on page 2 of Output 9.1 to the right of the heading "Corrected Total." Under the heading "DF," you can see that 17 degrees of freedom are associated with the total variance estimate. This goes in Table 9.1 under the heading *df*, to the right of the heading, "Total." Under the "Sum of Squares" heading on the output, you find that the total sum of squares is 2362.00. Note that this appears in Table 9.1 in the column headed *SS*.

Place a note at the bottom of the table to indicate the size of the total sample and your own version of the ANOVA summary table is now complete.

4. Review the Sample Means and the Results of the Multiple Comparison Procedure

Because the F statistic is significant, you reject the null hypothesis of no differences in population means and tentatively accept the alternative hypothesis that at least one of the means is different from at least one of the others. However, since you have three experimental conditions, you now have a new problem: which of these groups is significantly different from the others? You have performed a multiple comparison procedure called Tukey's HSD test to answer this question. The results of this analysis appear on page 3 of Output 9.1.

About halfway down this table, it indicates that "Alpha = 0.05." This means that if any of the groups are significantly different from the others, they will be different with a significance level of $p < .05$.

On the bottom half of the table, there are four columns of figures. The last column always represents your predictor variable (which, in this case, is REWARDGROUP). Below the REWARDGROUP heading, you find the values used to code the experimental conditions. In this case, you used the value of 3, 2, and 1, which coded the high-reward, mixed-reward, and low-reward conditions, respectively. To the left of the REWARDGROUP column is a column headed "N." This column indicates how many participants were in each condition. In the present case, there were six participants in each. To the left of the "N" column is a column headed "Mean," which provides the mean scores for the three groups on the criterion variable (the commitment variable in this study). You can see that the high-reward group had a mean commitment score of 33.33, the mixed-reward group had a mean score of 31.00, and the low-reward group had a mean score of 8.67.

Finally, to the left of the "Mean" column is the column headed "Tukey Grouping." It is this column that tells you which groups are significantly different from which. The rules to use here are simple:

- Two groups that are identified by the same letter are not significantly different from each other.
- Two groups that are identified by different letters are significantly different from each other.

For example, notice that group 3 is identified with the letter "A" in the grouping column while group 1 is identified with "B." This means that groups 3 and 1 are significantly different from each other on commitment. (Notice how different their means are.) Similarly, groups 2 and 1 are also identified by different letters, so they are also significantly different. However, groups 3 and 2 are both identified by "A." Therefore, groups 3 and 2 are not significantly different.

 Remember that you review the results of a multiple comparison procedure only if the F value from the preceding ANOVA summary table is statistically significant. If the F for the predictor variable is nonsignificant, then ignore the results of this Tukey HSD test.

5. Summarize the Analysis Results

When performing one between-subjects factor ANOVA, you can use the following format to summarize the results of your analysis:

A) Statement of the problem
B) Nature of the variables
C) Statistical test
D) Null hypothesis (H_0)
E) Alternative hypothesis (H_1)
F) Obtained statistic
G) Obtained probability (p) value
H) Conclusion regarding the null hypothesis
I) Magnitude of treatment effect

J) ANOVA summary table
K) Figure representing the results

An example summary follows:

A) Statement of the problem: The purpose of this study was to determine whether there was a difference among people in a high-reward relationship, people in a mixed-reward relationship, and people in a low-reward relationship with respect to their commitment to the relationship.

B) Nature of the variables: This analysis involved two variables. The predictor variable was type of rewards, which was measured on a nominal scale and could assume one of three values: a low-reward condition (coded as 1); a mixed-reward condition (coded as 2); and a high-reward condition (coded as 3). The criterion variable represented participants' commitment, which was measured on an interval/ratio scale.

C) Statistical test: One-way ANOVA, between-subjects design.

D) Null hypothesis (H_0): $M_1 = M_2 = M_3$. In the population, there is no difference between those in a high-reward relationship, those in a mixed-reward relationship, and those in a low-reward relationship with respect to their mean commitment scores.

E) Alternative hypothesis (H_1): In the population, there is a difference between at least two of the following three groups with respect to their mean commitment scores: those in a high-reward relationship; those in a mixed-reward relationship; and those in a low-reward relationship.

F) Obtained statistic: $F(2,15) = 122.12$.

G) Obtained probability (p) value: $p < .01$.

H) Conclusion regarding the null hypothesis: Reject the null hypothesis.

I) Magnitude of treatment effect: $R^2 = .94$ (large treatment effect).

J) ANOVA Summary Table: See Table 9.1.

K) Figure representing the results:

Figure 9.1 Mean Levels of Commitment Observed for Participants in Low-Reward, Mixed-Reward, and High-Reward Conditions (Significant Differences Observed)

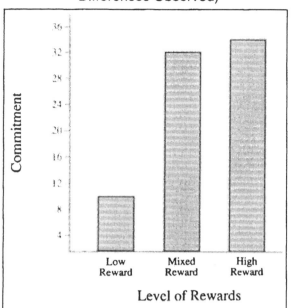

Formal Description of Results for a Paper

You could use the following approach to summarize this analysis for a research paper:

    Results were analyzed using a one-way ANOVA, between-subjects design. This analysis revealed a significant effect for type of rewards, $F(2,15) = 122.12$; $p < .01$. The magnitude of the treatment effect was computed as $R^2 = .94$ (large treatment effect). The sample means are displayed in Figure 9.1. Tukey's HSD test showed that participants in the high-reward and mixed-reward conditions scored significantly higher on commitment than did participants in the low-reward condition ($p < .05$). There were no significant differences between participants in the high-reward condition and participants in the mixed-reward condition.

Example with Nonsignificant Differences between Experimental Conditions

In this section, you repeat the preceding analysis, but this time you perform it on a dataset that was designed to provide nonsignificant results. This will help you learn how to interpret and summarize results when there are no significant between group differences.

Data for the New Analysis

Here is the dataset to be analyzed in this section of the chapter:

```
DATALINES;
1 15
1 17
1 10
1 19
1 16
1 18
2 14
2 17
2 16
2 18
2 11
2 12
3 17
3 20
3 14
3 16
3 16
3 17
;
RUN;
```

Remember that the first column of data (in column 1) of the preceding dataset represents each participant's score on REWARDGROUP (i.e., the type of rewards independent variable). The second column of data in columns 3 to 4 represents each participant's score on COMMITMENT (i.e., the dependent variable).

Results from the SAS Output

If the preceding dataset were analyzed using the PROC GLM program presented earlier, it would again produce four groupings of information, here presented as separate pages. The output from this analysis is reproduced as Output 9.2.

Output 9.2: Results of One-Way ANOVA: Nonsignificant Differences Observed

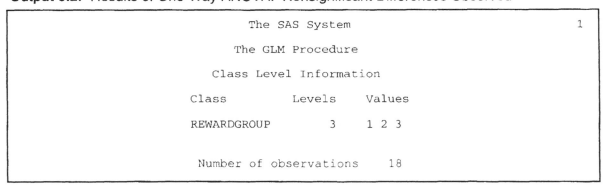

```
                        The SAS System                          1

                        The GLM Procedure

                    Class Level Information

            Class            Levels    Values

            REWARDGROUP          3      1 2 3


            Number of observations    18
```

(continued on the next page)

Output 9.2 *(continued)*

```
                                                                          2
                          The GLM Procedure

Dependent Variable: COMMITMENT

                                  Sum of
Source                  DF        Squares    Mean Square   F Value   Pr > F

Model                    2     12.1111111      6.0555556      0.83   0.4553

Error                   15    109.5000000      7.3000000

Corrected Total         17    121.6111111


           R-Square    Coeff Var    Root MSE    COMMITMENT Mean

           0.099589     17.18492    2.701851         15.72222


Source                  DF       Type I SS    Mean Square   F Value   Pr > F

REWARDGROUP              2     12.11111111     6.05555556      0.83   0.4553


Source                  DF     Type III SS    Mean Square   F Value   Pr > F

REWARDGROUP              2     12.11111111     6.05555556      0.83   0.4553
```

```
                                                                          3
                          The GLM Procedure

        Tukey's Studentized Range (HSD) Test for COMMITMENT

NOTE: This test controls the Type I experimentwise error rate, but it generally
      has a higher Type II error rate than REGWQ.

              Alpha                                   0.05
              Error Degrees of Freedom                  15
              Error Mean Square                         7.3
              Critical Value of Studentized Range   3.67338
              Minimum Significant Difference         4.0518
      Means with the same letter are not significantly different.


           Tukey Grouping        Mean      N    REWARDGROUP

                         A     16.667      6    3
                         A
                         A     15.833      6    1
                         A
                         A     14.667      6    2
```

(continued on the next page)

Output 9.2 *(continued)*

```
                                                                            4
                              The GLM Procedure

        Level of              ------------COMMITMENT-----------
        REWARDGROUP       N          Mean            Std Dev

        1                 6      15.8333333        3.18852108
        2                 6      14.6666667        2.80475786
        3                 6      16.6666667        1.96638416
```

Page 1 of Output 9.2 provides the class level information, which again shows that the class variable (REWARDGROUP) includes three levels. In addition, page 1 indicates that the analysis is again based on 18 observations.

Page 2 of Output 9.2 provides the analysis of variance table. Again, you are most interested in the information that appears at the very bottom of the page, in the section that provides information related to the Type III sum of squares. Under "F Value," you can see that this analysis resulted in an F statistic of only 0.83, which is quite small. Under "Pr > F," you can see that the probability value associated with this F statistic is .46. Because this p value is greater than .05, you fail to reject the null hypothesis and, instead, conclude that the type of rewards independent variable did not have a significant effect on rated commitment.

Page 3 of Output 9.2 provides the results of the Tukey HSD test, but you will not interpret these results since the F for the ANOVA was nonsignificant. It is, however, interesting to note that the Tukey test indicates that the three groups were not significantly different from each other. Notice how all three groups are identified by the same letter ("A").

Finally, page 4 of Output 9.2 provides the mean scores on commitment for the three groups. You can see that there is little difference between these means as all three groups display an average score on COMMITMENT between 14 and 17. This is what you would expect with nonsignificant differences. You can use the means from this table to prepare a figure that illustrates the results graphically.

Summarizing the Analysis Results

This section summarizes the present results according to the statistical interpretation format presented earlier. Since the same hypothesis is being tested, however, items A through E would be completed in exactly the same way as before. Therefore, only items F through J are presented here:

F) Obtained statistic: $F(2,15) = 0.83$.

G) Obtained probability (p) value: $p = .46$ (nonsignificant).

H) Conclusion regarding the null hypothesis: Fail to reject the null hypothesis.

I) Magnitude of treatment effect: $R^2 = .10$ (small treatment effect).

J) ANOVA summary table:

Table 9.2

ANOVA Summary Table for Study Investigating the Relationship between Type of Rewards and Commitment

Source	df	SS	MS	F	R^2
Type of rewards	2	12.11	6.06	0.83	.10
Within groups	15	109.50	7.30		
Total	17	121.61			

Note. N = 18.

K) Figure representing the results:

Figure 9.2 Mean Levels of Commitment Observed for Participants in Low-Reward, Mixed-Reward, and High-Reward Conditions (Nonsignificant Differences Observed)

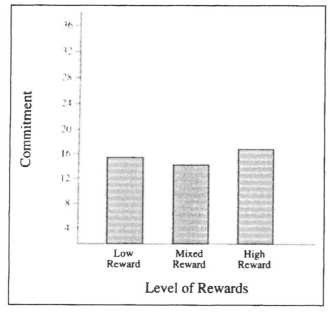

Formal Description of Results for a Paper

You could summarize the results from the present analysis in the following way for a published paper:

```
     Results were analyzed using a one-way ANOVA, between-
subjects design.  This analysis failed to reveal a significant
effect for type of rewards, F(2, 15) = 0.83, ns with a
negligible treatment effect of R² = .10.  The sample means are
displayed in Figure 9.2, which shows that the three
experimental groups demonstrated similar commitment scores.
```

Understanding the Meaning of the *F* Statistic

An earlier section mentioned that you obtain significant results in an analysis of variance when the *F* statistic produced in the ANOVA assumes a relatively large value. This section explains why this is so.

The meaning of the *F* statistic might be easier to understand if you think about it in terms of what results would be expected if the null hypothesis were true. If there really were no differences between groups with respect to their means, you would expect to obtain an *F* statistic somewhere close to 1.00. In the first analysis, the obtained *F* value was much larger than 1.00. While it is possible to obtain an *F* statistic of 122.12 when the means are equal, it is extremely unlikely. In fact, SAS calculated that the probability was less than 1 in 10,000 (i.e., the *p* value found on the output). Therefore, you rejected the null hypothesis of no population differences.

But why do you expect an *F* value of 1.00 when the population means are equal? And why do you expect an *F* value greater than 1.00 when the population means are unequal? To understand this, you need to understand how the *F* statistic is calculated. The formula for the *F* statistic is as follows:

$$F = \frac{MS_{\text{between groups}}}{MS_{\text{within groups}}}$$

The numerator in this ratio is $MS_{\text{between groups}}$, which represents the "mean square between groups." This is a measure of variability that is influenced by two sources of variation: error variability plus variability due to the differences in the population means. The denominator in this ratio is $MS_{\text{within groups}}$, which represents the "mean square within groups." This is a measure of variability that is influenced by only one source of variation: error variability. For this reason, the $MS_{\text{within groups}}$ is sometimes referred to as the MS_{error}.

Now you can see why the *F* statistic should be larger than 1.00 when there are differences in population means (i.e., when the null hypothesis is incorrect). Consider the following alternative formula for the *F* statistic:

$$F = \frac{MS_{\text{between groups}}}{MS_{\text{within groups}}} = \frac{\text{Error variability plus variability due to differences in the population means}}{\text{Error variability}}$$

If the means are different, your predictor variable (types of reward) has had an effect on commitment. Therefore, the numerator of the *F* ratio contains two sources of variation: both error variability plus variability due to the differences in the population means. The denominator ($MS_{\text{within groups}}$), however, was influenced by only one source of variance: error variability. It is clear that both the numerator and the denominator are influenced by error variability, but the numerator is affected by an additional source of variance, so the numerator should be larger than the denominator. Whenever the numerator is larger than the denominator in a ratio, the resulting quotient is greater than 1.00. That is why you expect to see an *F* value greater than 1.00 when there are differences between the means. And that is why you reject the null hypothesis of no population differences when you obtain a sufficiently large *F* statistic.

In the same way, you can also see why the *F* statistic should be approximately equal to 1.00 when there are no differences in mean values (i.e., when the null hypothesis is correct). Under these circumstances, your predictor variable (level of rewards) has no effect on commitment. Therefore, the $MS_{\text{between groups}}$ is not influenced by variability due to differences between means. Instead, it is influenced only by error variability. So, the formula for the *F* statistic reduces to the following:

$$F = \frac{MS_{\text{between groups}}}{MS_{\text{within groups}}} = \frac{\text{Error variability}}{\text{Error variability}}$$

Obviously, the number representing the numerator in this formula is going to be fairly close to the number representing the denominator. And whenever a numerator in a ratio is approximately equal to the denominator in that ratio, the resulting quotient will be fairly close to 1.00. Hence, when a predictor variable has no effect on the criterion variable, you expect an *F* statistic close to 1.00.

Using the LSMEANS Statement to Analyze Data from Unbalanced Designs

So far, the examples presented in this chapter are what are known as **balanced designs**, meaning that the same number of observations or participants have been assigned to each group or cell. Of the 18 children in the aggression study (both examples), six were assigned to each of the three treatment conditions (i.e., 0 grams of sugar, 20 grams of sugar, and 40 grams of sugar). When a research design is balanced, it is generally appropriate to use the MEANS statement with PROC GLM.

In contrast, a research design is said to be **unbalanced** if the number of observations per cell are unequal. In the aggression study, for instance, if 8 children were assigned to the first treatment condition (0 grams of sugar at lunch) and 5 children were assigned to both of the remaining cells (i.e., 20 grams of sugar and 40 grams of sugar), the design would then be unbalanced.

When analyzing data from an unbalanced research design, it is preferable to use the **LSMEANS statement** in your program rather than the MEANS statement. This is because the LSMEANS statement estimates the marginal means over a balanced population. In other words, LSMEANS estimates what the marginal means would be if you did have equal cell sizes. The marginal means estimated by the LSMEANS statement are less likely to be biased. ("LSMEANS" stands for "least-squares" means.)

Writing the LSMEANS Statements

Below is the syntax for the PROC GLM step of a SAS program that uses the LSMEANS statement rather than the MEANS statement:

```
PROC GLM DATA=filename;
   CLASS     predictor-variable;
   MODEL     criterion-variable = predictor-variable;
   LSMEANS   predictor-variable / PDIFF ADJUST=TUKEY;
   LSMEANS   predictor-variable;
RUN;
```

The preceding syntax is very similar to the syntax that used the MEANS statement presented earlier in this chapter. The primary difference is that the two MEANS statements have been replaced with two LSMEANS statements. In addition, the PDIFF command requests SAS to print p values for the significance tests related to the multiple comparison procedure. (These p values tell you whether or not there are significant differences between least-squares means for the different levels of the predictor variable.) The ADJUST=TUKEY command requests a multiple comparison adjustment for the p values for differences between least-squares means.

The Actual SAS Statements

Below are the actual statements that you would include in a SAS program to request a one-way ANOVA using LSMEANS rather than the MEANS statement. These statements could be used to analyze data from the studies described in this chapter. As both our examples have equal numbers of participants in each cell, the SAS output will not be computed here (and the associated output would be virtually identical given that both are balanced designs).

```
PROC GLM DATA=D1;
   CLASS     REWARDGROUP;
   MODEL     COMMITMENT = REWARDGROUP;
   LSMEANS   REWARDGROUP / PDIFF ADJUST=TUKEY;
   LSMEANS   REWARDGROUP;
RUN;
```

Discussion of the use of the LSMEANS and the associated SAS program is provided for reference should you encounter instances in which you have unequal numbers of observations.

Conclusion

The one-way analysis of variance is one of the most flexible and widely used procedures in the social sciences. It allows you to test for significant differences between groups and its applicability is not limited to studies with two groups (as the *t* test is limited). However, an even more flexible procedure is the *factorial* analysis of variance. Factorial ANOVA allows you to analyze data from studies in which more than one independent variable is manipulated. Not only does this allow you to test more than one hypothesis in a single study but it also allows you to investigate an entirely different type of effect called an "interaction." Factorial analysis of variance is introduced in the following chapter.

Assumptions Underlying One-Way ANOVA with One Between-Subjects Factor

- **Level of measurement.** The criterion variable should be assessed on an interval- or ratio-level of measurement. The predictor variable should be a nominal-level variable (a categorical variable) that includes two or more categories.

- **Independent observations.** A given observation should not be dependent on any other observation in any group. (For a more detailed explanation of this assumption, see Chapter 8.)

- **Random sampling.** Scores on the criterion variable should represent a random sample drawn from the populations of interest.

- **Normal distributions.** The distribution of observed values for the criterion variable (all groups) should approximate normal distributions.

- **Homogeneity of variance.** The populations represented by the various groups should have equal variances on the criterion variable. If the number of participants in the largest group is no more than 1.5 times greater than the number in the smallest group, then the test is robust against violations of the homogeneity assumption (Stevens, 2002).

References

SAS Institute Inc. (2004). *SAS/STAT 9.1 user's guide.* Cary, NC: Author.

Stevens, J. (2002). *Applied multivariate statistics for the social sciences* (4th ed.).
     Mahwah, NJ: Lawrence Erlbaum Associates.

Factorial ANOVA with Two Between-Subjects Factors

> **Overview.** This chapter shows how to enter data and prepare SAS programs to perform a two-way ANOVA using the GLM procedure. This chapter focuses on factorial designs with two between-subjects factors, meaning that each participant is exposed to only one condition under each independent variable. Guidelines are provided for interpreting results that do not display a significant interaction and separate guidelines are provided for interpreting results that do display a significant interaction. For significant interactions, you see how to display the interaction in a figure and how to perform tests for simple effects.

Introduction to Factorial Designs

The preceding chapter described a simple experiment in which you manipulated a single independent variable, type of rewards. Because there was a single independent variable in that study, it was analyzed using a one-way ANOVA.

But imagine that there are actually two independent variables that you want to manipulate. In this situation, you might think that it would be necessary to conduct two separate experiments, one for each independent variable; this would be incorrect. In many cases, you can manipulate both independent variables in a single study.

The research design used in such a study is called a factorial design. In a **factorial design**, two or more independent variables are manipulated in a single study so that the treatment conditions represent all possible combinations of the various levels of the independent variables.

In theory, a factorial design might include any number of independent variables. In practice, however, it generally becomes impractical to use many more than two or three. This chapter illustrates factorial designs that include just two independent variables and thus can be analyzed using a two-way ANOVA. More specifically, this chapter deals with studies that include two predictor variables, both measured on a nominal scale, as well as a single criterion variable assessed on an interval or ratio scale.

The Aggression Study

To illustrate the concept of factorial design, imagine that you are interested in conducting a study that investigates aggression in eight-year-old children. Here, aggression is defined as any verbal or behavioral act performed with the intention of harm. You want to test two hypotheses:

- Boys will display higher levels of aggression than girls.

- The amount of sugar consumed will have a positive effect on levels of aggression.

You perform a single investigation to test these two hypotheses. The hypothesis that you are most interested in is the second hypothesis, that consuming sugar will cause children to behave more aggressively. You will test this hypothesis by manipulating the amount of sugar that a group of school children consume at lunchtime. Each day for two weeks, one group of children will receive a lunch that contains no sugar at all (0 grams of sugar group). A second group will receive a lunch that contains a moderate amount of sugar

(20 grams), and a third group will receive a lunch that contains a large amount of sugar (40 grams). Each child will then be observed after lunch, and a pair of judges will tabulate the number of aggressive acts that the child commits. The total number of aggressive acts committed by each child over the two-week period will serve as the dependent variable in the study.

You begin with a sample of 60 children: 30 boys and 30 girls. The children are randomly assigned to treatment conditions in the following way:

- 20 children are assigned to the "0 grams of sugar" treatment condition;

- 20 children are assigned to the "20 grams of sugar" treatment condition;

- 20 children are assigned to the "40 grams of sugar" treatment condition.

In making these assignments, you ensure that there are equal numbers of boys and girls in each treatment condition. For example, you verify that, of the 20 children in the "0 grams" group, 10 are boys and 10 are girls.

The Factorial Design Matrix

The factorial design of this study is illustrated in Figure 10.1. You can see that this design is represented by a matrix that consists of two rows (running horizontally) and three columns (running vertically):

Figure 10.1 Experimental Design Used in Aggression Study

		Predictor A: Amount of Sugar Consumed		
		Level A1: 0 Grams	Level A2: 20 Grams	Level A3: 40 Grams
Predictor B: Participant Sex	Level B1: Males	10 Participants	10 Participants	10 Participants
	Level B2: Females	10 Participants	10 Participants	10 Participants

When an experimental design is represented in a matrix such as this, the matrix is easier to understand if you focus on just one aspect at a time. For example, consider just the three vertical columns of Figure 10.1. The three columns are headed "Predictor A: Amount of Sugar Consumed" so obviously, these columns represent the various levels of sugar consumption (independent variable). The first column represents the 20 participants in Level A1 (i.e., the participants who received 0 grams of sugar), the second column represents the 20 participants in Level A2 (i.e., 20 grams), and the last column represents the 20 participants in Level A3 (i.e., 40 grams).

Now consider just the two horizontal rows of Figure 10.1. These rows are headed "Predictor B: Participant Sex." The first row is headed "Level B1: Males," and this row represents the 30 male participants. The second row is headed "Level B2: Females," and represents the 30 female participants.

It is common to refer to a factorial design as a "r x c" design, in which "r" represents the number of rows in the matrix, and "c" represents the number of columns. The present study is an example of a 2 x 3 factorial design because it has two rows and three columns. If it included four levels of sugar consumption rather than three, it would be referred to as a 2 x 4 factorial design.

You can see that this matrix consists of six different cells. A **cell** is the location in the matrix where the row for one independent variable intersects with the column for a second independent variable. For example, look at the cell where the row named B1 (males) intersects with the column headed A1 (0 grams). The entry "10 Participants" appears in this cell, which means that there are 10 participants assigned to this particular combination of "treatments" under the two independent variables. More specifically, it means that there are 10 participants who are both (a) male and (b) given 0 grams of sugar. ("Treatments" appears in quotation marks in the preceding sentence because "participant sex" is obviously not a true independent variable that is manipulated by the researcher; it is merely a predictor variable.)

Now look at the cell in which the row labeled B2 (females) intersects with the column headed A2 (20 grams). Again, the cell contains the entry "10 Participants," which means that there is a different group of 10 children who experienced the treatments of (a) being female and (b) receiving 20 grams of sugar. You can see that there is a separate group of 10 children assigned to each of the six cells of the matrix. No participant appears in more than one cell.

Earlier, it was said that a factorial design involves two or more independent variables being manipulated so that the treatment conditions represent all possible combinations of the various levels of the independent variables. Figure 10.1 illustrates this concept. You can see that the six cells of the figure represent every possible combination of sex and amount of sugar consumed: males are observed under every level of sugar consumption; the same is true for females.

Some Possible Results from a Factorial ANOVA

Factorial designs are popular in social science research for a variety of reasons. One reason is that they allow you to test for several different types of effects in a single computation. The nature of these effects are illustrated later in this chapter.

First, however, it is important to note one drawback associated with factorial designs: they sometimes produce results that can be difficult to interpret compared to the results produced in a one-way ANOVA. Fortunately, this task of interpretation can be made much easier if you first prepare a figure that plots the results of the factorial study. This first section shows how this can be done.

Figure 10.2 presents one type of figure that is often used to illustrate the results of a factorial study. Notice that, with this figure, scores on the criterion variable (level of aggression displayed by the children) are plotted on the vertical axis. Remember that groups that appear higher on this vertical axis display higher mean levels of aggression.

Figure 10.2 A Significant Main Effect for Predictor A (Amount of Sugar Consumed) Only

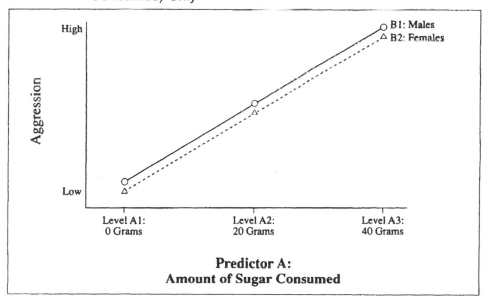

The three levels of predictor variable A (amount of sugar consumed) are plotted on the horizontal axis. The point at the left represents group A1 (who received 0 grams of sugar), the middle point represents group A2 (the 20-gram group), and the point at the right represents group A3 (the 40-gram group).

The two levels of predictor variable B (participant sex) are identified by drawing two different lines in the body of the figure itself. Specifically, the mean scores on aggression displayed by the males (level B1) are illustrated with small circles connected by a solid line, while the mean scores on aggression displayed by the females (level B2) are depicted by small triangles connected by a dashed line.

In summary, the important points to remember when interpreting the figures in this chapter are as follows:

- The possible scores on the criterion variable are represented on the vertical axis.
- The levels of predictor A are represented as points on the horizontal axis.
- The levels of predictor B are represented by drawing different lines within the figure itself.

With this foundation, you are ready to learn about the different types of effects that can be observed in a factorial design, and how these effects appear when they are plotted in this type of figure.

Significant Main Effects

When a predictor variable (or independent variable) in a factorial design displays a significant **main effect**, it means that there is a difference between at least two levels of that predictor variable with respect to mean scores on the criterion variable. In a one-way analysis of variance, there is essentially just one main effect: the main effect for the study's independent variable. In a factorial design, however, there is one main effect possible for each predictor variable examined in the study.

For example, the preceding study on aggression included two predictor variables: amount of sugar consumed and participant sex. This means that, in analyzing data from this investigation, it is possible to obtain any of the following outcomes related to main effects:

- a significant main effect for just predictor A (amount of sugar consumed);
- a significant main effect for just predictor B (participant sex);
- a significant main effect for both predictor A and B;
- no significant main effects for either predictor A or B.

This section describes these effects and illustrates how the effects appear when plotted in a figure.

A Significant Main Effect for Predictor A

Figure 10.2 shows a possible main effect for predictor A: "Amount of Sugar Consumed." (To save space, we henceforth refer to this as the "sugar consumption" variable.) Notice that a relatively low mean level of aggression was displayed by participants in the 0-gram condition of the sugar consumption. When you look above the heading "Level A1," you see that both the boys (represented with a small circle) and the girls (represented with a small triangle) display relatively low aggression scores. However, a somewhat higher level of aggression was exhibited by participants in the 20-gram condition: when you look above "Level A2," you can see that both boys and girls display a somewhat higher aggression level. Finally, an even higher level of aggression was exhibited by participants in the 40-gram condition: when you look above "Level A3," you can see that both boys and girls in this group display relatively high levels of aggression. In short, this trend shows that there is a main effect for the sugar consumption variable.

This leads to an important point. When a figure representing the results of a factorial study displays a significant main effect for predictor variable A, it demonstrates both of the following characteristics:

- The lines for the various groups are parallel.
- At least one line segment displays a relatively steep angle.

The first of the two conditions, that the lines should be parallel, ensures that the two predictor variables do not exhibit an interaction. This is important, because you normally will not interpret a significant main effect for a predictor variable if that predictor is involved in an interaction. In Figure 10.2, you can see that the lines for the two groups in the present study (the solid line for the boys and the dashed line for the girls) are parallel. This suggests that there probably is not an interaction between sex and sugar consumption in the present study. (The concept of interaction is explained in greater detail later in the section titled "A Significant Interaction.")

The second condition, that at least one line segment should display a relatively steep angle, can be understood by again referring to Figure 10.2. Notice that the line segment that begins at Level A1 (the 0-gram condition) and extends to Level A2 (the 20-gram condition) is not horizontal; instead, it displays an upward angle. This is because aggression scores for the 20-gram group were higher than aggression scores for the 0-gram group. When you obtain a significant effect for the predictor A variable, you should expect to see this type of angle. Similarly, you can see that the line segment that begins at A2 and continues to A3 also displays an upward angle, also consistent with a significant effect for the sugar consumption.

Remember that these guidelines are merely intended to help you understand what a main effect *looks like* when it is plotted as in Figure 10.2. To determine whether this main effect is *statistically significant*, it is necessary to review the results of the analysis of variance, to be discussed later.

A Significant Main Effect for Predictor B

You would expect to see a different pattern if the main effect for the other predictor variable (predictor B) were significant. Earlier, you learned that predictor A was represented by plotting three points on the horizontal axis. In contrast, you learned that predictor B was represented by drawing different lines within the body of the figure itself: one line for each level of predictor B. In the present study, predictor B is participant sex, so a solid line is used to represent mean scores for boys and a dashed line is used to represent mean scores for girls.

When predictor B is represented in a figure by plotting separate lines for its various levels, a significant main effect for that variable is evident when the figure displays both of the following:

- The lines for the various groups are parallel.
- At least two of the lines are relatively far apart.

For example, a main effect for predictor B in the current study is depicted by Figure 10.3. Consistent with the two preceding points, the lines in Figure 10.3 are parallel to one another (indicating that there is no interaction between sex and sugar consumption), and separated from one another. Regarding this separation, notice that (in general) boys tend to score higher on the measure of aggression compared to girls. Furthermore, notice that this tends to be true regardless of how much sugar participants consume. Figure 10.3 shows the general trend that you would expect when there is a main effect for predictor B only:

Figure 10.3 A Significant Main Effect for Predictor B (Participant Sex) Only

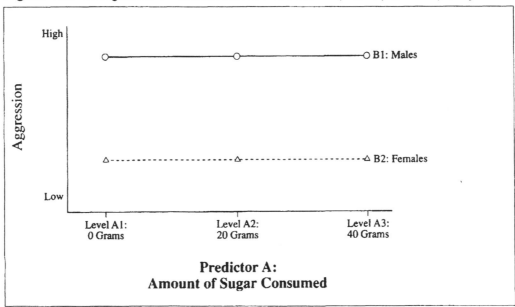

A Significant Main Effect for Both Predictor Variables

It is possible to obtain significant effects for both predictor A and predictor B in the same investigation. When there is a significant effect for both predictor variables, you should encounter each of the following:

- The lines for the various groups are parallel (indicating no interaction).
- At least one line segment displays a relatively steep angle (indicating a main effect for predictor A).
- At least two of the lines are relatively separated from each other (indicating a main effect for predictor B).

Figure 10.4 shows what the figure might look like under these circumstances.

Figure 10.4 Significant Main Effects for Both Predictor A (Amount of Sugar Consumed) and Predictor B (Participant Sex)

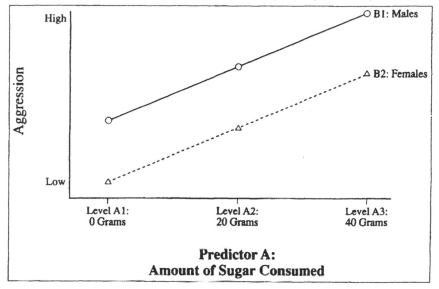

No Main Effects

Figure 10.5 shows what a figure might look like if there were main effects for neither predictor A nor predictor B. Notice that the lines are parallel (indicating no interaction), none of the line segments displays a relatively steep angle (indicating no main effect for predictor A), and that the lines are not separated (indicating no main effect for predictor B).

Figure 10.5 A Nonsignificant Interaction and Nonsignificant Main Effects

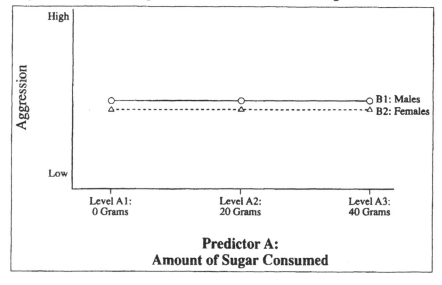

A Significant Interaction

An **interaction** can be defined in a number of ways. For example, with respect to experimental research (in which you manipulate independent variables), the following definition could be used:

> an interaction is a condition in which the effect of one independent variable on the dependent variable is different at different levels of a second independent variable.

On the other hand, when conducting nonexperimental research (in which you are simply measuring naturally occurring variables rather than manipulating independent variables), it could be defined in this way:

> an interaction is a condition in which the relationship between one predictor variable and the criterion is different at different levels of a second predictor variable.

These definitions are admittedly somewhat abstract at first glance. Once again, the concept of interaction is much easier to grasp when visually displayed. For example, Figure 10.6 displays a significant interaction between sugar consumption and participant sex in the present study. Notice that the lines for the two groups are no longer parallel: the line for the male participants now displays a somewhat steeper angle compared to the line for the females. This is the key characteristic of a figure that displays a significant interaction: lines that are not parallel.

Figure 10.6 A Significant Interaction between Predictor A (Amount of Sugar Consumed) and Predictor B (Participant Sex)

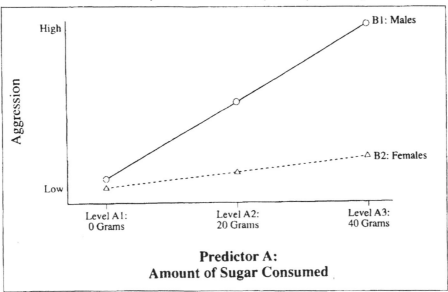

Notice how the relationships portrayed in Figure 10.6 are consistent with the definition for interaction: the relationship between one predictor variable (sugar consumption) and the criterion (aggression) is different at different levels of a second predictor variable (participant sex). More specifically, the figure shows that the relationship between sugar consumption and aggression is relatively *strong* for the male participants: consuming larger quantities of sugar results in dramatic increases in aggression among the male participants. (Notice that the boys who consumed 40 grams of sugar displayed much higher levels of aggression than the boys who consume 0 or 20 grams.) In contrast, the relationship between sugar consumption and aggression is relatively *weak* among the female participants. Notice that the line for the females is fairly "flat"; that is, there is little difference in aggression between the girls who consumed 0 grams versus 20 grams versus 40 grams of sugar.

Figure 10.6 shows why you would normally not interpret main effects when an interaction is significant. To understand why, consider this: would it make sense to say that there is a main effect for sugar consumption in the study illustrated by Figure 10.6? Probably not. It is clear that sugar consumption does seem to have an effect on aggression among boys, but the figure suggests that sugar consumption probably does not have any meaningful effect on aggression among girls. To say that there is a main effect for sugar consumption might mislead readers into believing that sugar causes aggression in all children in much the same way (which it apparently does not). Whether sugar fosters aggression appears to depend on the child's sex. In this situation, it would make more sense to, instead, do the following:

- Note that there was a significant interaction between sugar consumption and participant sex.

- Prepare a figure (like Figure 10.6) that illustrates the nature of the interaction.

- Test for simple effects.

Testing for **simple effects** is similar to testing for main effects but is done one group at a time. In the preceding analysis, for example, you would test for simple effects by first dividing your data into two groups. Data from the male participants would be separated from data from the female participants. With this done, you would perform an analysis to determine whether sugar consumption has a simple effect on aggression among just the male participants (and you would probably find that this simple effect is significant). Finally, you would perform a separate test to determine whether sugar consumption has a simple effect on aggression among just the female participants (and you would probably find that this simple effect is not significant). A later section shows how SAS can perform these tests for simple effects.

To summarize, an interaction means that the relationship between one predictor variable and the criterion is different at different levels of a second predictor variable. When an interaction is significant, you should interpret your results in terms of simple effects rather than main effects.

Example with a Nonsignificant Interaction

To illustrate how a two-way ANOVA is performed, let's modify the earlier study that examined the effect of rewards on commitment in romantic relationships. In Chapter 9, "One-Way ANOVA with One Between-Subjects Factor," only one independent variable was manipulated, and this variable was called "type of rewards." In this chapter, however, there are two independent variables: independent variable A will be "type of rewards" while independent variable B will be "type of costs."

In the study presented in Chapter 9, participants read descriptions of a fictitious "partner 10" and rated how committed they would be to a relationship with that person. One independent variable (called "type of rewards") was manipulated, and this variable consisted of three experimental conditions. Participants were assigned to either the low-, mixed-, or the high-reward conditions. This variable was manipulated by varying the description of partner 10 which was provided to participants:

- Participants in the low-reward condition were told to assume that this partner "...does not enjoy the same recreational activities that you enjoy, and is not very good looking."

- Participants in the mixed-reward condition were told to assume that "...sometimes this person seems to enjoy the same recreational activities that you enjoy, and sometimes s/he does not. Sometimes partner 10 seems to be very good looking, and sometimes s/he does not."

- Participants in the high-reward condition were told that this partner "...enjoys the same recreational activities that you enjoy, and is very good looking."

These same manipulations will be used in the present study. In a sample of 30 participants, one-third will be assigned to a low-reward condition, one-third to a mixed-reward condition, and one-third to a high-reward condition. This "type of reward" factor serves as independent variable A.

In this study, however, you also manipulate a second predictor variable (independent variable B) at the same time. The second independent variable will be called "type of costs" and it will consist of two experimental conditions. Half the participants (the "low-cost" group) will be told to assume that they are in a relationship that does not create significant personal hardships. Specifically, when they read the description of partner 10, it will state that "Partner 10 lives in the same neighborhood as you so it is easy to see him/her as often as you like." The other half of the participants (the "high-cost" group) will be told to imagine that they are in a relationship that does create significant personal hardships. When they read the description of partner 10, it will state that "Partner 10 lives in a distant city, so it is difficult and expensive to visit with him/her."

Your study now has two independent variables. Independent variable A is type of rewards and it has three levels (or conditions); independent variable B is type of costs and it has two levels. The factorial design used in the study is illustrated in Figure 10.7.

Figure 10.7 Experimental Design Used in Investment Model Study

		Type of Rewards		
		Low Reward (1)	Mixed Reward (2)	High Reward (3)
Type of Costs	Low Cost (1)	5 Participants (Cell 1-1)	5 Participants (Cell 1-2)	5 Participants (Cell 1-3)
	High Cost (2)	5 Participants (Cell 2-1)	5 Participants (Cell 2-2)	5 Participants (Cell 2-3)

This 2 x 3 matrix shows that there are two independent variables. One independent variable has two levels, and the other has three levels. There are a total of 30 participants who are divided into 6 cells or subgroups. Each cell contains five participants. Each horizontal row in the figure (running from left to right) represents a different level of the type of costs independent variable. There are 15 participants in the low-cost condition and 15 in the high-cost condition. Each vertical column in the figure (running from top to bottom) represents a different level of the type of rewards independent variable. There are 10 participants in the low-reward condition, 10 in the mixed-reward condition, and 10 in the high-reward condition.

It is strongly advised that you prepare a similar figure whenever you conduct a factorial study, as this will make it easier to enter your data and reduce the likelihood of input errors. Notice that the cell designator (the cell number, such as "1-1," "2-3," etc.) indicates which experimental condition to which a given participant was assigned under each of the two independent variables. The first number in the designator indicates which level under type of costs and the second number indicates level of rewards. For example, the designator "1-1" for cell 1-1 indicates that the participant was in condition 1 (the low-cost condition) under type of costs and in condition 1 (the low-reward condition) under type of rewards. This means that all five participants in this subgroup read that partner 10 lives in the same town as the participant (low-cost) but is not very good-looking (low-reward).

On the other hand, the designator "2-3" for cell 2-3 indicates that the participant was in condition 2 (the high-cost condition) under type of costs and was in condition 3 (the high-reward condition) under type of rewards. This means that all five participants in this subgroup read that partner 10 lives in a distant city (high-cost), but is very good-looking (high-reward). Similar reasoning applies in the interpretation of the designators for the remaining four cells in the figure.

When working with these cell designators, remember that the number for the row precedes and the number for the column. This means that the cell at the intersection of row 2 and column 1 is identified with the designator 2-1, not 1-2.

Writing the SAS Program

Creating the SAS Dataset

There were two predictor variables in this study (type of rewards and type of costs) and one criterion variable (commitment). This means that each participant has a score for three variables. First, with the variable REWGRP, you indicate which experimental condition this participant experienced under the type of rewards independent variable. Each participant is given a score of 1, 2, or 3 depending on the condition to which s/he is assigned. Second, with the variable COSTGRP, you indicate which condition the participants are assigned under the type of costs independent variable. Each participant is given a score of either 1 or 2 depending on the condition to which s/he is assigned. Finally, with the variable COMMIT, you will record participants' rated commitment to the relationship with partner 10. That is, you simply enter the number that is the sum of responses to the four questionnaire items that measure commitment to partner 10.

The data are entered so that there is one line of data for each participant. Assume that you enter your data using the following format:

Line	Column	Variable Name	Explanation
1	1	COSTGRP	Codes group membership, so that 1 = low-cost condition, and 2 = high-cost condition
	2	blank	
	3	REWGRP	Codes group membership, so that 1 = low-reward condition, 2 = mixed-reward condition, and 3 = high-reward condition
	4	blank	
	5-6	COMMIT	Commitment ratings obtained when participants rated partner 10

Remember that there are actually 30 participants in your study. However, to save space, assume for the moment that there are only 12 participants: two participants in cell 1-1; two participants in cell 1-2; and so forth. Here is what the dataset might look like if it were keyed according to the preceding format:

```
 1          1 1 08
 2          1 1 07
 3          1 2 13
 4          1 2 17
 5          1 3 36
 6          1 3 31
 7          2 1 09
 8          2 1 10
 9          2 2 15
10          2 2 18
11          2 3 35
12          2 3 29
```

Remember that the numbers on the far left are simply line numbers for reference; they are not actually a part of the dataset. In the dataset itself, the first vertical column of numbers (in column 1) codes the COSTGRP variable. Notice that participants who appear in lines 1 to 6 of the preceding dataset have a score of "1" for COSTGRP (indicating that they are in the low-cost condition.) In contrast, the participants on lines 7 to 12 of the preceding dataset have a score of "2" for COSTGRP indicating that they are in the high-cost condition.

The second column of numbers (in column 3) codes REWGRP. For example, the two participants from lines 1 to 2 of the preceding program have scores of "1" for REWGRP (indicating that they are in the low-reward condition), the two participants from lines 3 to 4 have scores of "2" on REWGRP (indicating that they are in the mixed-reward condition), and the two participants from lines 5 to 6 have scores of "3" for REWGRP (indicating that they are in the high-reward condition).

Finally, the last grouping in the dataset (columns 5 to 6) codes COMMIT (i.e., commitment scores from the participants). Assume that, with this variable, scores could range from 4 to 36 with higher scores representing higher levels of commitment to partner 10. You can see that the participant from line 1 had a score of "08" for COMMIT, the participant from line 2 had a score of "07," and so forth.

This approach to entering data for a two-way ANOVA is particularly useful because it provides a quick way to determine the cell (from the factorial design matrix) in which a given participant can be found. For example, the first two columns of data for the participant from line 1 provide the values "1" and "1" (indicating that this participant is in cell 1-1). Similarly, the first two columns for the participant from line 5 are "1" and "3" (indicating that this participant is in cell 1-3). The data in the preceding dataset are sorted so that participants in the same cell are grouped together; note that your data do not actually have to be sorted this way to be analyzed with PROC GLM.

Testing for Significant Effects with PROC GLM

Here is the general form for the PROC GLM step needed to perform a two-way ANOVA and follow it up with Tukey's HSD test. Enter your file name where *filename* appears.

```
PROC GLM DATA= filename;
   CLASS  predictorA  predictorB;
   MODEL  criterion-variable = predictorA  predictorB
          predictorA*predictorB;
   MEANS  predictorA  predictorB  / TUKEY;
   MEANS  predictorA  predictorB  predictorA*predictorB;
RUN;
```

This is the entire program, including the DATA step, needed to analyze some fictitious data from the preceding study:

```
1          DATA D1;
2             INPUT  #1  @1  COSTGRP  1.
3                        @3  REWGRP   1.
4                        @5  COMMIT   2.  ;
5          DATALINES;
6          1 1 08
7          1 1 13
8          1 1 07
9          1 1 14
10         1 1 07
11         1 2 19
12         1 2 17
13         1 2 20
14         1 2 25
15         1 2 16
16         1 3 31
17         1 3 24
18         1 3 37
19         1 3 30
20         1 3 32
21         2 1 09
22         2 1 14
23         2 1 05
24         2 1 14
25         2 1 16
26         2 2 22
27         2 2 18
28         2 2 20
29         2 2 19
30         2 2 21
31         2 3 24
32         2 3 33
33         2 3 24
34         2 3 26
35         2 3 31
36         ;
37         RUN;
38
39         PROC GLM DATA=D1;
40            CLASS REWGRP COSTGRP;
41            MODEL COMMIT = REWGRP COSTGRP REWGRP*COSTGRP;
42            MEANS REWGRP COSTGRP / TUKEY;
43            MEANS REWGRP COSTGRP REWGRP*COSTGRP;
44         RUN;
```

Notes Regarding the SAS Program

The GLM procedure is requested on line 39. In line 40, both REWGRP and COSTGRP are listed as classification variables.

The MODEL statement on line 41 specifies COMMIT as the criterion variable and lists three predictor terms: REWGRP; COSTGRP; and REWGRP*COSTGRP. The last term (REWGRP*COSTGRP) is the interaction term in the model. You create this interaction term by typing the names of your two predictor variables, connected by an asterisk (and no blank spaces).

The program includes two MEANS statements. The MEANS statement on line 42 requests that the Tukey multiple comparison procedure be used with predictor variables REWGRP and COSTGRP. (Do not list the interaction term in this MEANS statement.) The MEANS statement on line 43 requests that means be printed for all levels of REWGRP and COSTGRP. You should include the interaction term in this MEANS statement to obtain means for each of the six cells from the study's factorial design.

Steps in Interpreting the Output

With LINESIZE=80 and PAGESIZE=60 in the OPTIONS statement (first line), the preceding program would produce five pages of output. The information that appears on each page is summarized here:

- Page 1 provides class level information and the number of observations in the dataset.

- Page 2 provides the ANOVA summary table from the GLM procedure.

- Page 3 provides the results of the Tukey multiple comparison procedure for the REWGRP independent variable.

- Page 4 provides the results of the Tukey multiple comparison procedure for the COSTGRP independent variable.

- Page 5 provides three tables that summarize the means observed for the various groups that constitute the study:

  o The first table provides the means observed at each level of REWGRP.

  o The second table provides the means observed at each level of COSTGRP.

  o The third table provides the means observed for each of the six cells in the study's factorial design.

The results produced by the preceding program are reproduced here as Output 10.1.

Output 10.1   Results of Two-Way ANOVA Performed on Investment Model Data, Nonsignificant Interaction

```
                          The SAS System                                1

                          The GLM Procedure

                      Class Level Information

              Class           Levels      Values

              REWGRP               3      1 2 3

              COSTGRP              2      1 2


              Number of observations      30
```

```
                          The GLM Procedure                             2

Dependent Variable: COMMIT

                                    Sum of
Source                    DF        Squares     Mean Square    F Value    Pr > F

Model                      5    1746.266667     349.253333      24.42    <.0001

Error                     24     343.200000      14.300000

Corrected Total           29    2089.466667


              R-Square     Coeff Var     Root MSE     COMMIT Mean

              0.835748     19.03457      3.781534       19.86667


Source                    DF      Type I SS     Mean Square    F Value    Pr > F

REWGRP                     2    1711.666667     855.833333      59.85    <.0001
COSTGRP                    1       0.533333       0.533333       0.04    0.8485
REWGRP*COSTGRP             2      34.066667      17.033333       1.19    0.3212


Source                    DF    Type III SS     Mean Square    F Value    Pr > F

REWGRP                     2    1711.666667     855.833333      59.85    <.0001
COSTGRP                    1       0.533333       0.533333       0.04    0.8485
REWGRP*COSTGRP             2      34.066667      17.033333       1.19    0.3212
```

(continued on the next page)

Output 10.1 *(continued)*

```
                        The GLM Procedure                            3

              Tukey's Studentized Range (HSD) Test for COMMIT

NOTE: This test controls the Type I experimentwise error rate, but it generally
      has a higher Type II error rate than REGWQ.


              Alpha                                    0.05
              Error Degrees of Freedom                   24
              Error Mean Square                        14.3
              Critical Value of Studentized Range    3.53170
              Minimum Significant Difference          4.2233


        Means with the same letter are not significantly different.


        Tukey Grouping          Mean       N      REWGRP

                     A        29.200      10      3

                     B        19.700      10      2

                     C        10.700      10      1
```

```
                        The GLM Procedure                            4

              Tukey's Studentized Range (HSD) Test for COMMIT

NOTE: This test controls the Type I experimentwise error rate, but it generally
      has a higher Type II error rate than REGWQ.


              Alpha                                    0.05
              Error Degrees of Freedom                   24
              Error Mean Square                        14.3
              Critical Value of Studentized Range    2.91880
              Minimum Significant Difference          2.8499


        Means with the same letter are not significantly different.


        Tukey Grouping          Mean       N      COSTGRP

                     A        20.000      15      1
                     A
                     A        19.733      15      2
```

(continued on the next page)

Output 10.1 *(continued)*

```
                              The GLM Procedure                          5


        Level of             ------------COMMIT-----------
        REWGRP        N            Mean            Std Dev

        1           10        10.7000000         3.88873016
        2           10        19.7000000         2.58413966
        3           10        29.2000000         4.49196814


        Level of             ------------COMMIT-----------
        COSTGRP       N            Mean            Std Dev

        1           15        20.0000000         9.59166305
        2           15        19.7333333         7.56369776


  Level of      Level of             ------------COMMIT-----------
  REWGRP        COSTGRP        N            Mean            Std Dev

  1             1             5         9.8000000         3.42052628
  1             2             5        11.6000000         4.50555213
  2             1             5        19.4000000         3.50713558
  2             2             5        20.0000000         1.58113883
  3             1             5        30.8000000         4.65832588
  3             2             5        27.6000000         4.15932687
```

The fictitious dataset analyzed in this section was designed so that the interaction term would be nonsignificant. When the interaction is nonsignificant, interpreting the results of a two-factor ANOVA is very similar to interpreting the results of a one-factor ANOVA. Therefore, many of the instructions provided in the chapter on one-way ANOVA (Chapter 9) will not be repeated here ("making sure page 1 looks right," etc.). Instead, this section focuses on some issues that are specific to a two-factor ANOVA. The steps to follow in interpreting the output are discussed next.

1.   Determine Whether the Interaction Term Is Statistically Significant

Two-factor ANOVA allows you to test for three types of effects: (a) the main effect of predictor A (type of rewards, in this case); (b) the main effect of predictor B (type of costs); and (c) the interaction of predictors A and B. Remember that you may interpret a main effect only if the interaction is nonsignificant. One of your first steps must be to determine whether the interaction is significant. You can do this by looking at the analysis of variance results, which appears on page 2 of Output 10.1. (As before, look in the section that provides the "Type III SS.") Toward the bottom left of page 2, you see the heading "Source," as you did with the one-way ANOVA. In this case, however, you see three entries under "Source," indicating that there are three sources of variation in your data: (a) REWGRP; (b) COSTGRP; and (c) REWGRP*COSTGRP. The last entry, REWGRP*COSTGRP, is the interaction term in which you are interested. If this interaction is significant, it suggests that the relationship between one predictor and the criterion variable is different at different levels of the second predictor. In the present case, the interaction is associated with 2 degrees of freedom, a value of approximately 34.07 for the Type III sum of squares, a mean square of 17.03, an *F* value of 1.19, and a corresponding *p* value of .32. Remember that you generally view a result as being statistically significant only if the *p* value is less than .05. Since this *p* value of .32 is larger than .05, you conclude that there is no interaction between the two independent variables. You are therefore free to proceed with your review of the two main effects.

2. Determine If Either of the Main Effects Is Statistically Significant

A two-way ANOVA allows you to test two null hypotheses concerning main effects. The null hypothesis for the type of rewards predictor can be stated as follows:

> In the population, there is no difference between the high-reward group, the mixed reward group, and the low-reward group with respect to scores on the commitment variable.

The *F* statistic to test this null hypothesis can again be found toward the bottom of page 2 in Output 10.1. To the right of the heading "REWGRP," you can see that this effect is associated with 2 degrees of freedom, a value of approximately 1711.67 for the Type III sum of squares (rounded to two decimal places), a mean square of 855.83, an *F* value of 59.85, and a *p* value less than 0.01. With such a small *p* value, you can reject the null hypothesis of no main effect for type of rewards. Later, you will review the results of the Tukey test to see which pairs of groups (low-reward, high-reward, etc.) significantly differ.

The null hypothesis for the type of costs independent variable can be stated in a similar fashion:

> In the population, there is no difference between the high-cost group and the low-cost group with respect to scores on the commitment variable.

The appropriate *F* statistic is once again found toward the bottom of page 2. To the right of "COSTGRP," you see that this effect is associated with 1 degree of freedom, a value of approximately 0.53 for the Type III sum of squares, a mean square of 0.53, an *F* value of 0.04, and a *p* value of .85. This *p* value is greater than .05; therefore, you cannot reject the null hypothesis and, instead, conclude that there is not a significant main effect for type of costs.

3. Prepare Your Own Version of the ANOVA Summary Table

Use the information from page 2 of Output 10.1 to prepare the ANOVA summary table for the analysis. That table is reproduced here as Table 10.1.

Table 10.1

ANOVA Summary Table for Study Investigating the Relationship
between Type of Rewards, Type of Costs, and Commitment
(Nonsignificant Interaction)

Source	df	SS	MS	F	R^2
Type of rewards (A)	2	1711.67	855.83	59.85 *	.82
Type of costs (B)	1	0.53	0.53	0.04	.00
A x B Interaction	2	34.07	17.03	1.19	.02
Within groups	24	343.20	14.30		
Total	29	2089.47			

Note: N = 30.

* $p < .01$

On page 2 of Output 10.1, to the right of "REWGRP*COSTGRP," you will find
information concerning the study's interaction term. In your own ANOVA summary
table (such as Table 10.1), this will appear on the line headed "A x B Interaction."
As was the case with one-way ANOVA, look to the right of the word "Error" to find
the information needed for the "Within Groups" line of your table, and look to the
right of "Corrected Total" to find the information needed for the "Total" line of your
table.

In the preceding chapter, you learned that R^2 is a measure of the strength of the
relationship between a predictor variable and a criterion variable. In a one-way
ANOVA, R^2 tells you what percent of variance in the criterion is accounted for by the
study's predictor variable. In a two-way ANOVA, R^2 indicates what percent of
variance is accounted for by each predictor variable, as well as by their interaction
term.

As is the case in Chapter 9, you will have to perform a few hand calculations to
compute these R^2 values. To calculate R^2 for a given effect, divide the Type III sum
of squares associated with that effect by the corrected total sum of squares. For
example, to calculate the R^2 value for the type of rewards main effect, you would take
the appropriate terms from page 2 of Output 10.1 (or Table 10.1) and substitute them
in the formula for R^2, as is done here:

$$R^2 = \frac{\text{TYPE III SS}}{\text{CORRECTED TOTAL SUM OF SQUARES}} = \frac{1711.67}{2089.47} \approx .82$$

4. Review the Sample Means and Results of the Multiple Comparison Procedures

If a given main effect is statistically significant, you can review the Tukey test results for that main effect. Do this in the same way you would if this had been a one-way ANOVA. The results of the Tukey test for the present analysis are presented on pages 3 and 4 of Output 10.1. In this analysis, the type of rewards effect was significant, and the Tukey results on page 3 of Output 10.1 show that all three experimental groups are significantly different from each other, with the high-reward group showing the highest commitment scores, the mixed-reward group the next highest, and the low-reward group the lowest. (Chapter 9 describes how to interpret the Tukey test.)

The results of the Tukey test for the type of costs main effect appear on page 4 of Output 10.1. However, the type of costs main effect is not statistically significant; so, the Tukey results for that effect will not be interpreted. Technically, it would not really have been necessary to review the results of this multiple comparison procedure even if the type of costs main effect had been significant. This is because COSTGRP consists of only two groups, which means that you would not need a multiple comparison procedure to determine which "pairs" of groups were significantly different.

5. Summarize the Results of Your Analyses

In performing a two-way ANOVA, you should use the same statistical interpretation format as was used with one-way ANOVA:

A) Statement of the problem
B) Nature of the variables
C) Statistical test
D) Null hypothesis (H_0)
E) Alternative hypothesis (H_1)
F) Obtained statistic
G) Obtained probability (p) value
H) Conclusion regarding the null hypothesis
I) Magnitude of treatment effect
J) ANOVA summary table
K) Figure representing the results

However, with two-way ANOVA, it would be possible to perform this summary three times, since it is possible to test three null hypotheses in this study:

- The hypothesis of no interaction (which would be stated: "In the population, there is no interaction between type of rewards and type of costs in the prediction of commitment scores.")

- The hypothesis of no main effect for predictor A (type of rewards).

- The hypothesis of no main effect for predictor B (type of costs).

Because you should be fairly familiar with this format at this time, it is not completed again at this point.

Formal Description of Results for a Paper

Below is one approach that could be used to summarize the results of the preceding analysis in a scholarly paper:

```
    Results were analyzed using two-way ANOVA, with two
between-subjects factors.  This analysis revealed a significant
main effect for type of rewards, F(2, 24) = 59.85; p < .01 and
large treatment effect, R² = .82.  The sample means are
displayed in Figure 1.  Tukey's HSD test indicated that
participants in the high-reward condition scored significantly
higher on commitment than did those in the mixed-reward
condition, who, in turn, scored significantly higher than
participants in the low-reward condition (p < .05).  The main
effect for type of costs proved to be nonsignificant, F(1, 24)
= 0.04; ns.  The interaction between type of rewards and type
of costs also proved to be nonsignificant F(2, 24) = 1.19; ns.
```

Example with a Significant Interaction

When the interaction term is statistically significant, it is necessary to follow a different procedure when interpreting the results. In most cases, this consists of plotting the interaction in a figure and determining which of the simple effects are significant. This section shows how to do this.

For example, assume that the preceding study with 30 participants is repeated, but that this time the analysis is performed on the following dataset:

```
  5           DATALINES;
  6           1 1 08
  7           1 1 13
  8           1 1 04
  9           1 1 11
 10           1 1 05
 11           1 2 17
 12           1 2 12
 13           1 2 20
 14           1 2 14
 15           1 2 16
 16           1 3 31
 17           1 3 24
 18           1 3 37
 19           1 3 30
 20           1 3 32
 21           2 1 16
 22           2 1 10
 23           2 1 25
 24           2 1 15
 25           2 1 14
 26           2 2 21
 27           2 2 13
 28           2 2 16
 29           2 2 12
```

```
30            2 2 17
31            2 3 18
32            2 3 23
33            2 3 19
34            2 3 18
35            2 3 18
36            ;
37            RUN;
```

Analyzing this dataset with the program presented earlier would again result in five pages of output, with the same type of information appearing on each page. The results of the analysis of this dataset are reproduced here as Output 10.2:

Output 10.2   Results of Two-Way ANOVA Performed on Investment Model Data, Significant Interaction

```
                        The SAS System                            1


                        The GLM Procedure

                      Class Level Information

            Class           Levels     Values

            REWGRP               3      1 2 3

            COSTGRP              2      1 2


                 Number of observations      30
```

```
                        The GLM Procedure                         2

Dependent Variable: COMMIT

                             Sum of
Source               DF      Squares     Mean Square   F Value   Pr > F

Model                 5   1370.966667    274.193333     17.60   <.0001

Error                24    374.000000     15.583333

Corrected Total      29   1744.966667


            R-Square    Coeff Var     Root MSE     COMMIT Mean

            0.785669    22.38699      3.947573      17.63333


Source               DF     Type I SS    Mean Square   F Value   Pr > F

REWGRP                2   882.4666667   441.2333333     28.31   <.0001
COSTGRP               1    12.0333333    12.0333333      0.77   0.3883
REWGRP*COSTGRP        2   476.4666667   238.2333333     15.29   <.0001
```

(continued on the next page)

Output 10.2 *(continued)*

Source	DF	Type III SS	Mean Square	F Value	Pr > F
REWGRP	2	882.4666667	441.2333333	28.31	<.0001
COSTGRP	1	12.0333333	12.0333333	0.77	0.3883
REWGRP*COSTGRP	2	476.4666667	238.2333333	15.29	<.0001

```
                        The GLM Procedure                            3

              Tukey's Studentized Range (HSD) Test for COMMIT

NOTE: This test controls the Type I experimentwise error rate, but it generally
      has a higher Type II error rate than REGWQ.


              Alpha                                    0.05
              Error Degrees of Freedom                   24
              Error Mean Square                     15.58333
              Critical Value of Studentized Range   3.53170
              Minimum Significant Difference          4.4087


      Means with the same letter are not significantly different.


        Tukey Grouping          Mean      N    REWGRP

                    A         25.000     10    3

                    B         15.800     10    2
                    B
                    B         12.100     10    1
```

```
                        The GLM Procedure                            4

              Tukey's Studentized Range (HSD) Test for COMMIT

NOTE: This test controls the Type I experimentwise error rate, but it generally
      has a higher Type II error rate than REGWQ.


              Alpha                                    0.05
              Error Degrees of Freedom                   24
              Error Mean Square                     15.58333
              Critical Value of Studentized Range   2.91880
              Minimum Significant Difference          2.975


      Means with the same letter are not significantly different.


        Tukey Grouping          Mean      N    COSTGRP

                    A         18.267     15    1
                    A
                    A         17.000     15    2
```

(continued on the next page)

Output 10.2 *(continued)*

```
                         The GLM Procedure                              5


          Level of                ------------COMMIT-----------
          REWGRP        N              Mean            Std Dev

            1          10          12.1000000        6.08184913
            2          10          15.8000000        3.11982906
            3          10          25.0000000        7.00793201


          Level of                ------------COMMIT-----------
          COSTGRP       N              Mean            Std Dev

            1          15          18.2666667       10.3679910
            2          15          17.0000000        4.0355563


  Level of      Level of              ------------COMMIT-----------
  REWGRP        COSTGRP       N           Mean            Std Dev

    1             1           5         8.2000000        3.83405790
    1             2           5        16.0000000        5.52268051
    2             1           5        15.8000000        3.03315018
    2             2           5        15.8000000        3.56370594
    3             1           5        30.8000000        4.65832588
    3             2           5        19.2000000        2.16794834
```

Steps in Interpreting the Output for the Omnibus Analysis

This most recent dataset was constructed to generate a significant interaction between type of rewards and type of costs. The steps to be followed in interpreting this interaction are as follows:

1. Determine If the Interaction Term Is Statistically Significant

The analysis of variance table for the current analysis appears on page 2 of Output 10.2. As before, interpret the information that appears toward the bottom of the page, in the section that provides the Type III sum of squares. The interaction term is the REWGRP*COSTGRP term that appears as the last line on output page 2. You must first determine whether this interaction term is significant before reviewing any other results.

The last line on output page 2 shows that the REWGRP*COSTGRP interaction term displays an F value of 15.29 which, with 2 and 24 degrees of freedom, is associated with a p value less than 0.01. The interaction therefore is clearly significant. Because of this, you do not interpret the main effects for REWGRP or COSTGRP; instead, you plot the interaction and test for simple effects.

2. Plot the Interaction

Interactions are easiest to understand when they are represented in a figure that plots the means for each cell in the study's factorial design. The design used in the current study involved a total of six cells. (This design was presented in Figure 10.7.) The mean commitment scores for these six cells are presented on page 5 of Output 10.2.

There are actually three tables of means on page 5 of this output. The first table provides the means for each level of REWGRP, the second provides the means for each level of COSTGRP, and the third provides the means for the six cells that represent every possible combination of REWGRP and COSTGRP. It is this third table that will be used to create the figure.

The first line of the third table provides the mean score displayed by the cell that is coded with a "1" on REWGRP and a "1" on COSTGRP. Given the values that are used in coding levels of REWGRP and COSTGRP, you therefore know that the first line provides the mean for the participants who experienced the "low-reward" condition under REWGRP and the "low-cost" condition under COSTGRP. Notice that this line shows that this subgroup displayed a mean COMMIT score of 8.20.

The second line of the table provides the mean score displayed by the cell that is coded with a "1" on REWGRP and a "2" on COSTGRP. Given the values that are used to code the variables, you therefore know that this second line provides data for the participants who experienced the "low-reward" condition under REWGRP and the "high-cost" condition under COSTGRP. Notice that this line shows that this subgroup displayed a mean COMMIT score of 16.00. The remaining lines in the table can be read in the same way.

It is very difficult to understand a two-way interaction by simply reviewing cell means from a table such as the third table on output page 5. Interactions are much easier to understand if the means are instead plotted. Therefore, the cells means from Output 10.2 are plotted in Figure 10.8.

Figure 10.8 Mean Levels of Commitment as a Function of the Interaction between Type of Rewards and Type of Costs

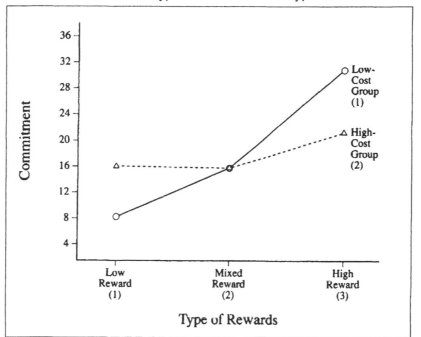

The vertical axis in Figure 10.8 is labeled "Commitment," to convey that this figure plots mean commitment scores. Notice that scores can range from a low of 4 to a high of 36.

The two different lines that appear in Figure 10.8 represent the different conditions under independent variable B (the type of costs independent variable). The solid line represents means commitment scores displayed by the low-cost group, and the dotted line represents mean commitment scores from the high-cost group.

The horizontal axis of the figure labels the three conditions under independent variable A (the type of rewards independent variable). For example, directly above the label "Low Reward," you see the mean scores displayed by participants in the low-reward condition. Notice that two mean scores are plotted: a small circle plots the mean commitment scores of those who are in the "low reward/low cost" cell; and a small triangle plots the mean commitment scores of participants who are in the "low reward/high cost" cell. The same system was used to plot mean scores for participants in the remaining cells.

When there is an interaction between two variables, it means that at least two of the line segments in the resulting figure are not parallel to each other. You can see that this is the case in Figure 10.8. Notice that the line for "Low-Cost Group" (the solid line) displays a steeper angle than is displayed by the line for "High-Cost Group" (the dashed line). This suggests that the type of rewards independent variable had a stronger effect on "Low-Cost Group," compared to the "High-Cost Group." Another way of stating this is to say that type of costs moderates the relationship between type of rewards and commitment.

This figure is also useful for again illustrating the definition of an **interaction**. Earlier, it was said that an interaction involves a situation in which the effect of one independent variable on the dependent variable is different at different levels of the second independent variable. In Figure 10.8, you can see that the effect of the rewards independent variable on commitment is stronger in the low-cost group than it is in the high-cost group.

Preparing a figure that plots an interaction (such as Figure 10.8) can be confusing. For this reason, this section presents structured, step-by-step guidelines to make the task easier. The following guidelines describe how to begin with the cells means from page 5 of Output 10.2, and transform them into the figure represented in Figure 10.8.

1. Label the vertical axis with the name of the criterion variable (in this case, COMMITMENT). On this same axis, provide some midpoints of scores that were possible. In Figure 10.8, this is done by using the midpoints of 4, 8, 12, and so forth.

2. Label the horizontal axis with the name of predictor A (*type of rewards*, in this case). On the same axis, provide one midpoint for each level of this independent variable and label these midpoints. For this variable, this is done by creating three midpoints labeled low-reward (1), mixed-reward (2), and high-reward (3).

3. Now, draw small circles on the figure to indicate the mean commitment scores of just those participants in the low-cost group. First, plot the mean for participants who are in the low-cost group under the type of costs predictor, and are also in the low-reward group under the type-of-rewards predictor. Go to the means provided on page 5 of Output 10.2 and find the entry for the group that is coded with a "1" under "COSTGRP" (1 stands for "low-cost") and is also coded with a "1" under "REWGRP" (1 stands for "low-reward"). It turns out that this is the first entry, and the mean COMMIT score for this subgroup is 8.20. Therefore, go to the figure and draw a small circle that is directly above the "low-reward" midpoint and directly to the right of where 8.20 would be on the COMMITMENT axis.

4. Next, plot the mean for participants who are in the low-cost/mixed-reward subgroup. On output page 5, find the entry for the subgroup that is coded with a "1" under "COSTGRP" and a "2" under "REWGRP." This is the third entry down, and their mean commitment score is 15.80. Therefore, draw a small circle that is directly above the "mixed-reward" midpoint and to the right of where 15.80 would be on the COMMITMENT axis. Finally, find the mean for the low-cost/high-reward subgroup, coded with a "1" under "COSTGRP" and a "3" under "REWGRP." This is the fifth entry in the table and the mean for this subgroup is 30.80. The circle representing this score goes above the high-reward midpoint on the figure. As a final step, draw a solid line connecting these three circles.

5. Now repeat this procedure. This time, you will draw small triangles to represent the scores of just those in the high-cost group. Remember that these are the subgroups coded with a "2" under "COSTGRP." The mean for the high-cost/low-reward group is 16.00, and it is the second entry in the table on page 5

of Output 10.2. To represent this score, a small triangle is drawn above the "low-reward" midpoint in the figure. The mean for the high-cost/mixed-reward subgroup is 15.80, and the mean for the high-cost/high-reward subgroup is 19.20. Note where triangles are drawn on the figure to represent these means. Once all three triangles are in place, they are connected with a dotted line. With this done, the figure is now complete.

Using this system, you will know that whenever you see a solid line connecting circles, you are looking at mean commitment scores from participants in the low-cost group, and whenever you see a dotted line connecting triangles, you are looking at mean commitment scores from the high-cost group.

Testing for Simple Effects

When there is a **simple effect** for independent variable A at a given level of independent variable B, it means that there is a significant relationship between independent variable A and the dependent variable at *that level* of independent variable B. As was stated earlier, the concept of a simple effect is perhaps easiest to understand by again referring to Figure 10.8.

First, consider the solid line (the line representing the low-cost group) in Figure 10.8. This line displays a relatively steep angle, suggesting that there might be a significant relationship between rewards and commitment for participants in the low-cost group. In other words, there might be a significant simple effect for rewards at the low-cost level of the "type of costs" independent variable.

Now, consider the dashed line in the same figure that represents the high-cost group. This line displays an angle that is less steep than the angle displayed by the solid line. It is impossible to tell by simply "eyeballing" the results in this way, but this might mean that there is not a simple effect for rewards at the high level of the "type of costs" independent variable.

Testing for simple effects is fairly straightforward. In simplified terms, this procedure involves dividing your sample into two groups (the low-cost group and the high-cost group): using only the low-cost group, performing a one-way ANOVA in which the predictor is type of rewards and the criterion is commitment; and using only the high-cost group, performing another one-way ANOVA where the predictor is type of rewards and the criterion is commitment.

In truth, the preceding description actually oversimplifies the process of testing for simple effects. The actual process is more complicated because you must use a special error term in computing the *F* tests of interest, rather than the *F* term that is provided with the one-way ANOVAs. The meaning of this is explained after first presenting the SAS program to test for simple effects.

Writing the SAS Program

The general form for the SAS program that will allow the testing of simple effects is shown here:

```
PROC GLM   DATA=dataset-name;
   CLASS   predictorA   predictorB;
   MODEL   criterion-variable = predictorA   predictorB
           predictorA*predictorB;
   MEANS   predictorA   predictorB   / TUKEY;
   MEANS   predictorA   predictorB   predictorA*predictorB;
RUN;

PROC SORT   DATA=dataset-name;
   BY predictorB;
RUN;

PROC GLM;
   CLASS   predictorA;
   MODEL   criterion-variable = predictorA;
   MEANS   predictorA / TUKEY;
   MEANS   predictorA;
   BY      predictorB;
RUN;
```

Here are the statements needed to test for simple effects with the present study:

```
1          PROC GLM DATA=D1;
2             CLASS REWGRP COSTGRP;
3             MODEL COMMIT = REWGRP COSTGRP REWGRP*COSTGRP;
4             MEANS REWGRP COSTGRP / TUKEY;
5             MEANS REWGRP COSTGRP REWGRP*COSTGRP;
6          RUN;
7
8          PROC SORT DATA=D1;
9             BY COSTGRP;
10         RUN;
11
12         PROC GLM DATA=D1;
13            CLASS REWGRP;
14            MODEL COMMIT = REWGRP;
15            MEANS REWGRP / TUKEY;
16            MEANS REWGRP;
17            BY COSTGRP;
18         RUN;
```

Lines 1 through 6 of the preceding program request the standard "omnibus" two-way ANOVA, identical to that performed earlier. Lines 8 through 10 request that SAS sort the data by the variable COSTGRP, so that all participants with a COSTGRP score of 1 are together and those with a COSTGRP score of 2 are together. Lines 12 through 18 ask for a standard one-way ANOVA with REWGRP as the predictor. However, line 17 asks SAS to perform this ANOVA once for each level of COSTGRP. That is, perform one ANOVA for participants with a COSTGRP score of 1 and a separate ANOVA for those with a COSTGRP score of 2.

Interpreting the Output

The preceding program would produce 13 pages of output and would actually perform three ANOVAs. The first would be the omnibus ANOVA, identical to the ANOVA that was reproduced as Output 10.2. The second ANOVA would result from the PROC GLM in which COMMIT was the dependent variable, REWGRP was the independent variable, and the analysis involved only participants with a score of "1" on COSTGRP (the low-cost group). Finally, the third ANOVA would result from the PROC GLM in which COMMIT was the dependent variable, REWGRP was the independent variable, and the analysis involved only participants with a score of "2" on COSTGRP (the high-cost group).

Because the omnibus ANOVA (based on all 30 participants) has already been presented in Output 10.2, it is not reproduced here. Instead, here is the output that was created by lines 8 through 18 of the preceding program (i.e., the two one-way ANOVAs). The information appearing in this part of the output is summarized below:

- Page 6 provides class-level information and the number of observations in the dataset for the low-cost subsample.

- Page 7 provides the ANOVA summary table for the low-cost subsample.

- Page 8 provides the results of the Tukey multiple comparison procedure for the low-cost subsample.

- Page 9 provides the mean commitment scores observed for the various levels of REWGRP for the low-cost subsample.

- Finally, pages 10 through 13 provide information corresponding to the information appearing on pages 6 through 9, except that the information on pages 10 through 13 is based on the analysis of data from the high-cost subsample.

Output 10.3   Results of One-Way ANOVA Procedures Necessary for Testing Simple Effects

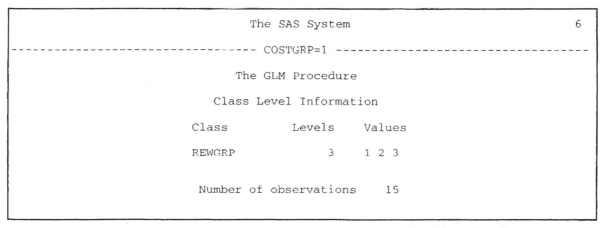

```
                              The SAS System                              6
--------------------------------- COSTGRP=1 -------------------------------

                             The GLM Procedure

                          Class Level Information

                   Class          Levels     Values

                   REWGRP              3      1 2 3


                   Number of observations      15
```

(continued on the next page)

Output 10.3 (continued)

```
                                                                         7
------------------------------- COSTGRP=1 -------------------------------
                         The GLM Procedure

Dependent Variable: COMMIT

                                  Sum of
 Source                   DF      Squares     Mean Square   F Value   Pr > F

 Model                     2   1322.533333     661.266667     43.50   <.0001
 Error                    12    182.400000      15.200000

 Corrected Total          14   1504.933333


          R-Square     Coeff Var      Root MSE     COMMIT Mean

          0.878799     21.34335      3.898718        18.26667


 Source                   DF     Type I SS     Mean Square   F Value   Pr > F

 REWGRP                    2   1322.533333     661.266667     43.50   <.0001


 Source                   DF    Type III SS    Mean Square   F Value   Pr > F

 REWGRP                    2   1322.533333     661.266667     43.50   <.0001
```

```
                                                                         8
------------------------------- COSTGRP=1 -------------------------------
                         The GLM Procedure

          Tukey's Studentized Range (HSD) Test for COMMIT

NOTE: This test controls the Type I experimentwise error rate, but it generally
      has a higher Type II error rate than REGWQ.


              Alpha                                   0.05
              Error Degrees of Freedom                  12
              Error Mean Square                        15.2
              Critical Value of Studentized Range   3.77278
              Minimum Significant Difference         6.5781


       Means with the same letter are not significantly different.


          Tukey Grouping        Mean      N    REWGRP

                     A        30.800      5    3

                     B        15.800      5    2

                     C         8.200      5    1
```

(continued on the next page)

Output 10.3 *(continued)*

```
                                                                           9
------------------------------- COSTGRP=1 -----------------------------------

                        The GLM Procedure


        Level of              ------------COMMIT-----------
        REWGRP         N           Mean           Std Dev

          1            5         8.2000000        3.83405790
          2            5        15.8000000        3.03315018
          3            5        30.8000000        4.65832588
```

```
                                                                          10
------------------------------ COSTGRP=2 ------------------------------------

                        The GLM Procedure


                     Class Level Information

             Class          Levels     Values

             REWGRP            3        1 2 3


            Number of observations     15
```

```
                                                                          11
------------------------------ COSTGRP=2-------------------------------------

                        The GLM Procedure

Dependent Variable: COMMIT
```

Source	DF	Sum of Squares	Mean Square	F Value	Pr > F
Model	2	36.4000000	18.2000000	1.14	0.3522
Error	12	191.6000000	15.9666667		
Corrected Total	14	228.0000000			

R-Square	Coeff Var	Root MSE	COMMIT Mean
0.159649	23.50489	3.995831	17.00000

Source	DF	Type I SS	Mean Square	F Value	Pr > F
REWGRP	2	36.40000000	18.20000000	1.14	0.3522

Source	DF	Type III SS	Mean Square	F Value	Pr > F
REWGRP	2	36.40000000	18.20000000	1.14	0.3522

(continued on the next page)

Output 10.3 *(continued)*

```
                                                                    12
------------------------------ COSTGRP=2 ------------------------------

                         The GLM Procedure

              Tukey's Studentized Range (HSD) Test for COMMIT

NOTE: This test controls the Type I experimentwise error rate, but it generally
         has a higher Type II error rate than REGWQ.


              Alpha                                    0.05
              Error Degrees of Freedom                   12
              Error Mean Square                     15.96667
              Critical Value of Studentized Range   3.77278
              Minimum Significant Difference         6.7419


         Means with the same letter are not significantly different.


         Tukey Grouping          Mean       N      REWGRP

                    A          19.200       5        3
                    A
                    A          16.000       5        1
                    A
                    A          15.800       5        2
```

```
                                                                    13
------------------------------ COSTGRP=2 ------------------------------

                         The GLM Procedure

         Level of              ------------COMMIT-----------
         REWGRP        N             Mean            Std Dev

            1          5        16.0000000        5.52268051
            2          5        15.8000000        3.56370594
            3          5        19.2000000        2.16794834
```

Page 7 of Output 10.3 provides the analysis of variance table from the ANOVA in which COMMIT served as the dependent variable, REWGRP was the independent variable, and the analysis involved only participants with a score of "1" on COSTGRP (the low-cost group). Output pages 6 through 9 are identified with the header "COSTGRP=1" at the top of each page, and this informs you that these pages provide results for only the low-cost group. Information from this output will be used to determine whether there was a simple effect for type of rewards at the low-cost level of the type of costs independent variable.

The last line of page 7 of Output 10.2 provides the F test for the ANOVA in which the independent variable was type of rewards and the analysis was based on only the low-cost participants. It can be seen that the F value of 43.50 is significant at $p < .01$. However, this is not the F value for the simple effect of interest. To compute the F for the simple effect, you must take the appropriate mean square from this table, and insert it in the following formula:

$$F = \frac{MS_{\text{simple effect}}}{\text{Omnibus Within-Groups } MS}$$

In the preceding formula, the $MS_{\text{simple effect}}$ is the mean square from the last line of page 7 of Output 10.2. In the present case, it can be seen that this mean square is 661.27; this value is now inserted in your formula:

$$F = \frac{661.27}{\text{Omnibus Within-Groups } MS}$$

But where do you find the "Omnibus Within-Groups MS" for the preceding formula? This comes from the omnibus two-way ANOVA that is based on the entire sample and included both type of rewards and type of costs as independent variables. The ANOVA summary table for this analysis is presented earlier on page 2 of Output 10.2. The omnibus MS is simply the "Error Mean Square" from Output 10.2 (middle of page 3). It can be seen that this Omnibus Within-Groups MS is 15.58; this term is now inserted in your formula:

$$F = \frac{661.27}{15.58}$$

With both values now in the formula, you can compute the F ratio that will indicate whether there is a significant simple effect for type of rewards in the low-cost group:

$$F = \frac{MS_{\text{simple effect}}}{\text{Omnibus Within-Group } MS} = \frac{661.27}{15.58} = 42.44$$

This is a large value of F, but is it large enough to reject the null hypothesis? To know this, you have to find the *critical value* of F for this analysis. This, in turn, requires knowing how many degrees of freedom are associated with this particular F test. In this case, you had 2 and 24 degrees of freedom for the numerator and denominator respectively. The "2" degrees of freedom for the numerator are the 2 that went into calculating the $MS_{\text{simple effect}}$. See the bottom of page 7 of Output 10.2, where the row for "REWGRP" intersects with the column for "DF," you see that there are 2 degrees of freedom associated with the type of rewards variable.

The "24" degrees of freedom for the denominator are the 24 degrees of freedom associated with the mean square error term from the omnibus ANOVA. These degrees of freedom can be found in the upper part of page 7 of Output 10.3. Look under the heading "DF" to the right of "Error."

Now that you have established the degrees of freedom associated with the analysis, you are free to find the critical value of F by referring to the F table in Appendix C at the back of this text. This table shows that, when $p = .05$ and there are 2 and 24 degrees of freedom, the critical value of F is 3.40. Your obtained value of F is 42.44, which is much larger than this critical value. Therefore, you can reject the null hypothesis and conclude that there is a simple effect for type of rewards at the low-cost level of the type of costs factor.

With this done, you now repeat the procedure to determine whether there is a simple effect for type of rewards with the high-cost group of participants. The $MS_{\text{simple effect}}$ necessary to perform this test may be found on page 11 of Output 10.3. (Notice that the top of each output page indicates "COSTGRP=2"; this informs you that the results on this page come from the high-cost group.)

To find the information necessary for this test, see the bottom of page 11 from Output 10.2 in the section for the Type III sum of squares. To find the $MS_{\text{simple effect}}$ for this subgroup, look below the heading "Mean Square." It can be seen that, for the high-reward participants, the $MS_{\text{simple effect}}$ equals 18.20. You can now insert this in your formula:

$$F = \frac{18.20}{\text{Omnibus Within-Groups } MS}$$

The error term for the preceding formula is the same error term used with the previous test. It is the Omnibus Within-Groups MS from Output 10.2. With both terms inserted in the formula, you can now compute the F ratio for the simple effect of type of reward in the high-cost group:

$$F = \frac{MS_{\text{simple effect}}}{\text{Omnibus Within-Groups } MS} = \frac{18.20}{15.58} = 1.17$$

This test is also associated with 2 and 24 degrees of freedom (same as the earlier test for a simple effect), so the same critical value of F described earlier (3.40) still applies. Your obtained value of F is 1.17, and this is not greater than the critical value of 3.40. Therefore, you fail to reject the null and conclude that there is not a significant simple effect for the type of rewards at the high-cost level of the type of costs factor.

Formal Description of Results for a Paper

Results from this analysis could be summarized in the following way for a scholarly paper:

```
     Results were analyzed using two-way ANOVA, with two
between-subject factors.  This revealed a significant Type of
Rewards x Type of Costs interaction, F(2, 24) = 15.29, p < .01,
R² = .27.  The nature of this interaction is displayed in
Figure 10.8.
     Subsequent analyses demonstrated that there was a simple
effect for type of rewards at the low-cost level of the type of
costs factor, F(2, 24) = 42.44, p < .05.  As evident from
Figure 10.8, the high-reward group displayed higher commitment
scores than the mixed-reward group, which, in turn,
demonstrated higher commitment scores than the low-reward
```

group. The simple effect for type of rewards at the high-cost
level of the type of costs factor was nonsignificant, $F(2, 24)$
= 1.17, *ns*.

Two Perspectives on the Interaction

The preceding example presents a potential simple effect for the type of rewards at two
different levels of the type of costs factor. If you had chosen, you could have also
investigated this interaction from a different perspective. That is, you could have studied
possible simple effects for the type of costs at three different levels of the type of rewards
factor.

To do this, you would have drawn the preceding figure so that the horizontal axis
represented the type of costs factor and had two midpoints (one for the low-cost group,
and one for the high-cost group). Within the body of the figure itself, there would have
been three lines, with one line representing the low-reward group, one representing the
mixed-reward group, and one representing the high-reward group.

In writing the SAS program, you would have sorted the dataset by the REWGRP
variable. Then you would have requested an ANOVA in which the criterion was
COMMIT and the predictor was COSTGRP. You would add a BY REWGRP statement
toward the end of the PROC GLM step and this would cause one ANOVA to be
performed for the low-reward group, one for the mixed-reward group, and one for the
high-reward group. The hand calculations to determine the simple effects would have
followed in a similar manner.

When a two-factor interaction is significant, you can always view it from these two
different perspectives. Furthermore, the interaction is often more interpretable (i.e.,
makes more sense) from one perspective than from the other. In many cases, therefore,
you should test for simple effects from both perspectives before interpreting the results.

Using the LSMEANS Statement to Analyze Data from
Unbalanced Designs

As discussed in Chapter 9, an experimental design is **balanced** if the same numbers of
observations (participants) appear in each cell of the design. For example, Figure 10.7
(presented earlier in this chapter) illustrates the research design used in the investment
model study. It shows that there are five participants in each cell of the design. When a
research design is balanced, it is generally appropriate to use the MEANS statement with
PROC GLM to request group means, multiple comparison procedures, and confidence
intervals.

In contrast, a research design is said to be **unbalanced** if some cells in the design contain
a different number of observations (participants) than other cells. For example, again
consider Figure 10.7 where the research contains six cells. If there were 20 participants
in one of the cells, but just five participants in each of the remaining five cells, the
research design would then be unbalanced.

When analyzing data from an unbalanced design, it is generally best to not use the MEANS statement. This is because (with unequal cell sizes), the MEANS statement can produce marginal means that are biased. When analyzing data from an unbalanced design, it is generally preferable to use the LSMEANS statement in your program, rather than the MEANS statement. This is because the LSMEANS statement estimates the marginal means over a balanced population. LSMEANS estimates what the marginal means would be if you did have equal cell sizes. In other words, the marginal means estimated by the LSMEANS statement are less likely to be biased.

Writing the LSMEANS Statements

The General Form

Below is the general form for the PROC step of a SAS program that uses the LSMEANS statement rather than the MEANS statement:

```
PROC GLM DATA = dataset-name;
    CLASS     predictorA  predictorB;
    MODEL     criterion-variable = predictorA  predictorB
              predictorA*predictorB;
    LSMEANS   predictorA  predictorB  predictorA*predictorB;
    LSMEANS   predictorA  predictorB  / PDIFF  ADJUST=TUKEY  ALPHA=alpha-
level;
RUN;
```

The preceding general form is very similar to the general form that used the MEANS statement earlier in this chapter. The primary difference is that the two MEANS statements is replaced with two LSMEANS statements. The first LSMEANS statement takes this form:

```
        LSMEANS  predictorA  predictorB  predictorA*predictorB;
```

You can see that this LSMEANS statement is identical to the earlier MEANS statement, except that "MEANS" is replaced with "LSMEANS."

The second LSMEANS statement is a bit more complex:

```
        LSMEANS  predictorA  predictorB  / PDIFF  ADJUST=TUKEY
                 ALPHA=alpha-level;
```

You can see that this second LSMEANS statement contains a forward slash, followed by a number of keywords for options. Here is what the key words request:

- **PDIFF** requests that SAS print p values for significance tests related to the multiple comparison procedure. These p values tell you whether or not there are significant differences between the least-squares means for the different levels under the two predictor variables.

- **ADJUST=TUKEY** requests a multiple comparison adjustment for the p values and confidence limits for the differences between the least-squares means. Including ADJUST=TUKEY requests an adjustment based on the Tukey HSD test. The adjustment can also be based on other multiple-comparison procedures.

- **ALPHA=** specifies the significance level to be used for the multiple comparison procedure and the confidence level to be used with the confidence limits. Specifying ALPHA=0.05 requests that the significance level (alpha) be set at .05 for the Tukey tests. If you had wanted alpha set at .01, you would have used the option ALPHA=0.01, and if you had wanted alpha set at .10, you would have used the option ALPHA=0.1.

The Actual SAS Statements

Below are the actual statements that you would include in a SAS program to request a factorial ANOVA using the LSMEANS statement rather than the MEANS statement. The following statements are appropriate to analyze data from the aggression study described in this chapter. Notice that alpha is set at .05 for the Tukey tests.

```
 1     PROC GLM DATA=D1;
 2         CLASS    REWGRP COSTGRP;
 3         MODEL    COMMIT = REWGRP COSTGRP REWGRP*COSTGRP;
 4         LSMEANS  REWGRP COSTGRP REWGRP*COSTGRP;
 5         LSMEANS  REWGRP COSTGRP / PDIFF  ADJUST=TUKEY  ALPHA=0.05;
 6     RUN;
 7
 8     PROC SORT DATA=D1;
 9         BY COSTGRP;
10     RUN;
11
12     PROC GLM DATA=D1;
13         CLASS REWGRP;
14         MODEL COMMIT = REWGRP;
15         LSMEANS REWGRP / PDIFF  ADJUST=TUKEY  ALPHA=0.05;
16         LSMEANS REWGRP;
17         BY COSTGRP;
18     RUN;
```

Output Produced by LSMEANS

The output produced by the LSMEANS statements is very similar to the output produced by the MEANS statements except that the means have been appropriately adjusted.

There are a few additional differences. For example, the MEANS statement prints means and standard deviations, while the LSMEANS statement prints only adjusted means. For the most part, however, if you have read the sections of this chapter that show how to interpret the results of the MEANS statements, you should have little difficulty interpreting the results produced by LSMEANS. Given that our example is a balanced design, the results obtained with the MEANS and LSMEANS statements will be virtually identical and not reproduced here. This discussion of when to use LSMEANS statements and the accompanying code is intended to provide you with an example should you be faced with unequal designs.

Conclusion

This chapter, as well as Chapter 9, deals with between-subjects investigations in which scores are obtained on only one criterion variable. Yet, social scientists often conduct studies that involve multiple criterion variables. Imagine, for example, that you are an industrial/organizational psychologist who hypothesizes that some organization-development intervention will positively affect several different aspects of job satisfaction: satisfaction with the work itself; satisfaction with supervision; and satisfaction with pay. When analyzing data from a field experiment that tests this hypothesis, it would be advantageous to use a multivariate statistic that allows you to test the effect of your manipulation on all three criterion variables simultaneously. One statistic that allows this type of test is the multivariate analysis of variance. This procedure is introduced in the following chapter.

Assumptions Underlying Factorial ANOVA with Two Between-Subjects Factors

- **Level of measurement.** The criterion variable should be assessed on an interval- or ratio-level of measurement. Both predictor variables should be nominal-level variables (i.e., categorical variables).

- **Independent observations.** A given observation should not be dependent on any other observation in another cell. (For a detailed explanation of this assumption, see Chapter 8, "*t* Tests: Independent Samples and Paired Samples.")

- **Random sampling.** Scores on the criterion variable should represent a random sample drawn from the populations of interest.

- **Normal distributions.** Each cell should be drawn from a normally distributed population. If each cell contains more than 30 participants, the test is robust against moderate departures from normality.

- **Homogeneity of variance.** The populations represented by the various cells should have equal variances on the criterion. If the number of participants in the largest cell is no more than 1.5 times greater than the number of participants in the smallest cell, the test is robust against violations of the homogeneity assumption.

Multivariate Analysis of Variance (MANOVA) with One Between-Subjects Factor

> **Overview.** This chapter shows how to enter data and prepare SAS programs that will perform a one-way multivariate analysis of variance (MANOVA) using the GLM procedure. You can think of MANOVA as an extension of ANOVA in that it allows for the inclusion of multiple criterion variables in a single computation. This chapter focuses on the between-subjects design in which each participant is assigned to only one condition under the independent variable. You are shown how to summarize both significant and nonsignificant MANOVA results.

Introduction: The Basics of Multivariate Analysis of Variance

Multivariate analysis of variance (MANOVA) with one between-subjects factor is appropriate when an analysis involves the following:

- a single predictor variable measured on a nominal scale;

- multiple criterion variables, each of which is measured on an interval or ratio scale.

MANOVA is similar to ANOVA in that it tests for significant differences between two or more groups of participants. There is an important difference between MANOVA and ANOVA, however: ANOVA is appropriate when the study involves just one criterion variable whereas MANOVA is appropriate when the study involves more than one criterion variable. With MANOVA, you can perform a single computation that determines whether there is a significant difference between treatment groups when compared simultaneously on all criterion variables.

The Aggression Study

To illustrate one possible use of MANOVA, consider the study on aggression that was introduced in Chapter 9, "One-Way ANOVA with One Between-Groups Factor." In that study, each of 60 children was assigned to one of the following treatment groups:

- a group that consumed 0 grams of sugar at lunch;

- a group that consumed 20 grams of sugar at lunch;

- a group that consumed 40 grams of sugar at lunch.

A pair of observers watched each child after lunch and recorded the number of aggressive acts displayed by that child. The total number of aggressive acts performed by a given child over a two-week period served as that child's score on the "aggression" dependent variable.

It was appropriate to analyze data from the preceding study with ANOVA because there was a single dependent variable: level of aggression. Now, consider how the study could be modified so that it would, instead, analyze the data using a multivariate procedure: MANOVA. Imagine that, as the researcher, you wanted to have more than one measure of aggression. After reviewing the literature, you believe that there are at least four different types of aggression that children might display:

- aggressive acts directed at children of the same sex;
- aggressive acts directed at children of the opposite sex;
- aggressive acts directed at teachers;
- aggressive acts directed at parents.

Assume that you now want to replicate your earlier study; this time, your observers will note the number of times each child displays an aggressive act in each of the four preceding categories. At the end of the two-week period, you will have scores for each child on each of the dependent variables listed above.

You now have a number of options as to how you will analyze your data. One option is to simply perform four ANOVAs (as described in Chapter 9). In each ANOVA, the independent variable would again be "amount of sugar consumed." In the first ANOVA, the dependent variable would be "number of aggressive acts directed at children of the same sex," in the second ANOVA, the dependent variable would be "number of aggressive acts directed at children of the opposite sex," and so forth.

However, a more appropriate alternative would be to perform a single computation that allows you to assess the effect of your independent variable on all four of the dependent variables simultaneously. This is what MANOVA allows. Performing a MANOVA will allow you to test the following null hypothesis:

In the population, there is no difference across the various treatment groups when they are compared simultaneously on the dependent variables.

Here is another way of stating this null hypothesis:

In the population, all treatment groups are equal on all dependent variables.

MANOVA produces a single statistic that allows you to test this null hypothesis. If the null hypothesis is rejected, it means that at least two of the treatment groups are significantly different with respect to at least one dependent variable. You can then perform follow-up tests to identify the pairs of groups that are significantly different, and the specific dependent variables on which they are different. In doing these follow-up tests, you might find that the groups differ on some dependent variables (e.g., "aggressive acts directed toward children of the same sex") but not on other dependent variables (e.g., "aggressive acts directed toward teachers").

A Multivariate Measure of Association

Chapter 9 introduced the R^2 statistic, a measure of association that is often computed when performing ANOVA. Values of R^2 can range from 0 to 1, with higher values indicating a stronger relationship between the predictor variable and the criterion variable in the study. If the study is a true experiment, you can view R^2 as an index of the magnitude of the treatment effect.

The current chapter introduces a *multivariate* measure of association: one that can be used when there are multiple criterion variables (as in MANOVA). This multivariate measure of association is called Wilks' lambda. Values of Wilks' lambda can range from 0 to 1, but the way that you interpret lambda is the opposite of the way that you interpret R^2. With lambda, small values (near 0) indicate a relatively strong relationship between the predictor variable and the multiple criterion variables (taken as a group), while larger values (near 1) indicate a relatively weak relationship. The *F* statistic that tests the significance of the relationship between the predictor and the multiple criterion variables is actually based on Wilks' lambda.

Overview of the Steps Followed in Performing a MANOVA

When you compute a multivariate analysis of variance with PROC GLM, the procedure produces two sets of results. First are the results from the univariate ANOVAs with one univariate ANOVA for each criterion variable in the analysis (here, **univariate** means "one criterion variable"). Each univariate ANOVA involves the same type of analysis that was described in Chapter 9. For example, with the present study, the predictor variable ("amount of sugar consumed") is the same in each analysis. However, one ANOVA will have "aggressive acts directed toward children of the same sex" as the criterion variable, one will have "aggressive acts directed toward children of the opposite sex" as the criterion, and so forth.

The second set of results produced when you request a MANOVA are the results of the multivariate analysis of variance (here, **multivariate** means "multiple criterion variables"). These multivariate results will include Wilks' lambda and the *F* statistic derived from Wilks' lambda.

As the preceding section suggests, there is a specific sequence of steps that you should follow when interpreting these results. First, you should review the multivariate *F* statistic derived from Wilks' lambda. If this multivariate *F* statistic is significant, you can reject the null hypothesis of no overall effect for the predictor variable. In other words, you can reject the null hypothesis that all groups are equal on all criterion variables. At that point, you can proceed to the univariate ANOVAs and interpret them.

When interpreting univariate ANOVAs, you first identify those criterion variables for which the univariate *F* statistic is significant. If the *F* statistic is significant for a given criterion variable, you can then proceed to interpret the results of the Tukey multiple comparison test to determine which pairs of groups significantly differ.

However, if the multivariate *F* statistic that is computed in the MANOVA is nonsignificant, this means that you cannot reject the null hypothesis that all groups have equal means on the criterion variables in the population. In most cases, your analysis should terminate at this point; you generally should not proceed to interpret the univariate ANOVAs, even if one or more of them displays a significant F statistic.

Similarly, even if the multivariate F statistic is significant, you should not interpret the results for any specific criterion variable that did not display a significant univariate F statistic. This is consistent with the general guidelines for univariate ANOVA presented in Chapter 9.

Example with Significant Differences between Experimental Conditions

To illustrate MANOVA, assume that you replicate the study that examined the effect of rewards on commitment in romantic relationships. This time, however, the study is modified so that you can obtain scores on three different dependent variables rather than just one.

It was hypothesized that the rewards people experience in romantic relationships have a causal effect on their commitment to those relationships. In Chapter 9, this hypothesis was tested by conducting a type of analogue experiment. All 18 participants in the experiment were asked to engage in similar tasks: to read the descriptions of 10 potential romantic "partners;" for each partner, imagine what it would be like to date this person and rate how committed they would be to a relationship with that person. For the first nine partners, each participant saw the same description. For partner 10, however, the different experimental groups were presented with a slightly different description of this person. For example, for the six participants in the high-reward condition, the last sentence in their description of partner 10 read as follows:

> This person enjoys the same recreational activities that you enjoy and is very good-looking.

However, for the six participants in the mixed-reward condition, the last part of the description of partner 10 read as follows:

> Sometimes, this person seems to enjoy the same recreational activities that you enjoy and sometimes s/he does not. Sometimes, partner 10 seems to be very good looking and sometimes s/he does not.

And for the six participants in the low-reward condition, the last sentence in the description of partner 10 read in this way:

> This person does not enjoy the same recreational activities that you enjoy and is not very good-looking.

In your current replication of the study from Chapter 9, you manipulate this "type of rewards" independent variable in exactly the same way. The only difference between the present study and the one described earlier involves the number of dependent variables measured. In the earlier study, you obtained data on just one criterion variable: commitment. **Commitment** was defined as the participants' rating of their commitment to remain in the relationship. Scores on this variable were created by summing participant responses to four questionnaire items. In the present study, however, you also obtain two additional dependent variables: participants' rating of their satisfaction with their relationship with partner 10 as well as their ratings of how long they intend to stay in the relationship with partner 10. Here, **satisfaction** is defined as the participants' emotional reaction to partner 10, whereas the **intention to stay** is defined as a rating of how long they intend to maintain the relationship with partner 10. Assume that satisfaction and intention to stay are measured with multiple items from a questionnaire, similar to those used to assess commitment.

You can see that you will obtain four scores for each participant in the study. One will simply be a score on a classification variable that indicates whether the participant is in the high-reward group, the mixed-reward group, or the low-reward group. The remaining three scores will be the participant's scores on the measures of commitment, satisfaction, and intention to stay. In conducting the MANOVA, you need to determine whether there is a significant relationship between the type of rewards predictor variable and the three criterion variables, taken as a group.

Writing the SAS Program

Creating the SAS Dataset

There is one predictor variable in this study, type of rewards. This variable could assume one of three values: participants were either in the high-reward group; the mixed-reward group; or the low-reward group. Since this variable simply codes group membership, it is measured on a nominal-scale. You give participants a score of "1" if they are in the low-reward condition, a "2" if they are in the mixed-reward condition, and a score of "3" if they are in the high-reward condition. You need a short SAS variable name for this variable, so call it REWGRP (for "reward group").

There are three criterion variables in this study. The first criterion variable, COMMITMENT, is given the SAS variable name, COMMIT. The second criterion, SATISFACTION, is called SATIS. The third, intention to stay, is called STAY. With each criterion variable, possible scores could range from a low of 4 to a high of 36.

This is the DATA step of the program that performs the MANOVA. It contains fictitious data from the preceding study:

```
 1          DATA D1;
 2             INPUT   #1   @1   REWGRP   1.
 3                          @3   COMMIT   2.
 4                          @6   SATIS    2.
 5                          @9   STAY     2. ;
 6          DATALINES;
 7          1 13 10 14
 8          1 06 10 08
 9          1 12 12 15
10          1 04 10 13
11          1 09 05 07
12          1 08 05 12
13          2 33 30 36
14          2 29 25 29
15          2 35 30 30
16          2 34 28 26
17          2 28 30 26
18          2 27 26 25
19          3 36 30 28
```

```
20          3 32 32 29
21          3 31 31 27
22          3 32 36 36
23          3 35 30 33
24          3 34 30 32
25          ;
26          RUN;
```

The data are entered so that there is one line of data for each participant. The format used in entering the preceding data is summarized in the following table:

Line	Column	Variable Name	Explanation
1	1	REWGRP	Codes group membership, so that 1 = low-reward condition 2 = mixed-reward condition 3 = high-reward condition
	2	blank	
	3-4	COMMIT	Commitment ratings obtained when participants rated partner 10
	5	blank	
	6-7	SATIS	Satisfaction ratings obtained when participants rated partner 10
	8	blank	
	9-10	STAY	Intention to stay ratings obtained when participants rated partner 10

Testing for Significant Effects with PROC GLM

In the following example, you write a SAS program to determine whether the overall MANOVA is significant and follow it with univariate ANOVAs and Tukey HSD tests. Remember that the results for all of these tests will appear in the output regardless of the significance of the multivariate test. However, you should interpret the univariate ANOVAs and Tukey tests only if the overall MANOVA is significant.

Here is the general form for the SAS program to perform a MANOVA followed by individual ANOVAs and Tukey's HSD tests:

```
PROC GLM  DATA=filename;
   CLASS  predictor-variable;
   MODEL  criterion-variables = predictor-variable;
   MEANS  predictor-variable / TUKEY;
   MEANS  predictor-variable;
   MANOVA  H = predictor-variable / MSTAT = exact;
RUN;
```

Here is the actual program (including the DATA step) that you would use to analyze data from the investment-model study just described:

```
1              DATA D1;
2                 INPUT  #1  @1  REWGRP  1.
3                            @3  COMMIT  2.
4                            @6  SATIS   2.
5                            @9  STAY    2. ;
6              DATALINES;
7              1 13 10 14
8              1 06 10 08
9              1 12 12 15
10             1 04 10 13
11             1 09 05 07
12             1 08 05 12
13             2 33 30 36
14             2 29 25 29
15             2 35 30 30
16             2 34 28 26
17             2 28 30 26
18             2 27 26 25
19             3 36 30 28
20             3 32 32 29
21             3 31 31 27
22             3 32 36 36
23             3 35 30 33
24             3 34 30 32
25             ;
26             RUN;
27
28             PROC GLM DATA=D1;
29                CLASS REWGRP;
30                MODEL COMMIT SATIS STAY = REWGRP;
31                MEANS REWGRP / TUKEY;
32                MEANS REWGRP;
33                MANOVA H = REWGRP / MSTAT = exact;
34             RUN;
```

Notes Regarding the SAS Program

The data are input in lines 6 through 26 of the preceding program, and the PROC GLM step requesting the MANOVA appears in lines 28 through 34. In the CLASS statement on line 29, you should specify the name of the nominal-scale predictor variable (REWGRP, in this case). The MODEL statement in line 30 is similar to that described in Chapter 9, except that all of your criterion variables should appear to the left of the "=" sign (COMMIT, SATIS, and STAY in this case). The name of the predictor variable should again appear to the right of the "=" sign.

The MEANS statement on line 31 requests that the Tukey multiple comparison procedure be performed for the REWGRP predictor variable. The second MEANS statement on line 32 merely requests that the means for the various treatment groups be printed.

The only new statement in the program is the MANOVA statement that appears on line 33. To the right of the term "H =" in this statement, you should specify the name of your study's predictor variable. The present program specifies "H = REWGRP."

Results from the SAS Output

Specifying LINESIZE=80 and PAGESIZE=60 in the OPTIONS statement (first line) causes the preceding program to produce nine pages of output. The following table indicates which specific results appear on each page:

Page	Information
1	Class level information
2	Univariate ANOVA results for COMMIT criterion
3	Univariate ANOVA results for SATIS criterion
4	Univariate ANOVA results for STAY criterion
5	Tukey HSD test results for COMMIT criterion
6	Tukey HSD test results for SATIS criterion
7	Tukey HSD test results for STAY criterion
8	Group means for each criterion variable
9	Results of the multivariate analysis: characteristic roots and vectors; Wilks' lambda; and results of the multivariate F test

The output produced by the preceding program is reproduced here as Output 11.1.

Output 11.1 Results of the Multivariate Analysis of Variance for the Investment Model Study, Significant Results

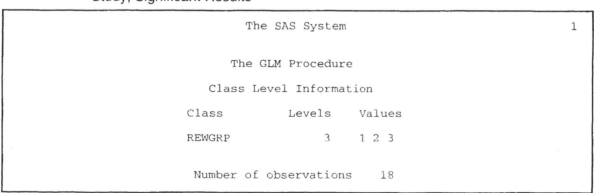

```
                        The SAS System                              1

                        The GLM Procedure

                     Class Level Information

            Class          Levels      Values

            REWGRP             3        1 2 3


            Number of observations      18
```

(continued on the next page)

Output 11.1 *(continued)*

```
                                                                    2
                        The GLM Procedure

Dependent Variable: COMMIT

                          Sum of
Source              DF    Squares      Mean Square   F Value   Pr > F

Model               2    2225.333333   1112.666667   122.12    <.0001

Error              15     136.666667      9.111111

Corrected Total    17    2362.000000


        R-Square    Coeff Var    Root MSE    COMMIT Mean

        0.942139    12.40464     3.018462     24.33333


Source              DF    Type I SS     Mean Square   F Value   Pr > F

REWGRP              2    2225.333333   1112.666667   122.12    <.0001


Source              DF    Type III SS   Mean Square   F Value   Pr > F

REWGRP              2    2225.333333   1112.666667   122.12    <.0001
```

```
                                                                    3
                        The GLM Procedure

Dependent Variable: SATIS

                          Sum of
Source              DF    Squares      Mean Square   F Value   Pr > F

Model               2    1825.444444    912.722222   143.11    <.0001

Error              15      95.666667      6.377778

Corrected Total    17    1921.111111


        R-Square    Coeff Var    Root MSE    SATIS Mean

        0.950202    11.08724     2.525426    22.77778


Source              DF    Type I SS     Mean Square   F Value   Pr > F

REWGRP              2    1825.444444    912.722222   143.11    <.0001


Source              DF    Type III SS   Mean Square   F Value   Pr > F

REWGRP              2    1825.444444    912.722222   143.11    <.0001
```

(continued on the next page)

Output 11.1 *(continued)*

```
                                                                          4
                              The GLM Procedure

Dependent Variable: STAY

                                     Sum of
     Source                 DF       Squares    Mean Square   F Value   Pr > F

     Model                   2    1346.333333    673.166667    51.61   <.0001

     Error                  15     195.666667     13.044444

     Corrected Total        17    1542.000000


              R-Square    Coeff Var    Root MSE    STAY Mean

              0.873109    15.26074     3.611709    23.66667


     Source                 DF      Type I SS    Mean Square   F Value   Pr > F

     REWGRP                  2    1346.333333    673.166667    51.61   <.0001


     Source                 DF     Type III SS   Mean Square   F Value   Pr > F

     REWGRP                  2    1346.333333    673.166667    51.61   <.0001
```

```
                                                                          5
                              The GLM Procedure

              Tukey's Studentized Range (HSD) Test for COMMIT

NOTE: This test controls the Type I experimentwise error rate, but it generally
      has a higher Type II error rate than REGWQ.


              Alpha                                     0.05
              Error Degrees of Freedom                    15
              Error Mean Square                       9.111111
              Critical Value of Studentized Range    3.67338
              Minimum Significant Difference          4.5266
         Means with the same letter are not significantly different.


         Tukey Grouping         Mean       N      REWGRP

                        A      33.333       6      3
                        A
                        A      31.000       6      2

                        B       8.667       6      1
```

(continued on the next page)

Output 11.1 *(continued)*

```
                                                                    6
                      The GLM Procedure

            Tukey's Studentized Range (HSD) Test for SATIS

NOTE: This test controls the Type I experimentwise error rate, but it generally
      has a higher Type II error rate than REGWQ.


            Alpha                                   0.05
            Error Degrees of Freedom                  15
            Error Mean Square                   6.377778
            Critical Value of Studentized Range  3.67338
            Minimum Significant Difference        3.7873


       Means with the same letter are not significantly different.


       Tukey Grouping           Mean       N    REWGRP

                     A        31.500       6    3
                     A
                     A        28.167       6    2

                     B         8.667       6    1
```

```
                                                                    7
                      The GLM Procedure

            Tukey's Studentized Range (HSD) Test for STAY

NOTE: This test controls the Type I experimentwise error rate, but it generally
      has a higher Type II error rate than REGWQ.


            Alpha                                   0.05
            Error Degrees of Freedom                  15
            Error Mean Square                   13.04444
            Critical Value of Studentized Range  3.67338
            Minimum Significant Difference        5.4163


       Means with the same letter are not significantly different.


       Tukey Grouping           Mean       N    REWGRP

                     A        30.833       6    3
                     A
                     A        28.667       6    2
                     B        11.500       6    1
```

(continued on the next page)

Output 11.1 *(continued)*

```
                                                                            8
                              The GLM Procedure

Level of          -----------COMMIT-----------    ------------SATIS-----------
REWGRP    N          Mean          Std Dev            Mean          Std Dev

1         6        8.6666667      3.44480285        8.6666667      2.94392029
2         6       31.0000000      3.40587727       28.1666667      2.22860195
3         6       33.3333333      1.96638416       31.5000000      2.34520788

          Level of          -------------STAY-----------
          REWGRP    N          Mean          Std Dev

          1         6       11.5000000      3.27108545
          2         6       28.6666667      4.08248290
          3         6       30.8333333      3.43025752
```

```
                                                                            9
                              The GLM Procedure
                      Multivariate Analysis of Variance

              Characteristic Roots and Vectors of: E Inverse * H, where
                     H = Type III SSCP Matrix for REWGRP
                          E = Error SSCP Matrix

Characteristic                  Characteristic Vector   V'EV-1
        Root      Percent          COMMIT          SATIS            STAY

  31.7043925       99.91         0.05732862      0.07825296      -0.01131767
   0.0276118        0.09        -0.05723078      0.08569473      -0.02621408
   0.0000000        0.00        -0.04299469     -0.02393025       0.08311775


              MANOVA Tests for the Hypothesis of No Overall REWGRP Effect
                     H = Type III SSCP Matrix for REWGRP
                          E = Error SSCP Matrix

                        S=2      M=0      N=5.5

          Statistic                        Value      P-Value

          Wilks' Lambda                  0.02975533    <.0001
          Pillai's Trace                 0.99629295    0.0013
          Hotelling-Lawley Trace        31.73200433    <.0001
          Roy's Greatest Root           31.70439252    <.0001
```

Interpretation of pages 1 through 8 (the univariate ANOVA results for the three criterion variables) would proceed in exactly the same manner as described in Chapter 9 in the section "Steps in Interpreting the Output." Because these steps have already been described, they are not covered again in this chapter. You are encouraged, however, to review these results to make sure that everything looks right prior to interpreting the multivariate results. You should make sure that all participants were included in the analyses, that the input statements were written correctly, and so forth. Again, remember that these univariate ANOVAs and Tukey tests should be interpreted only if the overall MANOVA is significant.

Steps in Interpreting the Output

1. Review Wilks' Lambda, the Multivariate *F* Statistic, and Its Associated *p* Value

Once you have reviewed the univariate statistics on pages 1 through 8 to verify that there are no obvious errors in entering data or in writing the program, you can review your MANOVA results on page 9 of the output.

You can state the multivariate null hypothesis for the present study as follows:

> In the population, there is no difference across groups for participants in the high-reward group, the mixed-reward group, and the low-reward group when compared simultaneously on commitment, satisfaction, and intention to stay in the relationship.

In other words: in the population, the three groups are equal on all criterion variables. Symbolically, you can represent the null hypothesis this way:

$$\begin{pmatrix} M_{11} \\ M_{21} \\ M_{31} \end{pmatrix} = \begin{pmatrix} M_{12} \\ M_{22} \\ M_{32} \end{pmatrix} = \begin{pmatrix} M_{13} \\ M_{23} \\ M_{33} \end{pmatrix}$$

Each "M" represents a sample mean for one of the treatment conditions on one of the criterion variables. With the preceding notation, the first number in each subscript identifies the criterion variable, and the second number identifies the experimental group. For example, M_{31} refers to the sample mean for the third criterion variable (the intention to stay) in the first experimental group (the low-reward group), whereas M_{12} refers to the sample mean for the first criterion variable (commitment) in the second experimental group (the mixed-reward group).

Below the heading "Statistic," the first subheading you find is "Wilks' Lambda." Where the row headed "Wilks' Lambda" intersects with the column headed "Value," you see the computed value of Wilks' lambda for this analysis. For the current analysis, lambda is approximately .03. Remember that small values (closer to 0) indicate a relatively strong relationship between the predictor variable and the multiple criterion variables, and so this obtained value of .03 indicates that there is a very strong relationship between REWGRP and the criterion variables. This is an exceptionally low value for lambda, much lower than you generally encounter in actual research.

This value for Wilks' lambda indicates that the relationship between the predictor and the criteria is strong, but is it statistically significant? To determine this, see where the row headed "Wilks' Lambda" intersects with the column headed "P-Value." You can see that this significance value is very small (i.e., $p < .01$). Because this p value is less than the threshold value of .05, you can reject the null hypothesis of no differences between groups. In other words, you conclude that there is a significant multivariate effect for type of rewards.

2. Summarize the Results of Your Analyses

This section shows how to summarize just the results of the multivariate test discussed in the previous section. When the multivariate test is significant, you can proceed to the univariate ANOVAs and summarize them using the format presented in Chapter 9.

You can use the following statistical interpretation format to summarize the multivariate results from a MANOVA:

- A) Statement of the problem
- B) Nature of the variables
- C) Statistical test
- D) Null hypothesis (H_0)
- E) Alternative hypothesis (H_1)
- F) Obtained statistic
- G) Obtained probability (p) value
- H) Conclusion regarding the null hypothesis

You could summarize the preceding analysis in the following way.

A) Statement of the problem: The purpose of this study was to determine whether there was a difference across groups of participants in a low-reward relationship, mixed-reward relationship, and high-reward relationship with respect to their commitment to the relationship, their satisfaction with the relationship, and their intention to stay in the relationship.

B) Nature of the variables: This analysis involved one predictor variable and three criterion variables. The predictor variable was "type of rewards," which was measured on a nominal-scale and could assume three values: a low-reward condition (coded as 1); a mixed-reward condition (coded as 2); and a high-reward condition (coded as 3). The three criterion variables were "commitment," "satisfaction," and "intention to stay." All were measured on an interval/ratio scale.

C) Statistical test: Wilks' lambda, derived through a one-way MANOVA, between-subjects design.

D) Null hypothesis (H_0):

$$\begin{pmatrix} M_{11} \\ M_{21} \\ M_{31} \end{pmatrix} = \begin{pmatrix} M_{12} \\ M_{22} \\ M_{32} \end{pmatrix} = \begin{pmatrix} M_{13} \\ M_{23} \\ M_{33} \end{pmatrix}$$

In the population, there is no difference across groups for participants in the high-reward group, the mixed-reward group, and the low-reward group when compared simultaneously on commitment, satisfaction, and intention to stay in the relationship.

E) Alternative hypothesis (H_1): In the population, there is a difference between participants in at least two of the conditions when they are compared simultaneously on commitment, satisfaction, and intention to stay.

F) Obtained statistic: Wilks' lambda = .03.

G) Obtained probability (*p*) value: $p < .01$.

H) Conclusion regarding the null hypothesis: Reject the null hypothesis.

Formal Description of Results for a Paper

You could summarize this analysis in the following way for a scholarly journal:

```
    Results were analyzed using one-way MANOVA, between-
subjects design.  This analysis revealed a significant
multivariate effect for type of rewards, Wilks' lambda = .03,
and p < .01.
```

Example with Nonsignificant Differences between Experimental Conditions

Assume that you perform a MANOVA a second time. Here, the analysis is performed on the following dataset:

```
DATALINES;
1 16 19 18
1 18 15 17
1 18 14 14
1 16 20 10
1 15 13 17
1 12 15 11
2 16 20 13
2 18 14 16
2 13 10 14
2 17 13 19
2 14 18 15
2 19 16 18
3 20 18 16
3 18 15 19
3 13 14 17
3 12 16 15
3 16 17 18
3 14 19 15
;
RUN;
```

Results from the SAS Output

The preceding dataset was constructed to provide nonsignificant results. Analyzed with the program presented earlier, it again results in nine pages of output, with the same type of information appearing on each page. To conserve space, however, Output 11.2 presents only the results from the multivariate analyses that appear on page 9 of the output:

Output 11.2 Results of the Multivariate Analysis of Variance for the Investment Model Study, Nonsignificant Results

```
                                                                           9
                          The GLM Procedure
                    Multivariate Analysis of Variance

            Characteristic Roots and Vectors of: E Inverse * H, where
                    H = Type III SSCP Matrix for REWGRP
                         E = Error SSCP Matrix

   Characteristic             Characteristic Vector  V'EV=1
          Root      Percent        COMMIT          SATIS          STAY

     0.31962691       88.15     -0.07731564     0.06272924     0.11858737
     0.04297906       11.85     -0.03447851     0.06784671    -0.03600286
     0.00000000        0.00      0.08317305     0.03925020     0.00373811


           MANOVA Tests for the Hypothesis of No Overall REWGRP Effect
                    H = Type III SSCP Matrix for REWGRP
                         E = Error SSCP Matrix

                      S=2      M=0      N=5.5

           Statistic                      Value     P-Value

           Wilks' Lambda               0.72656295    0.6147
           Pillai's Trace              0.28341803    0.6129
           Hotelling-Lawley Trace      0.36260596    0.6123
           Roy's Greatest Root         0.31962691    0.5796
```

To the right of the heading "Wilks' Lambda" in Output 11.2, you can see that this computation provided a lambda value of approximately .73. Because this is a relatively large number (i.e., closer to 1), it indicates that the relationship between type of rewards and the three criterion variables is substantially weaker than with the previous dataset.

Output 11.2 shows that this analysis was nonsignificant ($p = .61$). In short, the results of this analysis do not allow you to reject the null hypothesis of no group differences in the population. Because Wilks' Lambda is nonsignificant, your analysis would terminate at this point in most cases. You would not go on to interpret the univariate ANOVAs.

Summarizing the Results of the Analysis

Because this analysis tested the same null hypothesis that you tested earlier, you would prepare items A through E in the same manner described before. Therefore, this section presents only items F through H of the statistical interpretation format:

F) Obtained statistic: Wilks' lambda = .73.

G) Obtained probability (*p*) value: $p = .61$.

H) Conclusion regarding the null hypothesis: Fail to reject the null hypothesis.

Formal Description of Results for a Paper

You could summarize the results of this analysis for a scholarly paper in the following way:

```
   Results were analyzed using one-way MANOVA, between-
subjects design.  This analysis failed to reveal a significant
multivariate effect for type of rewards, Wilks' lambda = .73,
ns.
```

Conclusion

Chapters 9, 10, and 11 have focused on **between-group** research designs: designs in which each participant is assigned to just one treatment condition and provides data under that condition only.

In some situations, however, it is advantageous to conduct studies in which each participant provides data under *every* treatment condition. This is called a **repeated-measures** design and requires data analysis techniques that differ from the ones presented in the previous three chapters. The next chapter introduces procedures for analyzing data from the simplest type of repeated-measures design.

Assumptions Underlying Multivariate ANOVA with One Between-Subjects Factor

- **Level of measurement.** Each criterion variable should be assessed on an interval or ratio level of measurement. The predictor variable should be a nominal-level variable (i.e., a categorical variable).

- **Independent observations.** Across participants, a given observation should not be dependent on any other observation in any group. It is acceptable for the various criterion variables to be correlated with one another; however, a given participant's score on any criterion variable should not be affected by any other participant's score on any criterion variable. (For a more detailed explanation of this assumption, see Chapter 8, "*t* Tests: Independent Samples and Paired Samples.")

- **Random sampling.** Scores on the criterion variables should represent a random sample drawn from the populations of interest.

- **Multivariate normality.** In each group, scores on the various criterion variables should follow a multivariate normal distribution. Under conditions normally encountered in social science research, violations of this assumption have a small effect on the Type I error rate (i.e., the probability of incorrectly rejecting a true null hypothesis). On the other hand, when the data are platykurtic (form a relatively flat distribution), the power of the test might be significantly attenuated. (The power of the test is the probability of correctly rejecting a false null hypothesis.) Platykurtic distributions can be transformed to better approximate normality (see Stevens, 2002; or Tabachnick & Fidell, 2001).

- **Homogeneity of covariance matrices.** In the population, the criterion-variable covariance matrix for a given group should equal the covariance matrix for each remaining group. This is the multivariate extension of the "homogeneity of variance" assumption in univariate ANOVA. To illustrate, consider a simple example with just two groups and three criterion variables (V1, V2, and V3). To satisfy the homogeneity assumptions, the variance of V1 in group 1 must equal the variance of V1 in group 2. The same must be true for the variances of V2 and V3. In addition, the covariance between V1 and V2 in group 1 must equal the covariance between V1 and V2 in group 2. The same must be true for the remaining covariances (between V1 and V3 and between V2 and V3). It becomes evident that the number of corresponding elements that must be equal increases dramatically as the number of groups increases and/or as the number of criterion variables increases. For this reason, the homogeneity of covariance assumption is rarely satisfied in real-world research. Fortunately, the Type I error rate associated with MANOVA is relatively robust against typical violations of this assumption so long as the sample sizes are equal. The power of the test, however, tends to be attenuated when the homogeneity assumption is violated.

References

Stevens, J. (2002). *Applied multivariate statistics for the social sciences* (4th ed.). Mahwah, NJ: Lawrence Erlbaum Associates.

Tabachnick, B. G., & Fidell, L. S. (2001). *Using multivariate statistics* (4th ed.). New York: Harper Collins.

One-Way ANOVA with One Repeated-Measures Factor

> **Overview.** This chapter shows how to enter data and prepare SAS programs to perform one-way repeated measures ANOVA using the GLM procedure with the REPEATED statement. This chapter focuses on repeated-measures designs in which each participant is exposed to every condition under the independent variable. This chapter describes the necessary conditions for performing a valid repeated-measures ANOVA, discusses alternative analyses to use when the validity conditions are not met, and reviews strategies for minimizing sequence effects.

Introduction: What Is a Repeated-Measures Design?

A one-way repeated-measures ANOVA is appropriate when:

- the analysis involves a single predictor variable measured on a nominal scale;
- the analysis also involves a single criterion variable measured on an interval- or ratio-scale;
- each participant is exposed to each condition under the independent variable.

The **repeated-measures design** derives its name from the fact that each participant provides *repeated scores* on the criterion variable. Each participant is exposed to every treatment condition under the study's independent variable and provides scores on the criterion under each of these conditions. Perhaps the easiest way to understand the repeated-measures design is to contrast it with the **between-subjects** design in which each participant participates in only one treatment condition.

For example, Chapter 9, "One-Way ANOVA with One Between-Subjects Factor," presents a simple experiment that uses a between-subjects design. In that fictitious study, participants were randomly assigned to one of three experimental conditions. In each condition, they read descriptions of a number of fictitious romantic partners and rated their likely commitment to each partner. The purpose of that study was to determine whether the "level of rewards" associated with a given partner affected the participant's rated commitment to that partner. Therefore, the level of rewards was manipulated by varying the description of one specific partner (partner 10) that was presented to the three groups. The level of rewards was manipulated in this way:

- Participants in the low-reward condition read that partner 10 provides few rewards in the relationship.
- Participants in the mixed-reward condition read that partner 10 provides mixed rewards in the relationship.
- Participants in the high-reward condition read that partner 10 provides many rewards in the relationship.

After reading this description, each participant rated how committed he or she would probably be to partner 10.

This study was called a between-subjects study because the participants were divided into different treatment groups, and the independent variable was manipulated *between* these groups. In a between-subjects design, each participant is exposed to only one level of the independent variable. (In the present case, that means that a given participant read either the low-reward description, the mixed-reward description, or the high-reward description, but no participant read more than one of these descriptions of partner 10.)

In a repeated-measures design on the other hand, each participant is exposed to every level of the independent variable and provides scores on the dependent variable under each of these levels. For example, you could easily modify the preceding study so that it becomes a one-factor repeated-measures design. Imagine that you conduct your study with a single group of 20 participants rather than with three treatment groups. You ask each participant to go through a list of potential romantic partners and rate his or her commitment to each partner. Imagine further that all three versions of partner 10 appear somewhere in this list, and that a given participant responds to each of these versions individually.

For example, a given participant might be working her way through the list and the third potential partner she comes to happens to be the low-reward version of partner 10 (assume that you have renamed this fictitious partner to be "partner 3"). She rates her commitment to this partner and moves on to the next partner. Later, the 11th partner she comes to happens to be the mixed-reward version of partner 10 (now renamed "partner 11"). She rates her commitment and moves on. Finally, the 19th partner she comes to happens to be the high-reward version of partner 10 (now renamed as "partner 19"). She rates her commitment and moves on.

Your study now uses a repeated-measures design because each participant has been exposed to all three levels of the independent variable. To analyze these data, you would create three SAS variables to include the commitment ratings made under the three different conditions:

- One variable (perhaps named LOW) contains the commitment ratings made for the low-reward version of the fictitious partner.

- One variable (perhaps named MIXED) contains the commitment ratings made for the mixed-reward version of the fictitious partner.

- One variable (perhaps named HIGH) contains the commitment ratings made for the high-reward version of the fictitious partner.

To analyze your data, you would compare scores for each of the three variables. Perhaps you would hypothesize that the commitment score contained in HIGH would be significantly higher than the commitment scores contained in LOW or MIXED.

Make note of two cautions before moving on. First, remember that you need a special type of statistical procedure to analyze data from a repeated-measures study such as this. You should not analyze these data using the program for a one-way ANOVA with one between-subjects factor as was illustrated in Chapter 9. This chapter shows you the appropriate SAS program for analyzing repeated-measures data.

Second, the fictitious study described here was used merely to illustrate the nature of a repeated-measures research design; do not view it as an example of a *good* repeated-measures research design. (In fact, the preceding study suffers from several serious problems.) This is because repeated-measures studies are vulnerable to a number of problems that are not encountered with between-subjects designs. This means that you must be particularly concerned about design when your study includes a repeated-measures factor. Some of the problems associated with this design are discussed later in the section "Sequence Effects."

Example: Significant Differences in Investment Size across Time

To demonstrate the use of the repeated measures ANOVA, this chapter uses a new fictitious experiment that examines a different aspect of the investment model (Rusbult, 1980). Remember from earlier chapters that the investment model is a theory of interpersonal attraction that describes the variables that determine commitment to romantic relationships and other interpersonal associations.

Designing the Study

Some of the earlier chapters have described fictitious investigations of the investment model that involved the use of fictitious partners: written descriptions of potential romantic partners that the participants responded to as if they were real people. Assume that critics of the previous studies are very skeptical about the use of this analogue methodology and contend that investigations using fictitious partners do not generalize to how real individuals actually behave in the real world. To address these criticisms, this chapter presents a different study that could be used to evaluate aspects of the investment model while using actual couples.

This study focuses on the investment size construct from the investment model. **Investment size** refers to the amount of time, effort, and personal resources that an individual has put into his or her relationship with a romantic partner. People report heavy investments in a relationship when they have spent a good deal of time with their romantic partner, when they have a lot of shared activities or friends that they would lose if the relationship were to end, and so forth.

Assume that it is generally desirable for couples to believe that they have invested a good deal of time and effort in their relationships. Assume that this is desirable because research has shown that couples are more likely to stay together when they feel they have invested a great deal in their relationships.

The "Marriage Encounter" Intervention

Given that higher levels of perceived investment size are generally a good thing, assume that you are interested in finding interventions that are likely to increase perceived investments in a marriage. Specifically, you have read research indicating that a program called "marriage encounter" is likely to increase perceived investments. In a marriage encounter program, couples spend a weekend together under the guidance of counselors, sharing their feelings, learning to communicate, and engaging in exercises intended to strengthen their relationship.

Based on what you have read, you hypothesize that couples' perceived investments in their relationships will increase immediately after participation in a marriage encounter program. In other words, you hypothesize that, if couples are asked to rate how much they have invested in a relationship both before and immediately after the marriage encounter weekend, the "post" ratings will be significantly higher than the "pre" ratings. This is the *primary hypothesis* for your study.

However, being something of a skeptic, assume further that you do not expect these increased investment perceptions to endure. Specifically, you believe that if couples rate their perceived investments at a "follow-up" point two weeks after the marriage encounter weekend, these ratings will have declined to their initial "pre" levels that were observed just before the weekend. In other words, you hypothesize that there will not be a significant difference between investment ratings obtained before the weekend and those obtained three weeks after. This is the *secondary hypothesis* for your study.

To test these hypotheses, you conduct a study that uses a single-group experimental design with repeated measures. The design of this study is illustrated in Figure 12.1:

Figure 12.1 Single-Group Experimental Design with Repeated Measures

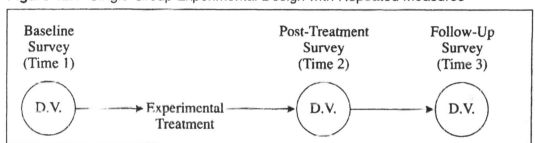

Specifically, you recruit 20 couples, all of whom are about to go through a marriage encounter program. The criterion variable in this study is *perceived investment*, and you measure this variable with a multiple-item questionnaire. Higher scores on this scale reflect higher levels of perceived investment.

You obtain investment ratings from each couple at three points in time. These points are illustrated by the three circles in Figure 12.1. (The "D.V." in each circle stands for "Dependent Variable.") Specifically:

- A "baseline survey" obtains investment scores at Time 1, just before the marriage encounter weekend.

- A "post-treatment survey" obtains investment scores at Time 2, immediately after the encounter weekend.

- A "follow-up survey" obtains investment scores at Time 3, three weeks after the encounter weekend.

Notice that, in Figure 12.1, an "Experimental Treatment" appears between Time 1 and Time 2. This treatment is the marriage encounter program.

Problems with Single-Group Studies

Before proceeding, be warned that the study described here uses a relatively weak research design. To understand why, remember the main hypothesis that you would like to test: the hypothesis that investment scores would increase significantly from Time 1 to Time 2 because of the marriage encounter program. Imagine for a moment that you obtain just these results when you analyze your data, that Time 2 investment scores are significantly higher than Time 1 scores. Does this provide strong evidence that the marriage encounter manipulation *caused* the increase in investment scores?

Not really. It would be very easy for someone to provide alternative explanations for the increase. It could be argued that investment scores naturally increase over time among married couples regardless of participation in a marriage encounter weekend. Perhaps an increase in scores was due to some television program that most of the couples saw and was not due to the marriage encounter. There is a long list of alternative explanations that could potentially be offered.

The point is simply that repeated-measures studies must be designed very carefully to avoid confounds and other problems, and this study was not designed very carefully. It is used here merely for illustration. A later section in this chapter ("Sequence Effects") discusses some of the problems associated with repeated-measures studies and reviews strategies for dealing with them. In addition, Chapter 13, "Factorial ANOVA with Repeated-Measures Factors and Between-Subjects Factors," shows how you can make the present single-group design much stronger through the addition of a control group.

Predicted Results

Remember that your primary hypothesis is that couples' ratings of investment in their relationships increases immediately following the encounter weekend. The secondary hypothesis is that this increase is a transient, or temporary, effect rather than a permanent change. These hypothesized results appear graphically in Figure 12.2. Notice that, in the figure, mean investment scores increase from Time 1 to Time 2, and then decrease again at Time 3.

Figure 12.2 Hypothesized Results for the Investment Model Study

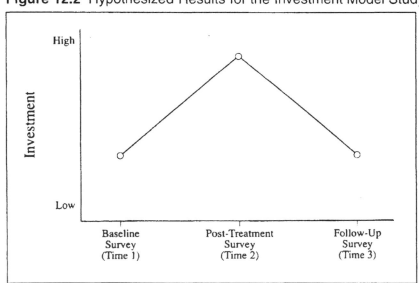

You test the primary hypothesis by comparing the investment scores obtained at post-treatment (Time 2) to the baseline scores (Time 1). By also comparing the follow-up scores (Time 3) to the baseline scores (Time 1), you learn whether any change observed at post-treatment is maintained. You can use this comparison to test the secondary hypothesis.

Assume that this study is exploratory because studies of this type have not previously been undertaken and you therefore are uncertain of what methodological problems might be encountered or the magnitude of possible changes. This example might serve as a pilot study to assist in the design of a more definitive study and the determination of an appropriate sample size. (A follow-up study based on the pilot data from this project is presented in Chapter 13.)

Writing the SAS Program

Twenty couples participated in this study. To keep things simple, imagine that only the investment ratings made by the wife in each couple were actually analyzed. This means that your sample actually consists of data from the 20 women who participated in the marriage encounter program.[1]

The criterion variable in this study is the size of the investments that participants (the wives) believe they have made in their relationships. Investment size was measured by a scale. Assume that these scores are measured on an *interval scale* (see Chapter 1, "Basic Concepts in Research and Data Analysis").

Creating the SAS Dataset

Enter the investment scores obtained at the three different points in time as three separate variables in the SAS dataset. You create a new variable named PRE to include investment scores obtained at Time 1, a variable named POST to include investment scores obtained at Time 2, and a variable named FOLLOWUP to include investment scores obtained at Time 3. A variable name ID (for "identification number") denotes the participant number. The following program could be used to create a dataset named REP that contains data from the fictitious study:

```
1        DATA REP;
2            INPUT #1 @1    ID          2.
3                     @5    PRE         2.
4                     @9    POST        2.
5                     @13   FOLLOWUP    2. ;
6        DATALINES;
7        01   08   10   10
8        02   10   13   12
9        03   07   10   12
10       04   06   09   10
11       05   07   08   09
12       06   11   15   14
13       07   08   10   09
14       08   05   08   08
```

[1] One assumption of this statistical procedure is independence of observations; thus, only one spouse per couple can be considered with use of one-way ANOVA with one repeated-measures factor (Kashy & Snyder, 1995). See "Assumptions Underlying the One-Way ANOVA with One Repeated-Measures Factor" at the end of this chapter.

```
15      09   12   11   12
16      10   09   12   12
17      11   10   14   13
18      12   07   12   11
19      13   08   08   09
20      14   13   14   14
21      15   11   11   12
22      16   07   08   07
23      17   09   08   10
24      18   08   13   14
25      19   10   12   12
26      20   06   09   10
27      ;
28      RUN;
```

The actual data from the study appear in lines 7 through 26 of the preceding program. The first column of data (in columns 1 and 2) includes each participant's identification number. The next column of data (in columns 5 and 6) includes each participant's score on PRE (the investment ratings observed at Time 1). The next column of data (in columns 9 and 10) includes each participant's score on POST (the investment ratings observed at Time 2). Finally, the last column of data (in columns 13 and 14) includes each participant's score on FOLLOWUP (investment ratings from Time 3).

Obtaining Descriptive Statistics with PROC MEANS

After inputting the dataset, you should perform a PROC MEANS to obtain descriptive statistics for the three investment variables. This serves two important purposes. First, scanning the sample size, minimum value, and maximum value for each variable provides an opportunity to check for obvious data entry errors. Secondly, you need the means and standard deviations for the variables to interpret significant differences found in the ANOVA. Performing this separate PROC MEANS is necessary because the means for within-subjects variables are not routinely included in the output of the PROC GLM program that performs the repeated-measures ANOVA.

NOTE

Here are the lines that you can add to the preceding program to obtain simple descriptive statistics for the study's variables:

```
29      PROC MEANS   DATA=REP;
30      RUN;
```

The output from this MEANS procedure appears in Output 12.1. After you review the columns headed "N," "Minimum," and "Maximum" to verify that there are no obvious errors in data entry, you should review the average investment model scores that appear under the heading "Mean." Where the row headed "PRE" intersects with the column headed "Mean," you can see that the mean investment score observed at Time 1 was 8.60. In the row headed "POST," you can see that this average investment score had increased to 10.75 by Time 2. Finally, the row headed "FOLLOWUP" shows that the mean score had increased to 11.00 by Time 3. Figure 12.3 plots these means graphically:

Output 12.1 Results of PROC MEANS, Investment Model Study

Variable	N	Mean	Std Dev	Minimum	Maximum
ID	20	10.5000000	5.9160798	1.0000000	20.0000000
PRE	20	8.6000000	2.1373865	5.0000000	13.0000000
POST	20	10.7500000	2.2912878	8.0000000	15.0000000
FOLLOWUP	20	11.0000000	2.0261449	7.0000000	14.0000000

Figure 12.3 Actual Results from the Investment Model Study

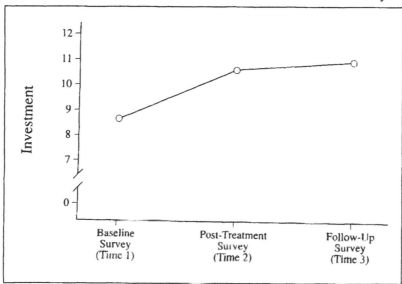

A review of the means presented in Figure 12.3 suggests that you might receive support for your study's primary hypothesis, but apparently will not receive support for the study's secondary hypothesis. Notice that mean investment scores seem to increase from the baseline survey (Time 1) to the post-treatment survey (Time 2). If subsequent analyses show that this increase is statistically significant, this finding would be consistent with your primary hypothesis that perceived investment would increase immediately following the marriage encounter weekend.

However, notice that the mean investment scores remain at a relatively high level at the follow-up survey, three weeks following the program (at Time 3). This trend is inconsistent with your secondary hypothesis that the increase in perceived investment would be short-lived.

At this point, of course, you are only "eyeballing" the data, and as yet it is not clear whether any of the differences that appear in Figure 12.3 are statistically significant. To determine this, you must analyze the data using a repeated-measures ANOVA. The next section shows how to do this.

Testing for Significant Effects with PROC GLM

The general form for the SAS program to perform a one-way repeated-measures ANOVA is as follows:

```
PROC GLM  DATA=filename;
   MODEL  trial1  trial2  trial3... trialn =  / NOUNI;
   REPEATED  trial-variable-name  #levels  CONTRAST (level#) /
SUMMARY;
RUN;
```

The actual SAS program that you need to analyze the dataset above is as follows:

```
1       PROC GLM  DATA=REP;
2          MODEL PRE POST FOLLOWUP =  / NOUNI;
3          REPEATED TIME 3 CONTRAST (1) / SUMMARY;
4       RUN;
```

Notes regarding the SAS Program

The repeated-measures ANOVA is similar to the between-subjects ANOVA in that both procedures can be performed with PROC GLM. However, notice that you write the MODEL statement differently in a repeated-measures analysis. In the repeated-measures ANOVA, the names of the variables that contain scores on the criterion variable appear to the left of the equals sign in the MODEL statement. In the general form of the program provided here, these variables are represented as "trial1 trial2 trial3... trial*n*". In the present study, the variable PRE contains investment scores obtained at Time 1, POST contains investment scores obtained at Time 2, and FOLLOWUP contains investment scores obtained at Time 3. Therefore, PRE, POST, and FOLLOWUP appear to the left of the equals sign in the MODEL statement.

It should be clear that you need a different criterion variable to include scores obtained under each level of the independent variable. This means that the number of variables appearing to the left of the equal sign in the MODEL statement should equal the number of levels under your independent variable. The present study had three levels under the repeated-measures independent variable (Time 1, Time 2, and Time 3), so three variables appear to the left of the equal sign (PRE, POST, and FOLLOWUP).

In a between-subjects study, the name of the predictor variable would normally appear to the right of the equals sign in the MODEL statement. Since there is no between-subjects factor in the present study, no variable name appears to the right of the equals sign in the preceding program. Yet, the MODEL statement does include a slash (which signals that options are being requested) followed by the NOUNI option. The NOUNI option suppresses the printing of output relevant to certain univariate analyses that are of no interest with this design. Without the NOUNI option, the SAS program computes and prints a univariate *F* test for each of the three levels of the trial variable. These tests are not of interest when analyzing within-subjects effects.

The REPEATED statement appears in line 3 of the preceding program. The general form of the program indicates that the first entry in the REPEATED statement should be the "trial-variable-name." This is a name that you supply to refer to your repeated-measures factor. In the present study, the three levels of your repeated-measures factor were "Time 1," "Time 2," and "Time 3." It therefore makes sense to give this repeated-measures factor the name TIME. (Notice where the name TIME appears on line 3 of the preceding program.) Obviously, you could have used any variable name of your choosing, provided that it complies with the usual naming rules for SAS variables.

The general form of the program shows that, to the right of the trial variable name, you are to provide the "#levels." This means that you should specify a number that represents the number of levels that appear under your repeated-measures factor. In the present study, the repeated-measures factor included three levels (Time 1, Time 2, and Time 3) and so the number "3" was entered next to TIME in the REPEATED statement.

The next entry in the REPEATED statement is the "CONTRAST (level#)" option. In a repeated-measures analysis, **contrasts** are planned comparisons between different levels of the repeated-measures variable. The CONTRAST option allows you to choose the types of contrasts that will be made. The number that you specify in the place of "level#" identifies the specific level of the repeated-measures factor against which the other levels will be compared. The preceding program specifies "CONTRAST (1)." A "1" appears in the parentheses with this option and this means that level 1 under the repeated-measures factor will be contrasted with level 2 and with level 3. In concrete terms, this requests that the mean investment scores obtained at Time 1 will be contrasted with those obtained at Time 2, and with those obtained at Time 3. Unless otherwise instructed, PROC GLM automatically performs tests that contrast the last (*n*th) level of the repeated-measures variable with each of the preceding levels.

NOTE

The preceding CONTRAST command requests that both the post-treatment (Time 2) and follow-up scores (Time 3) be compared to the baseline score (Time 1). These are the appropriate contrasts for this analysis because these contrasts directly evaluate the two hypotheses of this study. However, remember that you should interpret these contrast tests only if the analysis of variance shows that the effect for TIME (the repeated-measures factor) is significant.

In addition to the planned contrasts previously described, there are a number of post hoc multiple-comparison tests that are available using SAS/STAT software such as the Tukey test and the Scheffe test. Many of these tests control for the experiment-wise probability of making a Type I error (i.e., incorrectly rejecting a correct null hypothesis). In Chapter 9, you learned how to use these tests to determine which pairs of groups are significantly different on the criterion variable. Technically, these tests can also be used in a repeated-measures ANOVA. However, these multiple comparison procedures cannot be applied to the repeated-measures variable when you use the REPEATED statement. Use of these procedures is limited to variables that appear in the MODEL statement. If using one of these tests is essential, then you can run the repeated-measures ANOVA according to another method in which you do not use the REPEATED statement. See the section "Use of Other Post Hoc Tests with the Repeated-Measures Variable" in Chapter 13 for a description of this method.

Finally, the REPEATED statement ends with a slash and the SUMMARY option. This SUMMARY option requests the statistics from the contrasts to appear in the output.

In summary, the complete program (minus the DATA step) that provides means for the criterion variable and performs the repeated-measures ANOVA is as follows:

```
1    PROC MEANS  DATA=REP;
2    RUN;
3
4    PROC GLM  DATA=REP;
5       MODEL PRE POST FOLLOWUP =  / NOUNI;
6       REPEATED TIME 3 CONTRAST (1) / SUMMARY;
7    RUN;
```

Results from the SAS Output

With LINESIZE=80 and PAGESIZE=60 in the OPTIONS statement (first line), the preceding program (including both PROC MEANS and PROC GLM) produces five pages of output. The information that appears on each page is summarized here:

- Page 1 (not shown) provides means and other descriptive statistics for the criterion variables (reproduced earlier as Output 12.1).

- Page 2 displays the number of observations included in the analyses performed by PROC GLM.

- Page 3 provides the multivariate significance test for the repeated-measures factor (TIME, in this case).

- Page 4 provides the univariate significance test for TIME along with information related to the error term used in this *F* test, and two estimates of the epsilon statistic (discussed later).

- Page 5 provides the results of the planned comparisons requested with the CONTRAST option.

Output 12.2 provides the results of the GLM procedure requested by the preceding program. (The results of PROC MEANS appeared earlier as Output 12.1 and are not reproduced here.)

Output 12.2 Results of PROC GLM, Investment Model Study

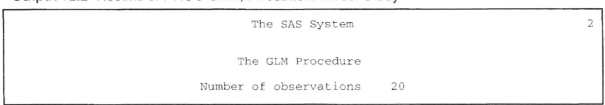

```
                      The SAS System                          2


                     The GLM Procedure

         Number of observations    20
```

(continued on the next page)

Output 12.2 *(continued)*

```
                                                                        3
                            The GLM Procedure
                  Repeated Measures Analysis of Variance

                 Repeated Measures Level Information

         Dependent Variable        PRE     POST FOLLOWUP

               Level of TIME         1       2       3


MANOVA Test Criteria and Exact F Statistics for the Hypothesis of no TIME Effect
               H = Type III SSCP Matrix for TIME
                   E = Error SSCP Matrix

                    S=1     M=0     N=8

Statistic                     Value    F Value   Num DF   Den DF   Pr > F

Wilks' Lambda              0.30600563    20.41       2        18    <.0001
Pillai's Trace            0.69399437    20.41       2        18    <.0001
Hotelling-Lawley Trace    2.26791373    20.41       2        18    <.0001

Roy's Greatest Root       2.26791373    20.41       2        18    <.0001
```

```
                                                                        4

                            The GLM Procedure
                  Repeated Measures Analysis of Variance
               Univariate Tests of Hypotheses for Within Subject Effects

Source                 DF     Type III SS     Mean Square    F Value    Pr > F

TIME                    2     69.63333333     34.81666667      30.28    <.0001
Error(TIME)            38     43.70000000      1.15000000

                                        Adj Pr > F
                    Source          G - G       H - F

                    TIME           <.0001      <.0001
                    Error(TIME)


                    Greenhouse-Geisser Epsilon      0.7593
                    Huynh-Feldt Epsilon             0.8115
```

(continued on the next page)

Output 12.2 *(continued)*

```
                                                                          5
                          The GLM Procedure
                  Repeated Measures Analysis of Variance
                  Analysis of Variance of Contrast Variables

TIME_N represents the contrast between the nth level of TIME and the 1st
Contrast Variable: TIME_2


  Source                    DF     Type III SS    Mean Square   F Value   Pr > F

  Mean                       1     92.45000000    92.45000000    29.01    <.0001
  Error                     19     60.55000000     3.18684211


Contrast Variable: TIME_3


  Source                    DF     Type III SS    Mean Square   F Value   Pr > F

  Mean                       1     115.2000000    115.2000000    43.09    <.0001
  Error                     19      50.8000000      2.6736842
```

Steps in Interpreting the Output

1. Make Sure That Everything Looks Right

First, check the number of observations listed on page 2 of the output to make certain that data from all participants were included in the analysis. If any data are missing, then all data for that participant will automatically be dropped from the analysis. Next, check the number of levels assigned to your criterion variable on page 3 of the output. (The criterion variable is called "Dependent Variable" in the output.)

2. Review the Appropriate *F* Statistic and Its Associated Probability Value

The first step in actually interpreting the results of the analysis is to review the *F* value and associated statistics on page 4 ("Univariate Tests of Hypotheses for Within Subject Effects") of the printout. The *F* value for the trial variable, as it was named in the REPEATED statement in the SAS program, appears here. This *F* value tells you whether to reject the null hypothesis. In the study, the null hypothesis is that, in the population, there is no difference in investment scores obtained at the three points in time. This hypothesis can be represented symbolically in this way:

$$H_0: \quad T1 = T2 = T3$$

where T1 is the mean investment score obtained at baseline, T2 is the mean score following a marriage encounter weekend, and T3 is the mean score two weeks after the weekend.

To find the *F* ratio that tests this null hypothesis, look on page 4 of Output 12.2 in the section headed "Source: TIME." Under the heading "F Value," you can see that the *F* statistic for the TIME effect is 30.28. There are 2 and 38 degrees of freedom for this *F* test. (Find these degrees of freedom on the left side of the output page under the heading "DF.")

Notice that there are actually three p values associated with this F test. The unadjusted p value appears under the heading "Pr > F." Two adjusted p values appear under the heading "Adj Pr > F." The use of these adjusted p values is explained in "Modified Univariate Tests," later in this chapter.

For the sake of simplicity, consider only the unadjusted p value at this point. You can see that this unadjusted p value (under "Pr > F") is very small at less than .01. Because this p value is less than .05, you can reject your null hypothesis of no differences in mean levels of commitment in the population. In other words, you conclude that there is an effect for your repeated-measures variable, TIME.

You might have noticed that page 3 of Output 12.2 provides a "MANOVA" test for the TIME effect. For the moment, ignore these results and focus only on the univariate test for the TIME effect. The multivariate (MANOVA) approach to conducting a repeated-measures analysis is discussed later in the section "Further Notes on Repeated Measures Analyses."

3. Prepare Your Own Version of the ANOVA Summary Table

As with other ANOVAs, the next step is to formulate the summary table. This is somewhat easier with this analysis as most of the values that you need for the table appear on page 4 of Output 12.2. Table 12.1 illustrates the ANOVA summary table for the analysis just completed:

Table 12.1

ANOVA Summary Table for Study of Investment Using a Repeated Measures Design

Source	df	SS	MS	F
Between Subjects	19	--	--	
Within Subjects	40	113.3		
Treatment	2	69.6	34.82	30.28 *
Residual	38	43.7	1.15	
TOTAL	59			

Note: $N = 20$.

* $p < .01$

4. Review the Results of the Contrasts

The significant effect for TIME revealed on page 4 of the output simply tells you that investment scores obtained at one point in time are significantly different from scores obtained at some other point in time. You still do not know which scores significantly differ. To determine this, you must consult the planned comparisons requested by the CONTRAST option. These appear on page 5 of Output 12.2.

Remember that you specified (in the SAS program with the CONTRAST option) that the mean from trial 1 (baseline) would be compared to each of the other two trials. Results of the analysis comparing trial 1 to trial 2 appear on the upper part of output page 5, below the heading "Contrast Variable: TIME_2." It is this comparison that directly evaluates your primary hypothesis that investment increases significantly at post-treatment (Time 2) compared to baseline (Time 1). The F ratio for this contrast appears where the "MEAN" row intersects the column headed "F Value." You can see that the obtained F value is 29.01 which, with 1 and 19 degrees of freedom, is significant at $p <$ .01. This tells you that there was, in fact, a significant increase in investment scores immediately following the marriage encounter program. This is what you might have expected, after you reviewed the mean scores plotted earlier in Figure 12.3. This finding supports your primary hypothesis that investment scores would display a significant increase from Time 1 to Time 2.

The other contrast appears on the bottom half of page 5 of Output 12.2, beneath the heading "Contrast Variable: TIME_3." This contrast compares baseline scores to the follow-up scores (taken three weeks after the program). The F value for this contrast is also statistically significant ($F[1, 19] = 43.09$, $p < .01$), indicating that the two treatment values are different. Inspection of the means in Figure 12.3 shows that the follow-up mean is greater than the baseline mean. Your second hypothesis, that the increase in investment would not be sustained, is therefore not supported. The increase in perceived investment size is maintained two weeks after the treatment.

Summarizing the Results of the Analysis

The standard statistical interpretation format used for between-subjects ANOVA (as described in Chapter 9) is also appropriate for the repeated-measures design. The outline of this format appears again here:

A) Statement of the problem
B) Nature of the variables
C) Statistical test
D) Null hypothesis (H_0)
E) Alternative hypothesis (H_1)
F) Obtained statistic
G) Obtained probability (p) value
H) Conclusion regarding the null hypothesis
I) ANOVA summary table
J) Figure representing the results

Since most sections of this format appeared in Chapter 9 or in this chapter (e.g., the ANOVA summary table), they are not repeated here; instead, the formal description of the results follows.

Formal Description of Results for a Paper

Mean investment size scores across the three trials are displayed in Figure 12.3. Results were analyzed using one-way analysis of variance (ANOVA), repeated-measures design revealing a significant effect for the treatment, $F(2,38) = 30.28$, $p < .01$. Contrasts showed that the baseline measure was significantly lower than the post-treatment trial, $F(1,19) = 29.01$; $p < .01$, and the follow-up trial, $F(1,19) = 43.09$, $p < .01$.

Further Notes on Repeated-Measures Analyses

Advantages of the Repeated-Measures Design versus the Between-Subjects Design

An alternative to the repeated-measures design is a between-subjects design, as described in Chapter 9. For example, you could have followed a between-subjects design in which two groups would be measured at only one point in time: immediately following the weekend program (at Time 2, from Figure 12.3). In this between-subjects study, one group of couples would attend the weekend encounter, and the other group would not attend. If you conducted the study well, you could then attribute any differences in the group means to the weekend experience.

With both the repeated-measures design and the between-subjects design, the sums of squares and mean squares are computed in similar ways. However, an advantage of the repeated-measures design is that each participant serves as his or her own control. Since each participant serves in each treatment condition, variability in scores due to individual differences between participants will not be a factor in determining the size of the treatment effect. Between-subjects variability is removed from the error term in computing the F test (see Table 12.1). This computation usually allows for a more sensitive test of treatment effects because the between-subjects variance is typically much larger than the within-subjects variance. This is true because multiple observations from the same participant tend to be positively correlated, even when you obtain measures across time.

An additional advantage with repeated measures is increased efficiency since this design requires only half the number of participants that you would need in a between-subjects design. This might be an important consideration when the targeted study population is limited. These statistical and practical differences illustrate the importance of carefully planning your analyses when designing an experiment.

Weaknesses of the One-Way Repeated Measures Design

The primary limitation of the present design is the lack of a control group. Since all participants receive the treatment in this design, there is no comparison that you can make to evaluate if any observed changes are truly the result of the experimental manipulation. In the study just described, for example, increases in investment scores might occur because of time spent together and have nothing to do with the specific program activities during the weekend (i.e., the treatment). Chapter 13, "Factorial

ANOVA with Repeated-Measures Factors and Between-Subjects Factors," shows how you can remedy this weakness with the addition of an appropriate control group.

Another potential problem with this type of design is that participants might be affected by a treatment in a way that changes responses to subsequent measures. This problem is called a **sequence effect**.

Sequence Effects

An important consideration in a repeated measures design is the potential for certain experimental confounds. In particular, the experimenter must control for order effects and carry-over effects.

Order Effects

Order effects result when the ordinal position of the treatments biases participant responses. Consider this example. An experimenter studying perception requires participants to perform a reaction-time task in each of three conditions. Participants must depress a button on a response pad while waiting for a signal, and after receiving the signal they must depress a different button. The dependent variable is reaction time and the independent variable is the type of signal. The independent variable has three levels: in condition 1, the signal is a flash of light; in condition 2, it is an audio tone; in condition 3, it is both the light and tone simultaneously. Each test session consists of 50 trials. A mean reaction time is computed for each session. Assume that these conditions are presented in the same order for all participants (e.g., in the morning, before lunch, and after lunch).

The problem with this research design is that reaction-time scores might be adversely affected after lunch by fatigue. Responses to condition 3 might be more a measure of fatigue than a true treatment effect. For example, suppose the experiment yields mean scores for 10 participants as shown in Table 12.2. It appears that presentation of both signals (tone and light) causes a delayed reaction time compared to the other two treatments (tone alone or light alone).

Table 12.2

Mean Reaction Time (msecs) for All Treatment Conditions

Type of Signal		
Tone	Light	Tone and Light
650	650	1125

An alternative explanation for the preceding results is that a fatigue effect in the early afternoon is causing the longer reaction times. According to this interpretation, you would expect each of the three treatments to have longer reaction times if presented during the early afternoon period. If you collected the data as described, there is no way to determine which explanation is correct.

To control for this problem, you must vary the treatment order. This technique, called **counterbalancing**, is used to present the conditions in different orders to different participants. A counterbalanced research design appears in Table 12.3:

Table 12.3

Counterbalanced Presentation of Treatment Conditions to Control for Sequence Effects

Treatment Order

	AM	Late AM	Early PM
Participant 1	Tone	Light	Both
Participant 2	Both	Tone	Light
Participant 3	Light	Both	Tone
Participant 4	Tone	Both	Light
Participant 5	Light	Tone	Both
Participant 6	Both	Light	Tone

Note that in Table 12.3, each treatment occurs an equal number of times at each point of measurement. To achieve complete counterbalancing, you must use each combination of treatment sequences for an equal number of participants so that possible error due to order effects is dispersed evenly across treatment conditions.

 Complete counterbalancing becomes impractical as the number of treatment conditions increases. For example, there are only 6 possible sequences of 3 treatments, but this increases to 24 sequences with 4 treatments, 120 sequences with 5 treatments, and so forth. Obviously, complete counterbalancing typically is feasible only if the independent variable can assume a relatively small number of values.

Carryover Effects

Carryover effects occur when an effect from one treatment changes (carries over to) participants' responses in the following treatment condition. For example, suppose you are investigating the hypnotic effect of three different drugs. Drug 1 is given on night 1, and sleep onset latency is measured by electroencephalogram. The same measure is collected on nights 2 and 3, when drugs 2 and 3 are administered. If drug 1 has a long half-life (e.g., flurazepam), then it may still exert an effect on sleep latency at night 2. This carryover effect would make it impossible to accurately assess the effect of drug 2.

To avoid potential carryover effects, the experimenter might decide to separate the experimental conditions by one week. Counterbalancing also provides some control over carryover effects. If all treatment combinations can be given, then each treatment will be followed (and preceded) by each other treatment with equal frequency.

However, carryover effects are not as likely to even out with counterbalancing as are order effects. The advantage of counterbalancing might lie more in enabling the experimenter to measure the extent of carryover effects and to make appropriate adjustments to the analysis.

Ideally, careful consideration of experimental design will allow the experimenter to avoid significant carryover effects in a study. This is another consideration when choosing between a repeated-measures and a between-subjects design. In the study of drug effects described above, the investigator can avoid any possible carryover effects by using a between-subjects design. In that design, each participant would receive only one of the drug treatments.

Validity Conditions for the Univariate One-Way ANOVA, Repeated-Measures Design

The analysis described previously was conducted as a conventional univariate repeated-measures ANOVA. This analysis is valid only if certain assumptions about the data hold true. Two assumptions for this test are normality and homogeneity of covariance. Statisticians have pointed out that, in many instances, data collected under real-world conditions do not meet the second assumption.

The violation of homogeneity of covariance is particularly problematic for repeated measures designs as compared to between-subjects designs. In the case of between-subjects designs, the analysis still produces a robust *F* test even when this assumption is not met (provided sample sizes are equal). On the other hand, a violation of this assumption with a repeated measures design leads to an increased probability of a Type I error (i.e., rejection of a true null hypothesis). Therefore, you must take greater care when analyzing repeated measures designs to either prove that the assumptions of the test have been met or alter the analysis to account for the effects of the violation.

A discussion of the validity conditions for the univariate ANOVA and alternative approaches to the analysis of repeated measures data appears next. This section merely introduces some of the relevant issues and is not intended as an exhaustive description of the statistical problems inherent with use of repeated measures analyses. For a more detailed treatment, see Barcikowski and Robey (1984) or LaTour and Miniard (1983).

The assumptions underlying a valid application of the conventional *F* test are that criterion scores from the experimental treatments have a multivariate normal distribution in the population (i.e., normality) and that the common covariance matrix has a spherical pattern (i.e., homogeneity of covariance). Normality is impossible to prove but becomes more likely as sample size increases.

Homogeneity of covariance refers to the covariance between participants for any two treatments. One way to conceptualize this is that participants should have the same rankings in scores for all pairs of levels of the independent variable. For example, if there are three treatment (T_x) conditions, the covariance between T_x1 and T_x2 should be comparable to the covariance between T_x2 and T_x3, and to the covariance between T_x1 and T_x3.

If there are only two levels of the independent variable, then this assumption is automatically satisfied because there is only one covariance value.

Homogeneity of covariance is sufficient for test validity, but a less-specific type of covariance pattern (i.e., sphericity) is a necessary validity condition (Huynh & Mandeville, 1979; Rouanet & Lepine, 1970). SAS performs a test for sphericity if you request the PRINTE option in the REPEATED statement. The general form for this option is as follows:

```
REPEATED trial-variable-name #levels CONTRAST (level#) / SUMMARY
             PRINTE;
```

The REPEATED statement from the actual program is as follows:

```
3           REPEATED TIME 3 CONTRAST (1) / SUMMARY  PRINTE;
```

Revising the program presented earlier so that it includes the PRINTE option results in seven pages of output. The results requested by the PRINTE option would appear on pages 3 and 4. Those pages appear here as Output 12.3:

Output 12.3 Results of Test for Sphericity Requested by PRINTE Option

```
                                                                      3
                        The GLM Procedure
                Repeated Measures Analysis of Variance

                Repeated Measures Level Information

        Dependent Variable       PRE      POST FOLLOWUP

            Level of TIME          1        2        3


    Partial Correlation Coefficients from the Error SSCP Matrix / Prob > |r|

            DF = 19            PRE           POST        FOLLOWUP

            PRE            1.000000      0.677055        0.692736
                                          0.0010          0.0007

            POST           0.677055      1.000000        0.895619
                            0.0010                        <.0001

FOLLOWUP        0.692736      0.895619      1.000000
                0.0007          <.0001

                    E = Error SSCP Matrix

    TIME_N represents the contrast between the nth level of TIME and the 1st

                           TIME_2        TIME_3

                TIME_2      60.55         45.80
                TIME_3      45.80         50.80
```

(continued on the next page)

Output 12.3 *(continued)*

```
                                                                        4
                          The GLM Procedure
                  Repeated Measures Analysis of Variance

         Partial Correlation Coefficients from the Error SSCP Matrix of the
           Variables Defined by the Specified Transformation / Prob > |r|

                  DF = 19          TIME_2            TIME_3

              TIME_2            1.000000          0.825803
                                                  <.0001

              TIME_3            0.825803          1.000000
                                <.0001

                          Sphericity Tests

                                  Mauchly's
         Variables                DF    Criterion    Chi-Square    Pr > ChiSq

      Transformed Variates        2    0.3156106    20.758429       <.0001
      Orthogonal Components        2    0.6830428     6.8615597      0.0324

MANOVA Test Criteria and Exact F Statistics for the Hypothesis of no TIME Effect
                  H = Type III SSCP Matrix for TIME
                      E = Error SSCP Matrix

                  S=1     M=0     N=8

Statistic                    Value     F Value    Num DF    Den DF    Pr > F

Wilks' Lambda             0.30600563    20.41        2        18      <.0001
Pillai's Trace            0.69399437    20.41        2        18      <.0001
Hotelling-Lawley Trace    2.26791373    20.41        2        18      <.0001
Roy's Greatest Root       2.26791373    20.41        2        18      <.0001
```

The test performed with orthogonal components is the test of interest. In Output 12.3, this test appears toward the middle of output page 4 under "Sphericity Tests," to the right of "Orthogonal Components," under the heading "Mauchly's Criterion."

In Output 12.3, this test is significant at approximately $p = .03$. Technically, a significant p value indicates that the data display significant non-sphericity and thus departure from the homogeneity of covariance assumption. Be warned, however, that this test is extremely sensitive and any deviation from sphericity results in a significant F test. However, you can compensate for small to moderate deviations from sphericity through use of an adjusted F test (discussion follows). If there is a severe departure from sphericity (e.g., $p < .01$), then another approach to the problem is to use multivariate tests.

Alternate Analyses

If the assumptions of the test are not met, then various alternatives to the conventional univariate ANOVA have been proposed. These alternatives consist of modifications of the univariate ANOVA or use of the multivariate ANOVA.

Modified Univariate Tests

As stated above, the primary concern when the sphericity pattern is not present is that the *F* test will be too liberal and lead to inappropriate rejection of the null hypothesis (i.e., Type I error). Therefore, modifications that are recommended to compensate for non-sphericity are generally aimed at making the test more conservative.

The currently accepted method to modify the *F* test to account for deviations from sphericity is to adjust the degrees of freedom associated with the *F* value. Several correction factors have been developed to accomplish this. The correction factor is named epsilon (Greenhouse & Geisser, 1959). Epsilon varies between a lower limit of $[1/(k-1)]$ and 1, depending on the degree to which the data deviate from a spherical pattern. The degrees of freedom are multiplied by the correction factor to yield a number that is either lower or unchanged. Therefore, a given test typically has fewer degrees of freedom and requires a greater *F* value to achieve a given *p* level. With epsilon at a value of 1 (meaning the assumption has been met), the degrees of freedom are unchanged. To the extent that sphericity is not present, epsilon is reduced and this further decreases the degrees of freedom to produce a more conservative test.

The computations for epsilon are performed routinely by SAS as part of the univariate analysis for repeated measures. (See page 4 of Output 12.2.) Although the exact procedure to follow is somewhat controversial and can depend on characteristics specific to a given dataset, there is some consensus for use of the following general guidelines:

- You should use the adjusted univariate test when the Greenhouse-Geisser (G-G) epsilon is greater than or equal to .75.

- If the G-G epsilon is less than .75, a multivariate analysis (MANOVA) provides a more powerful test.

A Greenhouse-Geisser (G-G) epsilon value of approximately .76 ($p < .05$) was previously reported in Output 12.2. According to the above guidelines, this value suggests use of the adjusted *F* test.

Multivariate ANOVA

As a repeated-measures design consists of within-subjects observations across treatment conditions, the individual treatment measures can be viewed as separate, correlated dependent variables. Thereby the dataset is easily conceptualized as multivariate even though the design is univariate, and it can be analyzed with multivariate statistics. In this type of analysis, each level of the repeated factor is treated as a separate variable. The SAS program described earlier in this chapter computes a multivariate ANOVA automatically, and includes the results in the output. The statistics for the multivariate test appear on page 3 of Output 12.2. SAS computes four test statistics, and the appropriate test to use depends on the characteristics of your dataset. For a review of the statistical literature on this topic, see Olson (1976).

The multivariate ANOVA has an advantage over the univariate test in that it requires no assumption of sphericity. Some statisticians recommend that the multivariate ANOVA be used frequently, if not routinely, with repeated measures designs (Davidson, 1972). This argument is made for several reasons.

In selecting a test statistic, it is always desirable to choose the most powerful test. A test is said to have power when it is able to correctly reject the null hypothesis. In many situations, the univariate ANOVA is more powerful than the MANOVA and is therefore the better choice. However, the test to determine if the assumption of sphericity has been met is not very powerful with small samples. Therefore, only when the n is large (i.e., 20 greater than the number of treatment levels or $n > k + 20$) does the test for sphericity have sufficient power. Remember that the multivariate test becomes just as powerful as the univariate test as the n grows larger. This, however, creates a sort of catch-22: with a small n, there is no certainty that the assumptions underlying the univariate ANOVA have been met; with a large n, the MANOVA is equal to, if not more powerful than, the univariate test.

Others argue that the univariate approach offers a more powerful test for many types of data and should not be so readily abandoned (Tabachnick & Fidell, 2001). In their view, the multivariate ANOVA should be reserved for those situations that cannot be analyzed with the univariate ANOVA.

With the current example, the sample is composed of 20 participants and thus falls below the sensitivity threshold for sphericity tests described above (i.e., $n < k + 20$). In other words, the decision to reject the null hypothesis of significant departure from the homogeneity of covariance assumption (i.e., Mauchly's Criterion of .68, $p < .05$) might not be correct. Fortunately, both the standard and adjusted F values (as well as the multivariate statistics) suggest rejection of the study's null hypothesis of no difference in investment scores across the three points in time.

Conclusion

You can use the one-way repeated-measures ANOVA to analyze data from studies in which each participant is exposed to every level of the independent variable. In some cases, this involves a single-group design in which repeated measurements are taken at different points in time. As this chapter has pointed out, these single-group repeated-measures studies often suffer from a number of problems. Fortunately, some of these problems can be rectified by adding a second group of participants, a control group, to the design. The next chapter shows how to analyze data from that type of study.

Assumptions Underlying the One-Way ANOVA with One Repeated-Measures Factor

Assumptions for the Multivariate Test

- **Level of measurement.** Repeated measures designs are so named because they normally involve obtaining repeated measures on some criterion variable from a single sample of participants. This criterion variable should be assessed on an interval- or ratio-level of measurement. The predictor variable should be a nominal-level variable (a categorical variable) that typically codes "time," "trial," "treatment," or some similar construct.

- **Independent observations.** A given participant's score in any one condition should not be affected by any other participant's score in any of the study's conditions. However, it is acceptable for a given participant's score in one condition to be dependent upon his or her own score in a different condition. This is another way of saying, for example, that it is acceptable for participants' scores in condition 1 to be correlated with their scores in condition 2 and condition 3.

- **Random sampling.** Scores on the criterion variable should represent a random sample drawn from the populations of interest.

- **Multivariate normality.** The measurements obtained from participants should follow a multivariate normal distribution. Under conditions normally encountered in social science research, violations of this assumption have only a very small effect on the Type I error rate (i.e., the probability of incorrectly rejecting a true null hypothesis).

Assumptions for the Univariate Test

The univariate test requires all of the preceding assumptions as well as the following assumption of sphericity:

- **Sphericity.** In order to understand the sphericity assumption, it is necessary to first understand the nature of the difference variables that are created in performing a repeated-measures ANOVA. Assume that a study is conducted in which each participant provides scores on the criterion variable under each of three conditions: the variable V1 includes scores from condition 1; V2 includes scores from condition 2; and V3 includes scores from condition 3.

  It is possible to create a difference variable (called D1) by subtracting participants' scores on V2 from their score on V1:

  ```
  D1 = V1 - V2
  ```

  Similarly, it is possible to create a separate difference variable by subtracting participants' scores on V3 from their score on V2:

  ```
  D2 = V2 - V3
  ```

  Difference variables are created in this way by subtracting participants' scores observed in adjacent conditions (e.g., V1 – V2, V2 – V3). In a given study, the number of difference variables created will be equal to $k-1$, where k = the number of conditions. The present study included three conditions and therefore will include two difference variables (D1 and D2).

  It is now possible to compute the variances of each of these difference variables as well as the covariances between the difference variables. You can arrange these values in a variance-covariance matrix. You review this matrix to determine whether its corresponding matrix in the population demonstrates sphericity.

  Two conditions must be satisfied for a variance-covariance matrix to demonstrate sphericity. First, each variance on the diagonal of the matrix should be equal to every other variance on the diagonal. In the present case, this means that the variance for D1 should equal the variance for D2. Second, each covariance off of the diagonal should equal zero. (This is analogous to saying that the correlations between the difference variables should be zero.) In the present case, this means that the covariance between D1 and D2 should be equal to zero.

PROC GLM performs a test for sphericity by requesting the PRINTE option in the REPEATED statement. When this test indicates that the sphericity assumption is satisfied, you may interpret the univariate test. When the test indicates that the sphericity assumption is not satisfied (as will often be the case), the situation becomes more complicated. The options available under these circumstances are discussed in detail in the section "Further Notes on Repeated-Measures Analysis," earlier in this chapter.

References

Barcikowski, R., & Robey, R. (1984). Decisions in single group repeated measures analysis: Statistical tests and three computer packages. *The American Statistician, 38,* 148-150.

Davidson, M. (1972). Univariate versus multivariate tests in repeated-measures experiments. *Psychological Bulletin, 77,* 446-452.

Greenhouse, S., & Geisser, S. (1959). On methods in the analysis of profile data. Psychometrika, 24, 95-112.

Huynh, H., & Mandeville, G. (1979). Validity conditions in repeated measures designs. *Psychological Bulletin, 86,* 964-973.

Kashy, D. A., & Snyder, D. K. (1995). Measurement and data analytic issues in couples research. *Psychological Assessment, 7,* 338-348.

LaTour, S., & Miniard, P. (1983). The misuse of repeated measures analysis in marketing research. *Journal of Marketing Research, 20,* 45-57.

Olson, C. (1976). On choosing a test statistic in multivariate analysis of variance. *Psychological Bulletin, 83,* 579-586.

Rouanet, H., & Lepine, D. (1970). Comparison between treatments in a repeated-measurement design: ANOVA and multivariate methods. *British Journal of Mathematical and Statistical Psychology, 23,* 147-163.

Rusbult, C. E. (1980). Commitment and satisfaction in romantic associations: A test of the investment model. *Journal of Experimental Social Psychology, 16,* 172-186.

Tabachnick, B. G., & Fidell, L. S. (2001). *Using multivariate statistics* (4th ed.). New York: Harper Collins.

Factorial ANOVA with Repeated-Measures Factors and Between-Subjects Factors

Overview. This chapter shows how to enter data and prepare SAS programs to perform a two-way mixed-design ANOVA using the REPEATED statement with the GLM procedure. Guidelines for the hierarchical interpretation of the analysis are provided. This chapter shows how to interpret a significant interaction between the repeated-measures factor and the between-subjects factor as well as how to interpret main effects in the absence of an interaction. It also provides an alternative method of performing the analysis that allows for use of a variety of SAS/STAT post-hoc multiple-comparisons tests.

Introduction: The Basics of Mixed-Design ANOVA

This chapter discusses analysis of variance procedures that are appropriate for studies that include both repeated-measures factors and between-subjects factors. These research designs are often referred to as **mixed designs**.

A mixed-design ANOVA is similar to the factorial ANOVA discussed in Chapter 10, "Factorial ANOVA with Two Between-Group Factors," in that the procedure discussed in Chapter 10 assumes that the criterion variable is assessed on an interval or ratio scale and that the predictor variables are assessed on a nominal scale. However, a mixed-design ANOVA differs from a factorial ANOVA with two between-group factors with respect to the nature of the predictor variables that are included in the analysis. Specifically, a mixed-designed ANOVA assumes that the analysis will include the following:

- at least one nominal-scale predictor variable that is a *between-subjects* factor (as in Chapter 9, "One-Way ANOVA with One Between-Subjects Factor");

- plus at least one nominal-scale predictor variable that is a *repeated-measures* factor (as in Chapter 12, "One-Way ANOVA with One Repeated-Measures Factor").

These are called mixed designs because they include a mix of between-subjects and repeated-measures factors.

Extension of the Single-Group Design

It is useful to think of a factorial mixed design as an extension of the single-group repeated-measures design presented in Chapter 12. That chapter stated that a one-factor repeated-measures design involves taking repeated measurements on the criterion variable from just one group of participants. The occasions at which measurements are taken are called **times** (or **trials**) and there is typically some type of experimental manipulation that occurs between two or more of these occasions. It is assumed that, if there is a change in scores on the criterion from one time to another, it is due to the experimental manipulation (though this assumption can easily be challenged, given the many possible confounds that are associated with this single-group design; more on this later).

Chapter 12 showed how to use a one-factor repeated-measures research design to test the effectiveness of a marriage encounter program to increase participants' perceptions of how much they have invested in their romantic relationships. The concept of investment size was drawn from Rusbult's investment model (Rusbult, 1980). Briefly, investment size refers to the amount of time and personal resources that an individual has put into the relationship with his or her romantic partner. In the preceding chapter, it was hypothesized that the participants' perceived investment in their relationships would increase immediately following the marriage encounter weekend.

Figure 13.1 illustrates the one-factor repeated-measures design that was used to test this hypothesis in Chapter 12. This figure contains three small circles with the initials "D.V." (for "dependent variable"). These circles represent the occasions on which the dependent or criterion variable was assessed. The criterion variable in the study was investment size. You can see that the survey assessing this criterion variable was administered at three points in time:

- A baseline survey (Time 1) was administered immediately before the marriage encounter weekend.

- A post-treatment survey (Time 2) was administered immediately after the marriage encounter weekend.

- A follow-up survey (Time 3) was administered two weeks after the marriage encounter weekend.

Figure 13.1 Single-Group Experimental Design with Repeated Measures

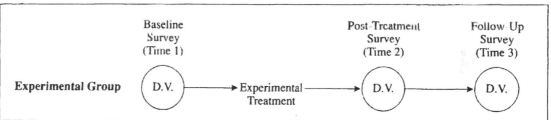

You could name the predictor variable in this study TIME; it contains three levels (Time 1, Time 2, and Time 3). Participants in the study provided data for each of these three levels of the predictor variable, which means that TIME is a repeated-measures factor.

Figure 13.1 shows that an experimental treatment (the marriage encounter program) was positioned between Times 1 and 2. You therefore hypothesized that the investment scores observed immediately after this treatment (at Time 2) would be significantly higher than scores observed just before this treatment (at Time 1).

However, Chapter 12 went to some lengths to point out that this single-group repeated-measures design can be a very weak research design, if used inappropriately. The fictitious single-group study that assessed the effectiveness of the marriage encounter program was described as an example of just such a weak design. The reason was simple: even if you obtained evidence consistent with your hypothesis (i.e., Time 2 scores were significantly higher than Time 1 scores), it would nonetheless provide very weak evidence to support your hypothesis. This is because there would be many alternative explanations for your findings, explanations that had nothing to do with the effectiveness of the marriage encounter program.

For example, if perceived investment increased from Time 1 to Time 2, someone could reasonably argue that this change was merely due to the passage of time, rather than the marriage encounter program. This argument would assert that the perception of investments naturally increases over time in all couples, regardless of whether they participate in a marriage encounter program. Because your study included just one group, you have no evidence showing that this alternative explanation is wrong.

As the preceding discussion suggests, data obtained from a single-group experimental design with repeated-measures generally provides relatively weak evidence of cause and effect. In order to minimize the weaknesses associated with this design, it is often necessary to expand it into a two-group experimental design with repeated measures (i.e., a mixed design).

Figure 13.2 Two-Group Experimental Design with Repeated Measures

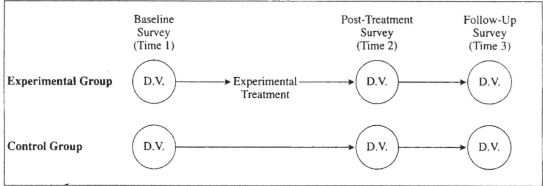

Figure 13.2 illustrates how to expand the current investment model study into a *two-group* design with repeated measures. The study would be conducted by starting with a single group of participants and randomly assigning each to one of two experimental conditions: an experimental group (which would go through the marriage encounter program); or a control group (which would not go through the program).

You would administer the measure of perceived investments to both groups at three points in time:

- a baseline survey (Time 1) immediately before the experimental group went through the marriage encounter program;

- a post-treatment survey (Time 2) immediately after the experimental group went through the program;

- a follow-up survey (Time 3) three weeks after the experimental group went through the program.

Advantages of the Two-Group Repeated-Measures Design

Including the control group makes this a much more rigorous study compared to the single-group design presented earlier. This is because the control group allows you to test the plausibility of many of the alternate explanations that can account for your study's results.

For example, consider a best-case scenario in which you obtain the results that appear in Figure 13.3. The fictitious results of Figure 13.3 show the mean investment scores displayed by the two groups at three points in time. The solid line depicts the mean investment scores for the experimental group (the group that experienced the marriage encounter training). This solid line shows that the experimental group displayed a relatively low level of perceived investment at Time 1 (mean investment score is approximately 7 on a 12-point scale), but that this increased substantively following the marriage encounter program at Times 2 and 3 (mean investment scores are approximately 11.5 and 11.5, respectively). These findings are consistent with your hypothesis that the marriage encounter program would positively affect perceptions of investment.

Figure 13.3 A Significant Interaction between TIME and GROUP

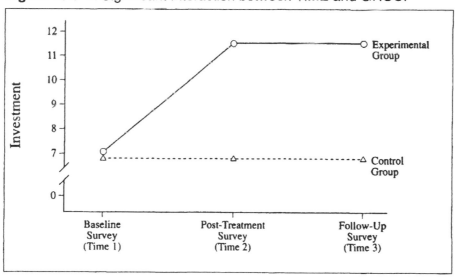

In contrast, now consider the dashed line of Figure 13.3, which represents the mean scores displayed by the control group (i.e., the group that did not receive the marriage encounter program). Notice that the mean investment scores for this group begin relatively low at Time 1 (at approximately 6.9 on the 12-point scale) and do not appear to increase at Times 2 or 3 (i.e., this broken line remains relatively flat). This finding is also consistent with your hypothesis; since the control group did not receive the training, you would not expect them to demonstrate large improvements in perceived investments over time.

The results of Figure 13.3 are consistent with your hypothesis regarding the effects of the program, and the presence of the control group allows you to discount many alternate explanations for the results. For example, it would no longer be tenable for someone to argue that increases in perceived investment were due simply to the passage of time (sometimes referred to as maturation effects). If that were the case, you would expect to see similar increases in perceived investments in the control group, yet these increases are not evident. Similarly, it is not tenable to say the post-treatment scores for the experimental group are higher than those for the control group because of some type of bias in assigning participants to conditions: Figure 13.3 shows that the two groups demonstrated similar levels of perceived investment at the baseline survey or Time 1 (before the experimental treatment). Because this two-group design protects you from these alternative explanations (as well as some others), it is generally preferable to the single-group procedure discussed in Chapter 12.

Random Assignment to Groups

It is useful at this point to review the important role played by randomization in a mixed-design study. In the experiment just described, it is essential that couples who volunteer for the study be randomly assigned to either the treatment or control groups. In other words, they must be *assigned* to conditions by you (the researcher), and these assignments must be made in a completely *random* or unsystematic manner (e.g., coin toss or from a table of random numbers).

The importance of random assignment becomes clear when you consider what might happen if participants are not randomly assigned to conditions. For example, imagine that the experimental group in this study consisted only of people who voluntarily chose to participate in the marriage encounter weekend, and that the control group consisted only of couples who chose not to participate in the program. Under these conditions, it would be reasonable to argue that a selection bias could affect the outcome of the study. **Selection biases** can occur any time that participants are assigned to groups in a nonrandom manner. When this happens, the results of the experiment can reflect pre-existing differences rather than treatment effects. For example, in the study just described, couples who volunteer to participate in a marriage encounter weekend might be different in some way from couples who do not volunteer. The volunteering couples might have different concerns about their relationships, different goals, or different views about themselves that can influence their scores on the measure of investment. Such pre-existing differences could render any results from the study meaningless.

One way to rectify this problem is to ensure that you recruit all couples from a list of couples who have volunteered to participate in the marriage encounter weekend, and you then randomly assign all couples to treatment conditions. In this way, you are helping to ensure that couples in both groups are as similar as possible at the start of the study.

Studies with More Than Two Groups

To keep things simple, this chapter focuses on only *two-group* repeated measures designs. It should be obvious, however, that it is possible to expand these studies to include more than two groups under the between-subjects factors. For example, the preceding study included only two groups: a control group and an experimental group. It would have been relatively easy to expand the study so that it included three groups of participants:

- the control group;
- experimental group 1 (which would experience the marriage encounter program);
- experimental group 2 (which would experience traditional marriage counseling).

Obviously, the addition of the third group would allow you to test additional hypotheses. At any rate, the procedures for analyzing the data from such a design would be a logical extension of the two-group procedures described in this chapter.

Some Possible Results from a Two-Way Mixed-Design ANOVA

Before showing how to analyze data from a mixed design, it is instructive to first review some of the results that are possible with this design. This will help illustrate the power of this design and will lay a foundation for concepts to follow.

A Significant Interaction

In Chapter 10, "Factorial ANOVA with Two Between-Group Factors," you learned that a **significant interaction** means that the relationship between one predictor variable and the criterion is different at different levels of the second predictor variable. (With experimental research, the corresponding definition is "the effect of one independent variable on the criterion variable is different at different levels of the second independent variable."). To better understand this definition, refer back to Figure 13.3, which illustrates the interaction between TIME and GROUP.

In this study, TIME is the variable that coded the repeated-measures factor (Time 1 scores versus Time 2 scores versus Time 3 scores). GROUP, on the other hand, is the variable that coded the between-subjects factor (experimental group versus control group). When there is a significant interaction between a repeated-measures factor and a between-groups factor, it means that the relationship between the repeated-measures factor and the criterion is different for the different groups coded under the between-subjects factor.

This is illustrated by the two lines in Figure 13.3. To understand this interaction, begin by focusing on just the solid line in the figure, which illustrates the relationship between TIME and investment scores for just the experimental group. Notice that the mean for the experimental group is relatively low at Time 1 but substantively higher at Times 2 and 3. This shows that there is a relationship between TIME and perceived investment for the experimental group.

Next, focus only on the dashed line that illustrates the relationship between TIME and perceived investment for just the control group. Notice that this line is flat (i.e., there is little change from Time 1 to Time 2 to Time 3) indicating that there is no relationship between TIME and investment size for the control group.

Combined, these results illustrate the definition for an **interaction**: the relationship between one predictor variable (TIME) and the criterion variable (investment size) is different at different levels of the second predictor variable.

You determine whether an interaction is significant by consulting the appropriate statistical test in your SAS output. However, it is also sometimes possible to identify a likely interaction by reviewing a figure that illustrates group means. For example, consider the solid line and the dashed line of Figure 13.3, particularly the first segment of each line that goes from Time 1 to Time 2. Notice that the solid line (for the experimental group) is not parallel to the dashed line (for the control group), the solid line has a much steeper angle. This is the hallmark of an interaction: nonparallel lines. Whenever a figure shows that a line segment for one group is not parallel to the corresponding line segment for a different group, it might mean that there is an interaction between the repeated-measures factor and the between-subjects factor.

In some (but not all) studies that employ a mixed design, your central hypothesis might require that there be a significant interaction between the repeated-measures variable and the between-subjects variable. For example, in the present study that assesses the effectiveness of the marriage encounter program, a significant interaction would be required to show that the experimental group displays a greater increase in investment size compared to the control group.

Significant Main Effects

If a given predictor variable is not involved in any significant interactions, you are free to determine whether or not that variable displays a significant main effect. When a predictor variable displays a significant **main effect**, it means that there is a difference between at least two of the levels of that predictor variable with respect to scores on the criterion variable.

The number of main effects that are possible in a study is equal to the number of predictor variables. The present investment model study includes two predictor variables; so, two main effects are possible: one for the repeated-measures factor (TIME); and one for the between-subjects factor (GROUP). These main effects can take a variety of different forms.

A Significant Main Effect for TIME

For example, Figure 13.4 shows one possible main effect for the TIME variable: an increasing linear time trend. Notice that the investment scores at Time 2 are higher than the scores at Time 1, and that the scores at Time 3 are somewhat higher than the scores at Time 2. Whenever the values of a predictor variable are plotted on the horizontal axis of a figure, a significant main effect for that variable is indicated when the line segments display a relatively steep angle.

Figure 13.4 A Significant Main Effect for TIME Only

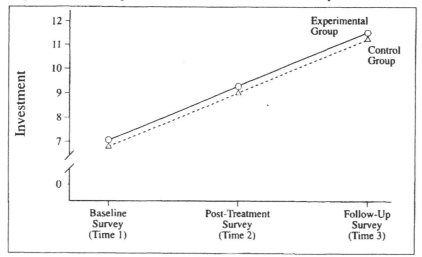

In Figure 13.4, the values of TIME are plotted on the horizontal axis. (These values are identified as "Baseline Survey, Time 1," "Post-Treatment Survey, Time 2," and "Follow-Up Survey, Time 3.") There is a relatively steep angle in the line that goes from Time 1 to Time 2, and also a relatively steep angle in the line that goes from Time 2 to Time 3. These results will typically suggest a significant main effect. Of course, you always check the appropriate statistical test in the SAS output to verify that the main effect is, in fact, significant.

Remember that the main effect for TIME is averaged over the two groups in the study. In the present case, this means that there is an overall main effect for TIME after collapsing the experimental and control groups (i.e., considering both as combined grouping).

A Significant Main Effect for GROUP

You would expect to see a different pattern in a figure if the main effect for the other predictor variable was significant. Earlier, it was stated that values of the predictor variable TIME are plotted as three separate points on the horizontal axis of the figure. In contrast, the values of the predictor variable GROUP are coded by drawing separate lines for the two groups under this variable: the experimental group is represented with a solid line, and the control group is represented with a dashed line.

When a predictor variable (such as GROUP) is represented in a figure by plotting separate lines for its various levels, a significant main effect for that variable is evident when at least two of the lines are relatively separate from each other. For example, consider Figure 13.5.

Figure 13.5 A Significant Main Effect for GROUP Only

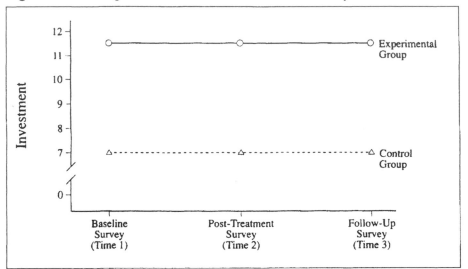

Begin your review of Figure 13.5 by first determining which effects are probably *not* significant. You can see that all line segments for the experimental group are parallel to their corresponding segments for the control group; this tells you that there is probably not a significant interaction between TIME and GROUP. Next, you can see that none of the line segments in Figure 13.5 display a relatively steep angle; this tells you that there is probably not a significant main effect for TIME.

However, notice that the solid line that represents the experimental group appears to be separated from the dashed line that represents the control group. Now, look at the individual data points. At Time 1, the experimental group displays a mean investment score that appears to be much higher than the one displayed by the control group. This same pattern of differences appears at Time 2 and Time 3. Combined, these are the results that you would expect to see if there were a significant main effect for GROUP. Figure 13.5 suggests that the experimental group consistently demonstrated higher investment scores than the control group.

A Significant Main Effect for Both Predictor Variables

It is possible to obtain significant main effects for both predictor variables simultaneously. Such an outcome appears in Figure 13.6. Notice that the line segments display a relatively steep angle (indicative of a main effect for TIME) and the lines for the two groups are also relatively separated (indicative of a main effect for GROUP).

Figure 13.6 Significant Main Effects for Both TIME and GROUP

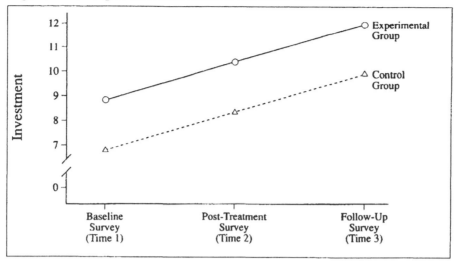

Nonsignificant Interaction, Nonsignificant Main Effects

Of course, there is no law that says that *any* of your effects have to be significant (as every researcher knows all too well!). This is evident in Figure 13.7. In that figure, notice that the lines for the two groups are parallel suggesting a probable nonsignificant interaction. There is also no angle to the line, suggesting that the main effect for TIME is nonsignificant. Finally, the line for the experimental group is not really separated from the line for the control group; this suggests that the main effect for GROUP is likewise nonsignificant.

Figure 13.7 A Nonsignificant Interaction and Nonsignificant Main Effects

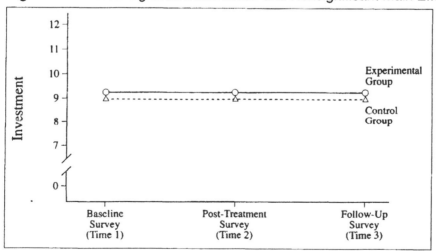

Problems with the Mixed-Design ANOVA

An earlier section indicated that the proper use of a control group in a mixed-design investigation can help remedy some of the weaknesses associated with the single-group repeated-measures design. However, there are other problems that can affect a repeated-measures analysis even when it includes a control group.

For example, Chapter 12, "One-Way ANOVA with One Repeated-Measures Factor," describes a number of sequence effects that can confound a study with a repeated-measures factor. Specifically, repeated-measures investigations often suffer from **order effects** (effects that occur when the ordinal position of treatments introduces response biases) as well as **carryover effects** (effects that occur when the effect from one treatment changes, or carries over, to participant responses in the following treatment conditions).

You should always be sensitive to the possibility of sequence effects when conducting research with any type of repeated-measures design. In some cases, it is possible to successfully deal with these problems through the proper use of counterbalancing, spacing of trials, additional control groups, or other strategies. Some of these approaches are discussed in the "Sequence Effects" section of Chapter 12.

Example with a Nonsignificant Interaction

The fictitious example presented here is a follow-up to the pilot study described in Chapter 12. The results of that pilot study suggest that scores on a measure of perceived investment significantly increase following a marriage encounter weekend. However (as described earlier), you could argue that those investment scores merely increased as a function of time, and not because of the experimental manipulation. In other words, the investment scores could have increased simply because the couples spent time together and this would have occurred with any activity, not just a marriage encounter experience.

To address concerns about this possible confound (as well as some others inherent in the one-group design), replicate the study as a two-group design. An experimental group again takes part in the marriage encounter program, and a control group does not. Repeated measures on perceived investment are obtained from both groups. (The design of this study is illustrated in Figure 13.2.) Remember that this two-group design is generally considered to be superior to the design described in Chapter 12; you would use it in place of that design in most instances.

Your primary hypothesis is that the experimental group will show a greater increase in investment scores at post-treatment and follow-up than the control group. This result would be confirmed by a significant GROUP × TIME interaction in the ANOVA. You also hypothesize (based on the results obtained from the pilot project described in Chapter 12) that the increased investment scores in the treatment group will still be found at follow-up. To determine whether the results are consistent with these hypotheses, it is necessary to check group means, review the results from the "omnibus" test of the interaction, and consult a number of post-hoc tests to be described later.

Writing the SAS Program

The criterion variable in this study is perceived **investment size**: the amount of time and effort that participants believe they have invested in their relationships (their marriages). Perceived investment will be assessed with a survey, and cumulative investment scores will be calculated so that higher scores indicate greater levels of investment. Assume that these scores are assessed on an interval scale and that responses to the scale have been shown to demonstrate acceptable psychometric properties (i.e., valid and reliable scale responses).

Creating the SAS Dataset

Because you will obtain investment scores at three points in time (Time 1, Time 2, and Time 3), you must create three SAS variables to contain these scores. Use the following approach:

- A SAS variable named PRE contains investment scores obtained at Time 1 (scores from the baseline survey).

- A SAS variable named POST contains investment scores obtained at Time 2 (from the post-treatment survey).

- A SAS variable named FOLLOWUP contains investment scores obtained at Time 3 (from the follow-up survey).

You need a number of additional variables to complete the analysis. Create the variable named PARTICIPANT to denote identification numbers (values will range from 01 through *n*, where *n* = the number of participants). In addition, create a variable named GROUP to designate membership in either the experimental or control groups. Code this variable so that a value of 1 indicates that the participant is assigned to the control group and a value of 2 indicates assignment to the experimental group.

You can use the following program to create a SAS dataset called MIXED that contains fictitious data from the mixed-design study:

```
1       DATA MIXED;
2       INPUT #1 @1     PARTICIPANT  2.
3                @5     GROUP        1.
4                @10    PRE          2.
5                @15    POST         2.
6                @20    FOLLOWUP     2. ;
7       DATALINES;
8       01  1    08   10   10
9       02  1    10   13   12
10      03  1    07   10   12
11      04  1    06   09   10
12      05  1    07   08   09
13      06  1    11   15   14
14      07  1    08   10   09
15      08  1    05   08   08
16      09  1    12   11   12
17      10  1    09   12   12
18      11  2    10   14   13
19      12  2    07   12   11
20      13  2    08   08   09
21      14  2    13   14   14
22      15  2    11   11   12
23      16  2    07   08   07
```

```
24      17   2      09    08    10
25      18   2      08    13    14
26      19   2      10    12    12
27      20   2      06    09    10
28      ;
29   RUN;
```

Lines 8 through 27 of the preceding program include data from the study. The first column includes the participant number variable labeled PARTICIPANT. You can see that 20 participants provided data.

The second column of data (in column 5) indicates the condition or GROUP to which participants are assigned. You can see that data from participants in the control group appear on lines 8 through 17 (coded as 1), while the data from the participants in the experimental group appear in lines 18 through 27 (identified by 2).

The third column of data (in columns 10 and 11) lists reported investment scores obtained at Time 1. These scores are given the SAS variable name PRE. The fourth column of data (columns 15 and 16) lists values for the SAS variable POST (investment scores obtained at Time 2). Finally, the last column of data (columns 20 and 21) lists values for the SAS variable FOLLOWUP (investment scores obtained at Time 3).

Obtaining Descriptive Statistics with PROC MEANS

After entering the data, perform a PROC MEANS to obtain descriptive statistics from all variables. This serves two important purposes. First, scanning the n, minimum value, and maximum value for each variable provides an opportunity to check for obvious data entry errors. Second, you need the means and standard deviations for the variables to interpret any significant effects observed in the ANOVA results to be reviewed later. Using PROC MEANS is particularly important when analyzing data from a mixed-design study because the means for within-subjects variables are not routinely included in the output of PROC GLM when it is used to perform a mixed-design ANOVA.

You need means and other descriptive statistics for all three of your investment score variables: PRE; POST; and FOLLOWUP. You need the overall means (based on the complete sample) as well as the means by GROUP (i.e., you need the means on these three variables for the control group as well as the means for the experimental group). You can obtain all of this information by adding the following lines to the preceding SAS program:

```
1      PROC MEANS   DATA=MIXED;
2      RUN;
3
4      PROC SORT    DATA=MIXED;
5         BY GROUP;
6      RUN;
7
8      PROC MEANS   DATA=MIXED;
9         BY GROUP;
10     RUN;
```

Lines 1 and 2 of the preceding program request that PROC MEANS be performed on all variables for the complete sample. Lines 4 through 6 sort the dataset by the variable GROUP, and lines 8 through 10 request that the MEANS procedure be performed twice: once for the control group; and once for the experimental group.

The output produced by the previous program appears below as Output 13.1. Results of the first PROC MEANS (performed on the combined sample) appear on page 1 of this output. It is instructive to review the means from this page of the output to get a sense for any general trends in the data. Remember that the variables PRE, POST, and FOLLOWUP contain investment scores obtained at Times 1, 2, and 3, respectively. The PRE variable displays a mean score of 8.60, meaning that the average investment score was 8.60 just before the marriage encounter weekend. The mean score on POST shows that the average investment score increased to 10.75 immediately after the marriage encounter weekend, and the mean score on FOLLOWUP shows that investment scores averaged 11.00 two weeks following the program. These means seem to display a fairly large increase in investment scores from Time 1 to Time 2 suggesting that you might observe a significant effect for TIME when you review ANOVA results (presented later).

Output 13.1 Results of PROC MEANS

```
                              The SAS System                                    1

                           The MEANS Procedure

Variable       N         Mean         Std Dev        Minimum        Maximum
-------------------------------------------------------------------------------
PARTICIPANT    20   10.5000000      5.9160798      1.0000000     20.0000000
GROUP          20    1.5000000      0.5129892      1.0000000      2.0000000
PRE            20    8.6000000      2.1373865      5.0000000     13.0000000
POST           20   10.7500000      2.2912878      8.0000000     15.0000000
FOLLOWUP       20   11.0000000      2.0261449      7.0000000     14.0000000
-------------------------------------------------------------------------------
```

```
                                                                                2

-------------------------------- GROUP=1 --------------------------------------

                           The MEANS Procedure

Variable       N         Mean         Std Dev        Minimum        Maximum
-------------------------------------------------------------------------------
PARTICIPANT    10    5.5000000      3.0276504      1.0000000     10.0000000
PRE            10    8.3000000      2.2135944      5.0000000     12.0000000
POST           10   10.6000000      2.2211108      8.0000000     15.0000000
FOLLOWUP       10   10.8000000      1.8737959      8.0000000     14.0000000
-------------------------------------------------------------------------------

-------------------------------- GROUP=2 --------------------------------------

Variable       N         Mean         Std Dev        Minimum        Maximum
-------------------------------------------------------------------------------
PARTICIPANT    10   15.5000000      3.0276504     11.0000000     20.0000000
PRE            10    8.9000000      2.1317703      6.0000000     13.0000000
POST           10   10.9000000      2.4698178      8.0000000     14.0000000
FOLLOWUP       10   11.2000000      2.2509257      7.0000000     14.0000000
-------------------------------------------------------------------------------
```

It is important to remember that, in order for your hypotheses to be supported, it is not adequate to merely observe a significant effect for TIME; instead, it is necessary that you obtain a significant TIME × GROUP interaction. Specifically, you must find that any increase in investment scores over time is greater among participants in the experimental group than among those in the control group. To see whether such an interaction has occurred, it is necessary to prepare a figure that plots data for the two groups separately and consult the appropriate statistical analyses. You can prepare the necessary figure by referring to the group means that appear on page 2 of Output 13.1. (A later section presents the appropriate statistical analyses.)

Figure 13.8 Mean Investment Scores from Output 13.8

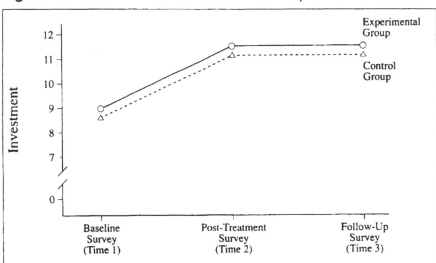

In Figure 13.8, the dashed line illustrates mean investment scores from the control group. These mean scores were obtained from the PRE, POST, and FOLLOWUP variables that appeared on the part of Output 13.1 labeled "GROUP=1." The solid line in the figure illustrates mean scores for the experimental group. These scores are obtained from the section of Output 13.1 labeled "GROUP=2." (For a review of how to prepare figures such as Figure 13.8 from a table of means, see Chapter 10.)

The general pattern of means plotted in Figure 13.8 does not suggest an interaction between TIME and GROUP. When two variables are involved in an interaction, the lines that represent the various groups tend not to be parallel to one another. So far, the lines for the control group and experimental group of Figure 13.8 do appear to be parallel. This might mean that the interaction is nonsignificant. However, the only way to be sure is to analyze the data and compute the appropriate statistical test. The next section shows how to do this.

Testing for Significant Effects with PROC GLM

The general form for the SAS program to perform a factorial ANOVA with one repeated-measures factor and one between-subjects factor is as follows:

```
PROC GLM   DATA=filename;
   CLASS   group-variable-name;
   MODEL   trial1  trial2  trial3... trialn = group-variable-name
           / NOUNI;
   REPEATED   trial-variable-name  #levels  CONTRAST (level#) /
SUMMARY;
RUN;
```

The actual SAS program needed to analyze this dataset is as follows:

```
1      PROC GLM   DATA=MIXED;
2         CLASS GROUP;
3         MODEL PRE POST FOLLOWUP = GROUP / NOUNI;
4         REPEATED TIME 3 CONTRAST (1) / SUMMARY;
5      RUN;
```

Notes Regarding the SAS Program

The analysis begins with the PROC GLM statement on line 1. The CLASS statement on line 2 identifies the variable that codes the between-subjects factor (i.e., the variable that codes the experimental group versus the control group). In this study, the between-subjects factor has the SAS variable name GROUP.

In the MODEL statement on line 3, the variables that contain the criterion variable scores are left of the equal sign. The number of variables equals the number of levels of the repeated-measures variable. These levels are represented as "trial1 trial2... trial*n*" in the general form. In this study, the repeated-measures variable has three levels (Time 1, Time 2, and Time 3), so the variables that represent these levels (PRE, POST, and FOLLOWUP) appear to the left of the equal sign.

Also in the MODEL statement, the variable that codes the between-subjects factor should be right of the equal sign. (This will be the same variable name listed in the CLASS statement.) In the present study, this between-subjects factor is GROUP.

The last entries in the MODEL statement are a slash (which indicates that options are to follow) and the NOUNI option. The NOUNI option suppresses the printing of certain univariate statistics that are of no interest in this analysis.

The REPEATED statement appears next in the program; in this statement, you must list a "trial-variable-name." That is, you must create a new variable name to represent your repeated-measures factor. In this study, the levels of the repeated-measures factor were Time 1, Time 2, and Time 3 (represented as PRE, POST, and FOLLOWUP in the MODEL statement). It therefore follows that an appropriate name for the repeated-measures variable in this study might be TIME. Note that variable name TIME appears as the first entry on the REPEATED statement on line 4 of the preceding program.

The general form for the program shows that the next entry in the REPEATED statement should be "#levels," a number that indicates how many levels are coded under the repeated-measures factor. In the present study, the repeated-measures variable has three levels (Time 1, Time 2, and Time 3), so the number 3 appears to the right of TIME in the REPEATED statement.

The next entry in the REPEATED statement is the "CONTRAST (level#)" option. In a repeated-measures analysis, **contrasts** are planned comparisons between different levels of the repeated-measures variable. The CONTRAST option allows you to choose the types of contrasts that will be made. The number that you specify in the place of "level#" identifies the specific level of the repeated-measures factor against which the other levels will be compared. The preceding program specifies "CONTRAST (1)." A "1" appears in the parentheses with this option, and this means that level 1 under the repeated-measures factor will be contrasted with level 2 and with level 3. In more concrete terms, this means that investment scores obtained at Time 1 will be contrasted with those obtained at Time 2, and with those obtained at Time 3. Unless otherwise instructed, PROC GLM automatically performs tests that contrast the last (*n*th) level of the repeated-measures variable with each of the preceding levels.

 The preceding CONTRAST command requests that both the post-treatment and follow-up scores be compared to the baseline score. However, remember that you should interpret these tests only if the interaction effect is not significant and the TIME effect is significant. If the interaction effect is significant, then other post-hoc procedures are necessary.

Finally, the REPEATED statement ends with a slash and the SUMMARY option. This SUMMARY option asks for the statistics from the contrasts to appear in the output.

Results from the SAS Output

With LINESIZE=80 and PAGESIZE=60 in the OPTIONS statement (first line), the preceding program would produce five pages of output. This output appears here as Output 13.2. The information that appears on each page is summarized here:

- Page 1 provides level information for the between-subjects factor (GROUP, in this case).
- Page 2 provides the following:
    - the level of information for the repeated-measures factor (TIME, in this case);
    - the results of the multivariate significance test for the main effect of TIME;
    - the results of the multivariate significance test for the TIME × GROUP interaction.
- Page 3 provides the results of the significance test for the main effect of GROUP.
- Page 4 provides the following:
    - the results of the univariate significance tests for the main effect for TIME and for the TIME × GROUP interaction;
    - information related to the error term in the univariate significance tests that involve the repeated-measures factor (the error degrees of freedom, the Type III error sum of squares, and the error mean square);
    - two estimates of the epsilon statistic.
- Page 5 provides the results of the planned comparisons requested with the CONTRAST option.

Output 13.2 Results of the Two-Way Mixed-Design ANOVA with a Nonsignificant Interaction

```
                        The SAS System                              1

                        The GLM Procedure

                     Class Level Information

            Class           Levels     Values

            GROUP              2        1 2


            Number of observations      20
```

```
                                                                    2
                        The GLM Procedure
                Repeated Measures Analysis of Variance

                Repeated Measures Level Information

        Dependent Variable        PRE     POST FOLLOWUP

            Level of TIME           1       2        3


MANOVA Test Criteria and Exact F Statistics for the Hypothesis of no TIME Effect
                H = Type III SSCP Matrix for TIME
                      E = Error SSCP Matrix

               S=1      M=0      N=7.5

Statistic                    Value    F Value   Num DF   Den DF   Pr > F

Wilks' Lambda             0.30518060   19.35      2        17     <.0001
Pillai's Trace            0.69481940   19.35      2        17     <.0001
Hotelling-Lawley Trace    2.27674828   19.35      2        17     <.0001
Roy's Greatest Root       2.27674828   19.35      2        17     <.0001


              MANOVA Test Criteria and Exact F Statistics
               for the Hypothesis of no TIME*GROUP Effect
              H = Type III SSCP Matrix for TIME*GROUP
                      E = Error SSCP Matrix

               S=1      M=0      N=7.5

Statistic                    Value    F Value   Num DF   Den DF   Pr > F

Wilks' Lambda             0.99234386    0.07      2        17     0.9368
Pillai's Trace            0.00765614    0.07      2        17     0.9368
Hotelling-Lawley Trace    0.00771521    0.07      2        17     0.9368
Roy's Greatest Root       0.00771521    0.07      2        17     0.9368
```

(continued on the next page)

Output 13.2 *(continued)*

```
                                                                      3
                        The GLM Procedure
                 Repeated Measures Analysis of Variance
            Tests of Hypotheses for Between Subjects Effects

Source             DF    Type III SS    Mean Square   F Value   Pr > F

GROUP               1     2.8166667      2.8166667      0.23     0.6355
Error              18   218.0333333     12.1129630
```

```
                                                                      4
                        The GLM Procedure
                 Repeated Measures Analysis of Variance
          Univariate Tests of Hypotheses for Within Subject Effects

Source             DF    Type III SS    Mean Square   F Value   Pr > F

TIME                2    69.63333333    34.81666667    28.84    <.0001
TIME*GROUP          2     0.23333333     0.11666667     0.10    0.9081
Error(TIME)        36    43.46666667     1.20740741

                                     Adj Pr > F
                Source            G - G       H - F

                TIME             <.0001      <.0001
                TIME*GROUP        0.8568      0.8818
                Error(TIME)


            Greenhouse-Geisser Epsilon       0.7605
            Huynh-Feldt Epsilon              0.8623
```

```
                                                                      5
                        The GLM Procedure
                 Repeated Measures Analysis of Variance
               Analysis of Variance of Contrast Variables

TIME_N represents the contrast between the nth level of TIME and the 1st

Contrast Variable: TIME_2


Source             DF    Type III SS    Mean Square   F Value   Pr > F

Mean                1    92.45000000    92.45000000    27.69    <.0001
GROUP               1     0.45000000     0.45000000     0.13    0.7178
Error              18    60.10000000     3.33888889


Contrast Variable: TIME_3


Source             DF    Type III SS    Mean Square   F Value   Pr > F

Mean                1   115.2000000    115.2000000     40.98    <.0001
GROUP               1     0.2000000      0.2000000      0.07    0.7927
Error              18    50.6000000      2.8111111
```

Steps in Interpreting the Output

1. Make Sure That Everything Looks Right

As always, review the output for possible signs of problems before interpreting the results. Most of these steps with a mixed-design ANOVA are similar to those used with between-subjects designs (e.g., check the number of observations listed on page 1 to make certain data from all participants were used in the analysis, and check the number of levels that appear under each predictor variable). Because most of these steps were already discussed in earlier chapters, they are not reviewed here.

2. Determine Whether the Interaction Term Is Statistically Significant

As discussed in Chapter 12, when an analysis included a repeated-measures factor, the SAS output includes results of univariate, modified univariate, and multivariate ANOVAs. The "Further Notes on Repeated-Measures Analyses" section from that chapter reviews some of the basic issues to consider when choosing between univariate versus multivariate statistics.

Interpretation of the current study is somewhat more complicated than the interpretation of the one-factor design described in Chapter 12. This is because the current mixed-design requires that you interpret the effects of both a between-subjects factor and an interaction in addition to the effect of the repeated-measures factor. To simplify matters, this chapter follows the same general procedure recommended in Chapter 10 in which you first check for interactions before proceeding to test for main effects and post-hoc analyses.

As discussed in Chapter 10, the first step in interpreting a two-factor ANOVA is to check the interaction effect. If the interaction effect is not statistically significant, then you can proceed with interpretation of main effects. You should interpret the univariate test of the interaction effect first; this univariate test appears on page 4 of Output 13.2. On the left side of page 4, find the heading "Source: TIME*GROUP"; the information relevant to the interaction appears in this section. Under the heading "F Value," you can see that the univariate F for the TIME $\times$ GROUP interaction is 0.10. With 2 and 36 degrees of freedom, this F has an associated p value of .91, and is clearly nonsignificant.

The next step is to check the multivariate test for the interaction. In most analyses, the univariate and multivariate tests yield tests with similar p values. The results of these two approaches are likely to vary only if the assumptions for the univariate test are grossly violated, or possibly if the Time variables are highly intercorrelated. The results of the multivariate test for the interaction appear on page 2 of Output 13.2. Find the section on this page headed "MANOVA Test Criteria and Exact F Statistics for the Hypothesis of no TIME*GROUP Effect." To the right of the heading "Wilks' Lambda," you can see that the MANOVA yielded an F value of 0.07 and a corresponding p value of approximately .94. Once again, you fail to reject the null hypothesis of a TIME*GROUP interaction.

The primary hypothesis for this study required a significant interaction but you now know that the present data do not support such an interaction. If you were actually conducting this research project, your analyses might terminate at this point. However, in order to illustrate additional steps in the analysis of data from a mixed-design, this section proceeds with the tests for main effects.

3. Determine If the Group Effect Is Statistically Significant

In this chapter, the term **group effect** is used to refer to the effect of the between-subjects factor. In the present study, the variable named GROUP represents this effect.

The group effect is of no real interest in the present investment-model investigation since support for the study's central hypothesis required a significant interaction. Nonetheless, it is still useful to plot group means and review the statistic for the group effect in order to validate the methodology used to assign participants to groups. For example, if the effect for GROUP proves to be significant and if the scores for one of the treatment groups are consistently higher than the corresponding scores for the other (particularly at Time 1), it could indicate that the two groups were not equivalent at the beginning of the study. This might suggest that there was some type of bias in the selection and assignment processes. Such a finding could invalidate any other results from the study.

The significance test for the group effect appears on page 3 of Output 13.2. You can see that the obtained F value for the GROUP effect is only 0.23 which, with 1 and 18 degrees of freedom, is nonsignificant (the p value for this F is quite large at approximately 0.64). This indicates that there was not an overall difference between experimental and control groups with respect to their mean investment scores. This finding is also illustrated by Figure 13.8, which shows that there is very little separation between the line for the experimental group and the line for the control group.

4. Determine If the Time Effect Is Statistically Significant

In this chapter, the term **time effect** is used to refer to the effect of the repeated-measures factor. Remember that the variable TIME was used to code this factor. A main effect for TIME would suggest that there was a significant difference between investment scores obtained at one time and the investment scores obtained at least one other time during the study (e.g., that the scores obtained at Time 2 are significantly higher than those obtained at Time 1).

The univariate test for the TIME effect appears on page 4 of Output 13.2, under the section headed "Source: TIME." Under the heading "F Value," you can see that the univariate F for the TIME effect is 28.84. With 2 and 36 degrees of freedom, this F is significant at $p < .01$.

The multivariate test for the TIME effect appears on page 2 of Output 13.2, under the heading "MANOVA Test Criteria and Exact F Statistics for the Hypothesis of no TIME Effect." To the right of "Wilks' Lambda," you can see that the multivariate F for the TIME effect is 19.35 which, with 2 and 17 degrees of freedom, is also significant at $p < .01$. Clearly, there is a significant effect for the TIME variable.

The group means plotted in Figure 13.8 are helpful in interpreting this TIME effect. The trend displayed by the means suggests that Time 2 scores can be significantly higher than Time 1 scores, and Time 3 scores can also be significantly higher than Time 1 scores. A later section shows how to interpret the results requested by the CONTRAST option to see if these differences are statistically significant.

5. Prepare Your Own Version of the ANOVA Summary Table

Table 13.1 summarizes the preceding analysis of variance:

Table 13.1

ANOVA Summary Table for Study Investigating Changes in Investments Following an Experimental Manipulation (Nonsignificant Interaction)

Source	df	SS	MS	F
Between Subjects	19	220.85		
Group (A)	1	2.82	2.82	0.23
Residual between	18	218.03	12.11	
Within Subjects	40	113.33		
Time (B)	2	69.63	34.82	28.84 *
A x B Interaction	2	0.23	0.12	0.10
Residual within	36	43.47	1.21	
Total	59	334.18		

Note: N = 20.

* $p < .01$

6. Review the Results of the Contrasts

The program presented earlier included the following REPEATED statement:

```
4          REPEATED TIME 3 CONTRAST (1) / SUMMARY;
```

The keyword CONTRAST in this statement is followed by a "1" in parentheses. This requests that level 1 under the TIME variable be contrasted with the other levels under TIME. In other words, it requests that investment scores observed at Time 1 be contrasted with investment scores at Time 2, and that Time 1 also be contrasted with Time 3.

The resulting contrast analyses appear on page 5 of Output 13.2. The analysis that compares Time 1 investment scores with Time 2 scores appears under the heading "Contrast Variable: TIME_2." The information of interest appears to the right of the heading "MEAN." You can see that this comparison results in an F ratio of 27.69, which is significant at $p < .01$. You can therefore reject the null hypothesis that there is no difference between Time 1 investment scores and Time 2 investment scores in the population. The nature of the means displayed in Figure 13.8 shows that Time 2 scores are significantly higher than Time 1 scores.

The analysis that compares Time 1 to Time 3 appears under the heading "Contrast Variable: TIME_3." With an *F* ratio of 40.98, it is clear that investment scores at Time 3 are also significantly higher than Time 1 scores ($p < .01$).

Summarizing the Results of the Analysis

The results of this mixed-design analysis could be summarized using the standard statistical interpretation format presented earlier in this text:

A) Statement of the problem
B) Nature of the variables
C) Statistical test
D) Null hypothesis (H_0)
E) Alternative hypothesis (H_1)
F) Obtained statistic
G) Obtained probability (*p*) value
H) Conclusion regarding the null hypothesis
I) ANOVA summary table
J) Figure representing the results

This summary could be performed three times: once for the null hypothesis of no interaction in the population; once for the null hypothesis of no main effect for GROUP; and once for the null hypothesis of no main effect for TIME. As this format has been previously described, it does not appear again here.

Formal Description of Results for a Paper

Below is one way that you could summarize the present results for a research paper. Notice that the names of the independent variables (time and group) are not capitalized, although the first letter of each variable's name is capitalized in the expression " Group × Time interaction."

```
     Results were analyzed using a two-way ANOVA with repeated
measures on one factor.  The Group x Time interaction was not
significant, F(2,36) = 0.10, ns  nor was the main effect for
group significant, F(1,18) = 0.23, ns.  However, this analysis
did reveal a significant effect for time, F(2,36) = 28.84, p <
.01.  Post-hoc contrasts found that investment scores at post-
treatment (F[1,18] = 27.69, p < .01) and follow-up (F[1,18] =
40.98, p < .01) were significantly higher than the scores
observed at baseline or Time 1.
```

Example with a Significant Interaction

Remember that your initial hypothesis was that the investment scores would increase more for the experimental group than for the control group and that, to be supported, this hypothesis required a significant TIME × GROUP interaction. This section presents the results of an analysis of a different set of fictitious data: participant responses that will provide the desired interaction. You will see that it is necessary to perform several follow-up analyses when you obtain a significant interaction in a mixed-design ANOVA. The dataset for this appears below. Otherwise, the SAS program is the same as the previous example and need not be repeated here.

```
 7  DATALINES;
 8  01  1    10   09   10
 9  02  1    12   13   12
10  03  1    10   10   10
11  04  1    09   09   09
12  05  1    08   08   08
13  06  1    12   13   12
14  07  1    10   09   10
15  08  1    08   09   10
16  09  1    11   12   11
17  10  1    12   11   12
18  11  2    10   14   13
19  12  2    07   12   11
20  13  2    08   09   11
21  14  2    13   14   14
22  15  2    11   11   12
23  16  2    07   08   08
24  17  2    09   08   10
25  18  2    08   13   14
26  19  2    11   12   12
27  20  2    06   09   10
28  ;
29  RUN;
```

Output 13.3 presents the results of an analysis in which the TIME × GROUP interaction is significant. The following section shows how to interpret the output and perform the necessary tests for simple effects.

Output 13.3 Results of Two-Way Mixed Design ANOVA with a Significant Interaction

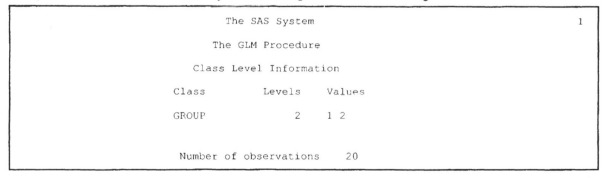

```
                        The SAS System                              1

                        The GLM Procedure

                     Class Level Information

          Class          Levels    Values

          GROUP              2      1 2


             Number of observations    20
```

(continued on the next page)

Output 13.2 *(continued)*

```
                                                                           2

                        The GLM Procedure
                Repeated Measures Analysis of Variance

                Repeated Measures Level Information

        Dependent Variable        PRE      POST FOLLOWUP

            Level of TIME           1        2        3


MANOVA Test Criteria and Exact F Statistics for the Hypothesis of no TIME Effect
                H = Type III SSCP Matrix for TIME
                      E = Error SSCP Matrix

                    S=1       M=0       N=7.5

Statistic                     Value    F Value    Num DF    Den DF    Pr > F

Wilks' Lambda              0.44416364    10.64         2        17    0.0010
Pillai's Trace             0.55583636    10.64         2        17    0.0010
Hotelling-Lawley Trace     1.25142248    10.64         2        17    0.0010
Roy's Greatest Root        1.25142248    10.64         2        17    0.0010


                MANOVA Test Criteria and Exact F Statistics
                for the Hypothesis of no TIME*GROUP Effect
                H = Type III SSCP Matrix for TIME*GROUP
                        E = Error SSCP Matrix

                    S=1       M=0       N=7.5

Statistic                     Value    F Value    Num DF    Den DF    Pr > F

Wilks' Lambda              0.56796607     6.47         2        17    0.0082
Pillai's Trace             0.43203393     6.47         2        17    0.0082
Hotelling-Lawley Trace     0.76066856     6.47         2        17    0.0082
Roy's Greatest Root        0.76066856     6.47         2        17    0.0082
```

```
                                                                           3
                        The GLM Procedure
                Repeated Measures Analysis of Variance
              Tests of Hypotheses for Between Subjects Effects

    Source            DF    Type III SS    Mean Square    F Value    Pr > F

    GROUP              1      0.2666667      0.2666667       0.03     0.8632
    Error             18    157.1333333      8.7296296
```

(continued on the next page)

Output 13.2 *(continued)*

```
                                                                         4
                           The GLM Procedure
                   Repeated Measures Analysis of Variance
           Univariate Tests of Hypotheses for Within Subject Effects

 Source               DF      Type III SS      Mean Square    F Value    Pr > F

 TIME                  2      21.73333333      10.86666667     12.28    <.0001
 TIME*GROUP            2      13.73333333       6.86666667      7.76     0.0016
 Error(TIME)          36      31.86666667       0.88518519

                                          Adj Pr > F
                  Source              G - G      H - F

                  TIME                0.0003     0.0002
                  TIME*GROUP          0.0037     0.0023
                  Error(TIME)


                  Greenhouse-Geisser Epsilon      0.7927
                  Huynh-Feldt Epsilon             0.9049
```

```
                                                                         5
                           The GLM Procedure
                   Repeated Measures Analysis of Variance
                 Analysis of Variance of Contrast Variables

TIME_N represents the contrast between the nth level of TIME and the 1st

Contrast Variable: TIME_2


 Source               DF      Type III SS      Mean Square    F Value    Pr > F

 Mean                  1      24.20000000      24.20000000      9.55     0.0063
 GROUP                 1      16.20000000      16.20000000      6.39     0.0210
 Error                18      45.60000000       2.53333333


Contrast Variable: TIME_3


 Source               DF      Type III SS      Mean Square    F Value    Pr > F

 Mean                  1      39.20000000      39.20000000     21.64     0.0002
 GROUP                 1      24.20000000      24.20000000     13.36     0.0018
 Error                18      32.60000000       1.81111111
```

(continued on the next page)

Output 13.2 *(continued)*

```
                                                                           6
---------------------------------- GROUP=1 ----------------------------------

                          The MEANS Procedure

 Variable       N          Mean        Std Dev       Minimum       Maximum
 ---------------------------------------------------------------------------
 PARTICIPANT   10     5.5000000      3.0276504     1.0000000    10.0000000
 PRE           10    10.2000000      1.5491933     8.0000000    12.0000000
 POST          10    10.4000000      1.7763883     8.0000000    13.0000000
 FOLLOWUP      10    10.5000000      1.1785113     9.0000000    12.0000000
 ---------------------------------------------------------------------------


---------------------------------- GROUP=2 ----------------------------------

 Variable       N          Mean        Std Dev       Minimum       Maximum
 ---------------------------------------------------------------------------
 PARTICIPANT   10    15.5000000      3.0276504    11.0000000    20.0000000
 PRE           10     9.0000000      2.2110832     6.0000000    13.0000000
 POST          10    11.0000000      2.3570226     8.0000000    14.0000000
 FOLLOWUP      10    11.5000000      1.9002924     8.0000000    14.0000000
 ---------------------------------------------------------------------------
```

Steps in Interpreting the Output

1. Determine Whether the Interaction Term Is Statistically Significant

After you scan the results in the usual manner to verify that there are no obvious errors in the program or dataset, check the appropriate statistics to see whether the interaction between the between-subjects factor and the repeated-measures factor is significant. The univariate results appear on page 4 of Output 13.3, in the section headed "Source: TIME*GROUP." The F value for this interaction term is 7.76. With 2 and 36 degrees of freedom, the unadjusted p value for this F is less than .01 as is the adjusted p value (Greenhouse-Geisser adjustment). (See "Modified Univariate Tests," in Chapter 12 for a discussion of when to report Greenhouse-Geisser adjusted p values.) The interaction is therefore significant.

This finding is consistent with the hypothesis that the relationship between the TIME variable and investment scores is different for the two experimental groups. To determine whether the experimental group shows greater increases in investments than the control group (as hypothesized), you must plot the interaction.

2. Plot the Interaction

To plot the interaction, you once again need to obtain mean investment scores for each of the two groups at each level of TIME. You can obtain these means by using PROC MEANS with a BY GROUP statement, as follows:

```
1      PROC SORT DATA=MIXED;
2         BY GROUP;
3      RUN;
4
5      PROC MEANS DATA=MIXED;
6         BY GROUP;
7      RUN;
```

Chapter 10 shows how you can use the results from this MEANS procedure to plot an interaction. The means from the present dataset appear on page 6 of Output 13.3, and are plotted in Figure 13.9.

Figure 13.9 Significant TIME x GROUP Interaction Obtained in the Analysis of the Investment Model Data

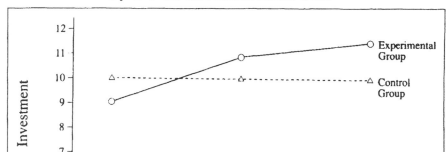

Notice that the line for the experimental group in Figure 13.9 (the solid line) is not perfectly parallel to the line for the control group (the dashed line). This is what you would expect with a significant TIME × GROUP interaction. You can see that the line for the control group is relatively flat. There does not appear to be much of an increase in perceived investment from Time 1 to Time 2 to Time 3. In contrast, notice the angle of the line for the experimental group. There is an apparent increase in investment size from Time 1 to Time 2, and another slight increase from Time 2 to Time 3.

These findings are consistent with the hypothesis that there would be a larger increase in investment scores in the experimental group than in the control group. To have confidence in this conclusion, you must test for simple effects.

3. Test for Simple Effects

When you test for **simple effects**, you determine whether there is a significant relationship between one predictor variable and the criterion for participants at one level of the second predictor variable. (The concept of simple effects first appeared in Chapter 10.)

For example, in the present study you might want to determine whether there is a significant relationship between TIME and the investment variable among those participants in the experimental group. If there is, you could state that there is a simple effect for TIME at the experimental group level of GROUP.

To understand the meaning of this, consider just the solid line in Figure 13.9 (i.e., the line that represents the experimental group). This line plots investment scores for the experimental group at three points in time. If there is a simple effect for TIME in the experimental group, it means that investment scores obtained at one point in time are significantly different from investment scores obtained at at least one other point in time. If the marriage encounter program really is effective, you would expect to see a simple effect for TIME in the experimental group (i.e., investment scores should improve over time in this group). You would not expect to see a simple effect for TIME in the control group since they did not experience the program.

Testing for simple effects following a mixed-design ANOVA is relatively straightforward. To test for the simple effects of the repeated-measures factor (TIME, in this case), it is necessary to do the following:

- divide the participants into subgroups based on their classification under the group variable (GROUP, in this case);

- perform a one-way ANOVA with one repeated-measures factor on just those participants in the first treatment group;

- repeat this repeated-measures ANOVA for each of the remaining subgroups.

For the present study, this means that you first divide participants into experimental and control groups. Next, using data from just the experimental group, you perform a one-way repeated measures ANOVA in which investment scores are the criterion and TIME is the predictor variable. You then repeat this analysis using data from the control group. You would compute and interpret these ANOVAs according to the procedures described in Chapter 12.

The first step in this process is to divide participants into the experimental and control groups. You do this with the following statements:

```
1        DATA CONTROL;
2            SET MIXED;
3            IF GROUP=1 THEN OUTPUT;
4        RUN;
5
6        DATA EXP;
7            SET MIXED;
8            IF GROUP=2 THEN OUTPUT;
9        RUN;
```

Lines 1 through 4 of the preceding program create a new dataset called CONTROL. It begins by setting the dataset equal to the MIXED dataset but retains only those participants whose score on the GROUP variable is equal to 1 (see line 3). Therefore, the dataset called CONTROL contains only data from the control group. In the same way, you can see that lines 6 through 9 create a new dataset called EXP that contain only data from participants in the experimental group (i.e., those with a score on GROUP equal to 2).

Later in the same program, you will add the lines to test the simple effect for TIME in the control group. The following lines accomplish this:

```
1       PROC GLM  DATA=CONTROL;
2          MODEL PRE POST FOLLOWUP =   / NOUNI;
3          REPEATED TIME 3 CONTRAST (1) / SUMMARY;
4       RUN;
```

In line 1 of the preceding program, the DATA=CONTROL option requests that the analysis be performed using the dataset that contains only responses from the control group. To test this simple effect in the experimental group, it is necessary only to change the dataset that is named in this line:

```
1       PROC GLM  DATA=EXP;
2          MODEL PRE POST FOLLOWUP =   / NOUNI;
3          REPEATED TIME 3 CONTRAST (1) / SUMMARY;
4       RUN;
```

Therefore, the complete program (minus data input) that creates the two new datasets and performs the two tests for simple effects is as follows:

```
1       DATA CONTROL;
2          SET MIXED;
3          IF GROUP=1 THEN OUTPUT;
4       RUN;
5
6       DATA EXP;
7          SET MIXED;
8          IF GROUP=2 THEN OUTPUT;
9       RUN;
10
11      PROC GLM  DATA=CONTROL;
12         MODEL PRE POST FOLLOWUP =   / NOUNI;
13         REPEATED TIME 3 CONTRAST (1) / SUMMARY;
14      RUN;
15
16      PROC GLM  DATA=EXP;
17         MODEL PRE POST FOLLOWUP =   / NOUNI;
18         REPEATED TIME 3 CONTRAST (1) / SUMMARY;
19      RUN;
```

The results of these two ANOVAs would constitute tests of the simple effects of TIME for each level of GROUP. See Chapter 12 for guidelines in interpreting the results. The results of this analysis for the current study are presented here as Output 13.4.

Output 13.4 Results of Tests for Simple Effects of TIME among Participants in the Control Group (Output Pages 1 to 4) and Participants in the Experimental Group (Output Pages 5 to 8)

```
                            The SAS System                              1

                            The GLM Procedure

                   Number of observations      10
```

```
                                                                        2
                            The GLM Procedure
                 Repeated Measures Analysis of Variance

                    Repeated Measures Level Information

          Dependent Variable        PRE     POST FOLLOWUP

              Level of TIME           1       2       3


MANOVA Test Criteria and Exact F Statistics for the Hypothesis of no TIME Effect
              H = Type III SSCP Matrix for TIME
                    E = Error SSCP Matrix

                    S=1      M=0      N=3

Statistic                   Value   F Value    Num DF    Den DF    Pr > F

Wilks' Lambda             0.80769231    0.95        2         8    0.4256
Pillai's Trace            0.19230769    0.95        2         8    0.4256
Hotelling-Lawley Trace    0.23809524    0.95        2         8    0.4256
Roy's Greatest Root       0.23809524    0.95        2         8    0.4256
```

```
                                                                        3
                            The GLM Procedure
                 Repeated Measures Analysis of Variance
          Univariate Tests of Hypotheses for Within Subject Effects

Source                DF    Type III SS    Mean Square   F Value   Pr > F

TIME                   2     0.46666667    0.23333333      0.76    0.4825
Error(TIME)           18     5.53333333    0.30740741

                                              Adj Pr > F
                    Source                  G - G      H - F

                    TIME                    0.4736     0.4825
                    Error(TIME)


                    Greenhouse-Geisser Epsilon    0.9212
                    Huynh-Feldt Epsilon           1.1474
```

(continued on the next page)

Output 13.4 *(continued)*

```
                                                                              4
                          The GLM Procedure
                 Repeated Measures Analysis of Variance
                 Analysis of Variance of Contrast Variables

TIME_N represents the contrast between the nth level of TIME and the 1st

Contrast Variable: TIME_2


 Source               DF     Type III SS    Mean Square    F Value    Pr > F

 Mean                  1      0.40000000     0.40000000       0.64    0.4433
 Error                 9      5.60000000     0.62222222


Contrast Variable: TIME_3


 Source               DF     Type III SS    Mean Square    F Value    Pr > F

 Mean                  1      0.90000000     0.90000000       1.98    0.1934
 Error                 9      4.10000000     0.45555556
```

```
                                                                              5
                          The GLM Procedure

                   Number of observations      10
```

```
                                                                              6
                          The GLM Procedure
                 Repeated Measures Analysis of Variance

                 Repeated Measures Level Information

         Dependent Variable        PRE      POST FOLLOWUP

            Level of TIME            1        2        3


MANOVA Test Criteria and Exact F Statistics for the Hypothesis of no TIME Effect
                  H = Type III SSCP Matrix for TIME
                      E = Error SSCP Matrix

                  S=1       M=0       N=3

Statistic                  Value    F Value    Num DF    Den DF    Pr > F

Wilks' Lambda           0.28777671     9.90        2         8    0.0069
Pillai's Trace          0.71222329     9.90        2         8    0.0069
Hotelling-Lawley Trace  2.47491639     9.90        2         8    0.0069
Roy's Greatest Root     2.47491639     9.90        2         8    0.0069
```

(continued on the next page)

Output 13.4 *(continued)*

```
                                                                          7
                            The GLM Procedure
                     Repeated Measures Analysis of Variance
              Univariate Tests of Hypotheses for Within Subject Effects

Source                       DF     Type III SS    Mean Square    F Value    Pr > F

TIME                          2     35.00000000    17.50000000     11.96     0.0005
Error(TIME)                  18     26.33333333     1.46296296

                                                  Adj Pr > F
                     Source                     G - G     H - F

                     TIME                       0.0024    0.0015
                     Error(TIME)


                     Greenhouse-Geisser Epsilon      0.7017
                     Huynh-Feldt Epsilon             0.7921
```

```
                                                                          8
                            The GLM Procedure
                     Repeated Measures Analysis of Variance
                     Analysis of Variance of Contrast Variables

TIME_N represents the contrast between the nth level of TIME and the 1st

Contrast Variable: TIME_2


Source                       DF     Type III SS    Mean Square    F Value    Pr > F

Mean                          1     40.00000000    40.00000000      9.00     0.0150
Error                         9     40.00000000     4.44444444


Contrast Variable: TIME_3


Source                       DF     Type III SS    Mean Square    F Value    Pr > F

Mean                          1     62.50000000    62.50000000     19.74     0.0016
Error                         9     28.50000000     3.16666667
```

The test of the simple effect of TIME at the control group level of GROUP appears on pages 1 through 4 of Output 13.4. The results of the univariate test appear on output page 3, which shows an F value of 0.76. With 2 and 18 degrees of freedom, the simple effect for TIME in the control group is nonsignificant ($p = .48$).

The test of the simple effect of TIME at the experimental group level of GROUP appears on pages 5 through 8 of Output 13.4. The results for the univariate test appear on output page 7 which shows an F value of 11.96. With 2 and 18 degrees of freedom, the simple effect for TIME in the experimental group is significant ($p < .01$).

Since the simple effect for TIME is significant in the experimental group, you can now proceed to interpret the results of the planned contrasts that appear on page 8 of Output 13.4. These contrasts show that there is a significant difference between scores obtained at Time 1 versus those obtained at Time 2, $F(1, 9) = 9.00, p < .05$, and that there was also a significant difference between scores obtained at Time 1 versus Time 3, $F(1, 9) = 19.74, p < .01$.

Although the preceding seems fairly straightforward, there are actually two complications in the interpretation of these results. First, you are now performing multiple tests (one test for each level under the group factor), and this means that the significance level you have initially chosen will have to be adjusted so that your experiment-wise probability of making a Type I error does not get out of hand. Looney and Stanley (1989) have recommended dividing the initial alpha (.05 in the present study) by the number of tests to be conducted, and using the result as the required significance level for each test (also referred to as a Bonferroni correction). For example, assume that in this investigation your alpha is initially set at .05 (as per convention). You are performing two tests for simple effects; so, this initial alpha must be divided by 2, resulting in a revised alpha of .025. When you conduct your tests for simple effects, you conclude that an effect is significant only if the p value that appears in the output is less than .025. This approach to adjusting alpha is viewed as a rigorous adjustment; you might want to consider alternative approaches for protecting the experiment-wise error rate.

The second complication involves the error term used in the analyses. With the preceding approach, the error term used in computing the F ratio for a simple effect is the mean square error from the one-way ANOVA based on data from just one group. An alternative approach is to use the mean square error from the "omnibus" test that includes the between-subjects factor, the repeated-measures factor, and the interaction term (from Output 13.3). This second approach has the advantage of using the same yardstick (the same mean square error) in the computation of both simple effects. Some authors recommend this procedure when the mean squares for the various groups are homogeneous. On the other hand, the approach just presented in which different error terms are used in the different tests has the advantage (or disadvantage, depending upon your perspective) of providing a slightly more conservative F test since it involves fewer degrees of freedom for the denominator. For a discussion of alternative approaches for testing simple effects, see Keppel (1982, pp. 428–431), and Winer (1971, pp. 527–528).

So far, this section has discussed simple effects for only the TIME factor. It is also possible to perform tests for simple effects for the GROUP factor. You can test three simple effects for GROUP in the present study. Specifically, you can determine whether there is a significant difference between the experimental group and the control group with respect to the following:

- investment scores obtained at Time 1;
- investment scores obtained at Time 2;
- investment scores obtained at Time 3.

To test a simple effect for GROUP, you simply perform a one-way ANOVA with one between-subjects factor. In this analysis, the criterion variable is the variable that includes investment scores at a specific point in time (such as Time 1). The predictor variable is GROUP.

For example, assume that you want to test the simple effect for GROUP with respect to investment scores taken at Time 1. The investment scores obtained at Time 1 are contained in a variable named PRE (as discussed earlier). The following program tests this simple effect:

```
1       PROC GLM DATA=MIXED;
2          CLASS GROUP;
3          MODEL PRE = GROUP ;
4       RUN;
```

To test the simple effect for GROUP with respect to investment scores observed at Time 2, it is only necessary to change the MODEL statement in the preceding program so that the variable PRE is replaced with the variable POST (Time 2 scores). Similarly, you can test the simple effect at Time 3 by using FOLLOWUP as the criterion variable. These lines perform these analyses:

```
5       PROC GLM DATA=MIXED;
6          CLASS GROUP;
7          MODEL POST = GROUP ;
8       RUN;
9
10      PROC GLM DATA=MIXED;
11         CLASS GROUP;
12         MODEL FOLLOWUP = GROUP ;
13      RUN;
```

The results produced by the preceding lines appear here as Output 13.5.

Output 13.5 Results of Tests for Simple Effects of GROUP for Data Gathered at Time 1 (Output pages 8 and 9), Time 2 (Output pages 10 and 11), and Time 3 (Output pages 12 and 13)

```
                                                              8
                    The GLM Procedure

                  Class Level Information

         Class           Levels     Values

         GROUP                2     1 2


         Number of observations     20
```

(continued on the next page)

Output 13.5 *(continued)*

```
                                                                      9
                        The GLM Procedure

Dependent Variable: PRE

                              Sum of
Source                 DF    Squares    Mean Square   F Value   Pr > F

Model                   1   7.20000000    7.20000000     1.98   0.1769

Error                  18  65.60000000    3.64444444

Corrected Total        19  72.80000000

            R-Square    Coeff Var     Root MSE     PRE Mean

            0.098901    19.88586      1.909043     9.600000

Source                 DF   Type I SS   Mean Square   F Value   Pr > F

GROUP                   1   7.20000000    7.20000000     1.98   0.1769

Source                 DF  Type III SS  Mean Square   F Value   Pr > F

GROUP                   1   7.20000000    7.20000000     1.98   0.1769
```

```
                                                                     10
                        The GLM Procedure

                     Class Level Information

          Class        Levels    Values

          GROUP           2       1 2

                Number of observations    20
```

```
                                                                     11
                        The GLM Procedure

Dependent Variable: POST

                              Sum of
Source                 DF    Squares    Mean Square   F Value   Pr > F

Model                   1   1.80000000    1.80000000     0.41   0.5284

Error                  18  78.40000000    4.35555556

Corrected Total        19  80.20000000

            R-Square    Coeff Var     Root MSE     POST Mean

            0.022444    19.50464      2.086997     10.70000
```

(continued on the next page)

Output 13.5 *(continued)*

Source	DF	Type I SS	Mean Square	F Value	Pr > F
GROUP	1	1.80000000	1.80000000	0.41	0.5284

Source	DF	Type III SS	Mean Square	F Value	Pr > F
GROUP	1	1.80000000	1.80000000	0.41	0.5284

12

The GLM Procedure

Class Level Information

Class	Levels	Values
GROUP	2	1 2

Number of observations   20

13

The GLM Procedure

Dependent Variable: FOLLOWUP

Source	DF	Sum of Squares	Mean Square	F Value	Pr > F
Model	1	5.00000000	5.00000000	2.00	0.1744
Error	18	45.00000000	2.50000000		
Corrected Total	19	50.00000000			

R-Square	Coeff Var	Root MSE	FOLLOWUP Mean
0.100000	14.37399	1.581139	11.00000

Source	DF	Type I SS	Mean Square	F Value	Pr > F
GROUP	1	5.00000000	5.00000000	2.00	0.1744

Source	DF	Type III SS	Mean Square	F Value	Pr > F
GROUP	1	5.00000000	5.00000000	2.00	0.1744

You can assess the simple effects for GROUP at the Time 1 level of TIME by reviewing pages 8 and 9 of Output 13.5. These pages show the results of the analysis in which PRE is the criterion variable. In the section for the Type III sum of squares at the bottom of output page 9, you can see that the *F* for this simple effect is only 1.98 which, with 1 and 18 degrees of freedom, is nonsignificant ($p = .18$). In other words, the simple effect for GROUP at the Time 1 level of TIME is nonsignificant.

By the same token, you can see that the simple effect for GROUP is also nonsignificant at Time 2 ($F = 0.41$, $p = .53$ from output page 11), and at Time 3 ($F = 2.00$, $p = .17$ from output page 13).

To step back and get the big picture concerning GROUP × TIME interaction discovered in this analysis, it becomes clear that the interaction is due to a simple effect for TIME at the experimental group level of GROUP. On the other hand, there is no evidence of a simple effect for GROUP at any level of TIME.

4. Prepare Your Own Version of the ANOVA Summary Table

Table 13.2 presents the ANOVA summary table from the analysis.

Table 13.2

ANOVA Summary Table for the Study Investigating the Change in Investment following an Experimental Manipulation (Significant Interaction)

Source	df	SS	MS	F
Between Subjects	19	157.40		
Group (A)	1	0.27	0.27	0.86
Residual between	18	157.13	8.73	
Within Subjects	40	67.33		
Time (B)	2	21.73	10.87	12.28 *
A x B Interaction	2	13.73	6.87	7.76 *
Residual within	36	31.87	0.86	
Total	59	224.73		

Note: N = 20.

* $p < .01$

Formal Description of Results for a Paper

Results were analyzed using a two-way ANOVA with repeated measures on one factor. The Group x Time interaction was significant, $F(2,36) = 7.76$, $p < .01$. Tests for simple effects showed that the mean investment scores for the control group displayed no significant differences across time, $F(2,18) = 0.76$, *ns*. However, the experimental group did display significant increases in perceived investment across time,

```
F(2,18) = 11.96, p < .01.  Post-hoc contrasts showed that the
experimental group has significantly higher scores at both
post-test, F(1,9) = 9.00; p < .05 and follow-up, F(1,9) =
19.74; p < .01 compared to baseline.
```

Use of Other Post-Hoc Tests with the Repeated-Measures Variable

In some instances, the contrasts described above will not be ideal, and you will need other multiple comparison tests for the interpretation of the analysis. For example, if there were ten levels of the repeated-measures factor, and if this factor were significant in the overall analysis, you might need to perform a multiple comparison procedure to determine which levels are different from each other. Although you could do this by using a large number of CONTRAST statements, this approach would be undesirable because each contrast is equivalent to a paired-samples *t* test. A multiple comparison procedure that controls for experiment-wise error rate would be preferable.

Several such tests are available in SAS/STAT for this situation (e.g., the Bonferroni test, the Scheffe test, the Tukey test). However, you cannot use these tests with the repeated-measures factor when you use the REPEATED statement. To utilize these tests, you must name the variable being analyzed in the MODEL statement. Fortunately, SAS can compute the ANOVA in a way that allows you to use these multiple comparison tests when analyzing a repeated factor. Essentially, this approach requires that the analysis be performed without use of the REPEATED statement. This procedure is somewhat more cumbersome than the approach previously described as it requires that the dataset be restructured. See Chapter 8 of Cody and Smith (1997) for ways to undertake such multiple comparison tests.

Conclusion

In summary, the two-way mixed design ANOVA is appropriate for a type of experimental design that is frequently used in social science research. In the most common form of this design, participants are divided on a group factor and provide repeated measures on the criterion variable at various time intervals. This analysis is easily handled by PROC GLM. Conducting the analysis with the REPEATED statement offers the advantage that the error terms for the various effects are automatically selected by the program, while conducting the analysis without the REPEATED statement allows you to request a number of post-hoc multiple comparison procedures.

Assumptions Underlying Factorial ANOVA with Repeated-Measures Factors and Between-Subjects Factors

All of the statistical assumptions associated with a one-way ANOVA, repeated-measures design are also required for a factorial ANOVA with repeated-measures factors and between-subjects factors. In addition, the latter design requires a homogeneity of covariances assumption for the multivariate test and the homogeneity assumption as well as two symmetry conditions for the univariate test. This section reviews the assumptions for the one-way repeated measures ANOVA and introduces the new assumptions for the factorial ANOVA, mixed design.

Assumptions for the Multivariate Test

- **Level of measurement.** The criterion variable should be assessed on an interval or ratio level of measurement. The predictor variables should both be nominal-level variables (i.e., categorical variables). One predictor codes the within-subjects variable, and the second codes the between-subjects variable.

- **Independent observations.** A given participant's score in any one condition should not be affected by any other participant's score in any of the study's conditions. However, it is acceptable for a given participant's score in one condition to be dependent upon his or her own score in a different condition (under the within-subjects predictor variable).

- **Random sampling.** Scores on the criterion variable should represent a random sample drawn from the populations of interest.

- **Multivariate normality.** The measurements obtained from participants should follow a multivariate normal distribution. Under conditions normally encountered in social science research, violations of this assumption have only a small effect on the Type I error rate (i.e., the probability of incorrectly rejecting a true null hypothesis).

- **Homogeneity of covariance matrices.** In the population, the criterion-variable covariance matrix for a given group (under the between-subject's predictor variable) should be equal to the covariance matrix for each of the remaining groups. This assumption was discussed in greater detail in Chapter 11, "Multivariate Analysis of Variance (MANOVA) with One Between-Subjects Factor."

Assumptions for the Univariate Test

The univariate test requires all of the preceding assumptions as well as the following assumptions of sphericity and symmetry:

- **Sphericity.** Sphericity is a characteristic of a difference-variable covariance matrix that is obtained when performing a repeated-measures ANOVA. (The concept of sphericity was discussed in greater detail in Chapter 12.) Briefly, two conditions must be satisfied for the covariance matrix to demonstrate sphericity. First, each variance on the diagonal of the matrix should be equal to every other variance on the diagonal. Second, each covariance off the diagonal should equal zero. (This is analogous to saying that the correlations between the difference variables should be zero.) Remember that, in a study with a between-subjects factor, there is a separate difference-variable covariance matrix for each group under the between-subjects variable.

- **Symmetry conditions.** There are two symmetry conditions, and the first of these is the sphericity condition just described. The second condition is that the difference-variable covariance matrices obtained for the various groups (under the between-subjects factor) should be equal to one another.

For example, assume that a researcher has conducted a study that includes a repeated-measures factor with three conditions and a between-subjects factor with two conditions. Participants assigned to condition 1 under the between-subjects factor are designated as "group 1" and those assigned to condition 2 are designated as "group 2." With this research design, one difference-variable covariance matrix is obtained for group 1 and a second for group 2. (The nature of these difference-variable covariance matrices was discussed in Chapter 12.) The symmetry conditions are met if both matrices demonstrate sphericity and each element in the matrix for group 1 is equal to its corresponding element in the matrix for group 2.

References

Cody, R. P., & Smith, J. K. (1997). *Applied statistics and the SAS programming language* (4th ed.). Upper Saddle River, NJ: Prentice Hall.

Keppel, G. (1982). *Design and analysis: A researcher's handbook* (2nd ed.). Englewood Cliffs, NJ: Prentice Hall.

Looney, S. & Stanley, W. (1989). Exploratory repeated measures analysis for two or more groups. *The American Statistician, 43,* 220-225.

Rusbult, C. E. (1980). Commitment and satisfaction in romantic associations: A test of the investment model. *Journal of Experimental Social Psychology, 16,* 172-186.

Winer, B. J. (1971). *Statistical principles in experimental design* (2nd ed.). New York: McGraw-Hill.

Multiple Regression

Overview. This chapter shows how to perform multiple regression analysis to investigate the relationship between a continuous criterion variable and multiple continuous predictor variables. It describes the different components of the multiple regression equation, and discusses the meaning of R^2 and other results from a multiple regression analysis. It shows how bivariate correlations, multiple regression coefficients, and uniqueness indices can be reviewed to assess the relative importance of predictor variables. Fictitious data are examined using PROC CORR and PROC REG to show how the analysis can be conducted and to illustrate how the results can be summarized in tables and in text.

Introduction: Answering Questions with Multiple Regression

Multiple regression is a highly flexible procedure that allows researchers to address many different types of research questions with many different types of data. Perhaps the most common multiple regression analysis involves a single continuous criterion variable measured on an interval or ratio scale and multiple continuous predictor variables also assessed on an interval or ratio scale.

For example, you might be interested in determining the relative importance of variables that are believed to predict adult income. To conduct your research, you obtain information for 1,000 Canadian adults. The criterion variable in your study is annual income for these participants. The predictor variables are age, years of education, and income of parents. In this study, the criterion variable as well as the predictor variables are each continuous and are all assessed on an interval or ratio scale. Because of this, multiple regression is an appropriate data analysis procedure.

Analysis with multiple regression allows you to answer a number of research questions. For example, it allows you to determine:

- whether there is a significant relationship between the criterion variable and the multiple predictor variables, when examined as a group;
- whether the multiple regression coefficient for a given predictor variable is statistically significant (this coefficient represents the amount of weight given to a specific predictor, while holding constant the other predictors);

- whether a given predictor accounts for a significant amount of variance in the criterion, beyond the variance accounted for by the other predictors.

By conducting the preceding analyses, you learn about the relative importance of the predictor variables included in your multiple regression equation. Researchers conducting nonexperimental research in the social sciences are often interested in learning about the relative importance of naturally occurring predictor variables. This chapter shows how to perform such analyses.

Because multiple regression is such a flexible procedure, there are many other types of regression analyses that are beyond the scope of this chapter. For example, in the study dealing with annual income that was discussed earlier, all predictor variables were continuous on an interval or ratio scale. Nominal (classification) variables might also be used as predictors in a multiple regression analysis provided that they have been appropriately transformed using dummy-coding or effect-coding. Because this chapter provides only an introduction to multiple regression, it does not cover circumstances in which nominal-scale variables are included as predictors. Also, this chapter covers only those situations in which a linear relationship exists between the predictor variables and the criterion; curvilinear and interactive relationships are not discussed.

Once you learn the basics of multiple regression from this chapter, you can learn more about advanced regression topics (e.g., dummy-coding nominal variables or testing curvilinear relationships) in Cohen, Cohen, West, and Aiken (2003), or Pedhazur (1982).

Multiple Regression versus ANOVA

Chapters 9 through 13 presented analysis of variance (ANOVA) procedures that are commonly used to analyze data from **experimental research**: research in which one or more categorical independent variables (such as "experimental condition") are manipulated to determine how they affect a study's dependent variable.

For example, imagine that you were interested in studying prosocial behavior: actions intended to help others. Examples of prosocial acts might include donating money to the poor, donating blood, doing volunteer work at a hospital, and so forth. You might have developed an experimental treatment that you believe will increase the likelihood that people will engage in prosocial acts. To investigate this, you conduct an experiment in which you manipulate the independent variable (e.g., half of your participants are given the experimental treatment and the other half are given a placebo treatment). You then assess your dependent variable, the number of prosocial acts that the participants later perform. It would be appropriate to analyze data from this study using one-way ANOVA, because you had a single criterion variable assessed on an interval/ratio scale (number of prosocial acts) as you had a single predictor variable measured on a nominal scale (experimental group).

Multiple regression is similar to ANOVA in at least one important respect: with both procedures, the criterion variable should be continuous and should be assessed on either an interval- or ratio-level of measurement. Chapter 1 indicates that a **continuous** variable is one that can assume a relatively large number of values. For example, the "number of prosocial acts performed over a six-month period" can be a continuous variable, provided that participants demonstrate a wide variety of scores (e.g., 0, 4, 10, 11, 20, 25, 30).

However, multiple regression also differs from ANOVA in some ways. The most important difference involves the nature of the predictor variables. When data are analyzed with ANOVA, the predictor variable is a categorical variable (i.e., a variable that simply codes group membership). In contrast, predictor variables in multiple regression are generally continuous.

As an illustration, assume that you conduct a study in which you administer a questionnaire to a group of participants to assess the number of prosocial acts each has performed. You can then proceed to obtain scores for the same participants on each of the following predictor variables:

- age;
- income;
- a questionnaire-type scale that assesses level of moral development.

Perhaps you have a hypothesis that the number of prosocial acts performed is causally determined by these three predictor variables, as illustrated by the model in Figure 14.1:

Figure 14.1 A Model of the Determinants of Prosocial Behavior

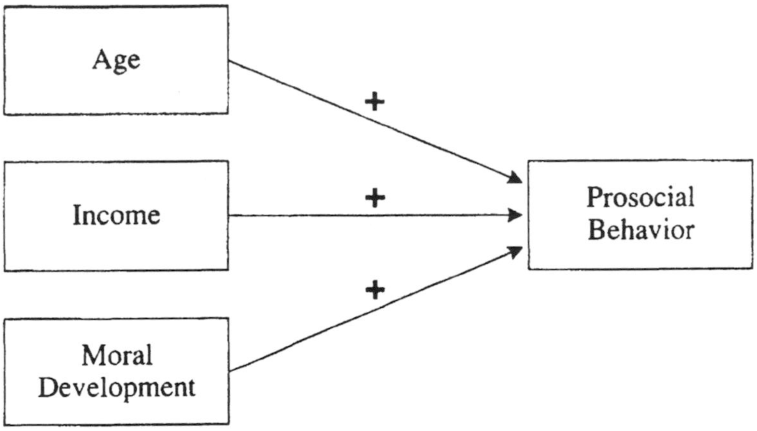

You can see that each predictor variable in your study (i.e., age, income, and moral development) is continuous, and is assessed on an interval or ratio scale. This, combined with the fact that your criterion variable is also a continuous variable assessed on an interval/ratio scale, means that you can analyze your data using multiple regression.

This is a most important distinction between the two statistical procedures. With ANOVA, the predictor variables are always categorical; with multiple regression, they are generally continuous ("generally" because categorical variables are sometimes used in multiple regression provided they have been dummy-coded or effect-coded). For more information, see Cohen, Cohen, West, and Aiken (2003) or Pedhazur (1982).

Multiple Regression and Naturally Occurring Variables

Multiple regression is particularly well-suited for studying the relationship between **naturally occurring** predictor and criterion variables; that is, variables that are not manipulated by the researcher, but are simply measured as they naturally occur. The preceding prosocial behavior study provides good examples of naturally occurring predictor variables: age; income; and level of moral development.

This is what makes multiple regression such an important tool in the social sciences. It allows researchers to study variables that cannot be experimentally manipulated. For example, assume that you have a hypothesis that domestic violence (an act of aggression against a domestic partner) is caused by the following:

- childhood trauma experienced by the abuser;
- substance abuse;
- low self-esteem.

The model illustrating this hypothesis appears in Figure 14.2. It is obvious that you would not experimentally manipulate the predictor variables of the model and later observe the participants as adults to see if the manipulation affected their propensity for domestic violence. However, it is possible to simply measure these variables as they naturally occur and determine whether they are related to one another in the predicted fashion. Multiple regression allows you to do this.

Figure 14.2 A Model of the Determinants of Domestic Violence

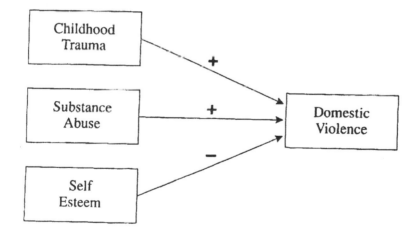

Does this mean that ANOVA is only for the analysis of manipulated predictor variables, while multiple regression is only for the analysis of naturally occurring variables? Not necessarily, because naturally occurring variables can be predictor variables in an ANOVA, provided they are categorical. For example, ANOVA can be used to determine whether participant sex (a naturally occurring predictor variable) is related to relationship commitment (a criterion variable). In addition, a categorical manipulated variable (such as "experimental condition") can be used as a predictor variable in multiple regression, provided that it has been dummy-coded or effect-coded. The main distinction to remember is this: with ANOVA, the predictor variables can only be categorical variables; with multiple regression, the predictor variables can be either categorical or continuous.

> **How large must my sample be?** Multiple regression is a large-sample procedure; unreliable results might be obtained if the sample does not include at least 100 observations, preferably 200. The greater the number of predictor variables included in the multiple regression equation, the greater the number of participants that will be necessary to obtain reliable results. Most experts recommend at least 15 to 30 participants per predictor variable. See Cohen (1992).

"Proving" Cause and Effect Relationships

Multiple regression can determine whether a given set of variables is useful for predicting a criterion variable. Among other things, this means that multiple regression can be used to determine the following:

- whether or not the relationship between the criterion variable and predictor variables (taken as a group) is statistically significant;

- how much variance in the criterion is accounted for by the predictors;

- which predictor variables are relatively important predictors of the criterion.

Although the preceding section often refers to causal models, it is nonetheless important to remember that the procedures discussed in this chapter do not provide evidence concerning cause and effect relationships between predictor variables and the criterion. For example, consider the causal model presented in Figure 14.2. Assume that naturally occurring data for these four variables are gathered and analyzed using multiple regression. Assume further that the results are significant: that multiple regression coefficients for all three of the predictor variables are significant and in the predicted direction. Even though these findings are consistent with your theoretical model, they do not "prove" that these predictor variables have a causal effect on domestic violence. Because the data are correlational, there is probably more than one way to interpret the observed relationships among the four variables. The most that you would be justified in saying is that your findings are *consistent* with the causal model portrayed in Figure 14.2. It would be inappropriate to say that the results "prove" that the model is correct.

Then why analyze correlational data with multiple regression at all? There are several reasons. Very often, researchers are not really interested in testing a causal model. Perhaps the purpose of the study is simply to understand relationships among a set of variables.

However, even when the research is based on a causal model, multiple regression can still be useful. For example, if you obtained correlational data relevant to the domestic violence model of Figure 14.2, analyzed it, and found that none of the multiple regression coefficients were significant, this would still be useful information because it showed that the model failed to survive a test (i.e., it failed to survive an analysis that investigated the predicted relationships between the variables).

If the model does survive an analysis with multiple regression (i.e., if significant results are obtained), this is useful as well. You can prepare a research report indicating that the results were consistent with the hypothesized model. In other words, the model survived an attempt at disconfirmation. If you are dealing with predictor variables that can be manipulated ethically, you might choose next to conduct an experiment to determine whether the predictors appear to have a causal effect on the criterion variable under controlled circumstances.

In summary, it is important to remember that multiple regression procedures discussed in this chapter provide virtually no evidence of cause and effect relationships. However, they are extremely useful for determining whether one set of variables can significantly and meaningfully predict variation in a criterion. Use of the term *prediction* in multiple regression should therefore not be confused with *causation*.

Background: Predicting a Criterion Variable from Multiple Predictors

The criterion (or predicted) variable in multiple regression is represented by the symbol Y, and is therefore often referred to as the "Y variable." The predictor variables are represented as X_1, X_2, X_3... X_n, and are referred to as the "X variables" (i.e., independent variables). The purpose of multiple regression is to understand the relationship between the Y variable and the X variables when taken as a group.

A Simple Predictive Equation

For example, consider again the model of the determinants of prosocial behavior presented in Figure 14.1. This model hypothesizes that the number of prosocial acts performed by an individual in a given period of time can be predicted by three variables:

- the participant's age;
- the participant's income;
- the participant's level of moral development.

Notice that each arrow (assumed causal path) in the figure is identified with either a "+"or a "–" sign. A plus sign indicates that you expect the relevant predictor variable to demonstrate a positive relationship with the criterion whereas a minus sign indicates that you expect the predictor to demonstrate a negative relationship. The nature of these signs in Figure 14.1 shows that you expect:

- a positive relationship between age and prosocial behavior, meaning that older participants will perform more prosocial acts;
- a positive relationship between income and prosocial behavior, meaning that more affluent people will perform more prosocial acts;
- a positive relationship between moral development and prosocial behavior, meaning that participants who score higher on the paper-and-pencil measure of moral development will perform more prosocial acts.

Assume that you have administered a questionnaire to a sample of 100 participants to assess their level of moral development. (Scores on this scale can range from 10 to 100 with higher scores reflecting higher levels of development.) From the same participants, you have also obtained information regarding their age and income. You now want to use this information to predict the number of prosocial acts that the participant will perform in the next six months. More specifically, you want to create a new variable, Y′, that represents your best guess of how many prosocial acts participants will perform. Y′ represents your prediction of each participant's standing on the criterion variable (as distinguished from Y, which is the participant's *actual* standing on the criterion variable).

Assume that the three predictor variables are positively related to prosocial behavior. To arrive at a prediction of how many prosocial behaviors in which participants will engage, one of your options is to simply add together the participants' scores on the three X variables and allow the sum to be your best guess of how many prosocial acts they will perform. You could do this using the following equation:

$$Y' = X_1 + X_2 + X_3$$

where:

Y' = the participants' predicted scores on "prosocial behavior";

X_1 = the participants' actual scores on "age";

X_2 = the participants' actual scores on "income" (in thousands);

X_3 = the participants' actual scores on the moral development scale.

To make this more concrete, consider the fictitious data presented in Table 14.1. This table presents scores for four of the study's participants on the three predictor variables.

Table 14.1

Fictitious Data, Prosocial Behavior Study

Participant	Age (X_1)	Income (in thousands) (X_2)	Moral Development (X_3)
1. Lars	19	15	12
2. Hiro	24	32	28
3. Karim	33	45	50
.			
.			
.			
100. Eamonn	55	60	95

To arrive at an estimate of the number of prosocial behaviors in which the first participant (Lars) will engage, you could insert his scores on the three X variables into the preceding equation:

$$Y' = X_1 + X_2 + X_3$$

$$Y' = 19 + 15 + 12$$

$$Y' = 46$$

So your best guess is that the first participant, Lars, will perform 46 prosocial acts in the next six months. By repeating this process for all participants, you could go on to compute their Y' scores in the same way. Table 14.2 presents the predicted scores on the prosocial behavior variable for some of the study's participants.

Table 14.2

Predicted Scores on the Prosocial Behavior Variable Using a Simple Predictive Equation

Participant	Predicted Scores On Prosocial Behavior (Y')	Age (X$_1$)	Income (in thousands) (X$_2$)	Moral Development (X$_3$)
1. Lars	46	19	15	12
2. Hiro	84	24	32	28
3. Karin	128	33	45	50
.				
.				
.				
100. Eamonn	338	55	60	95

Notice the general relationships between Y' and the X variables in Table 14.2. Because of the way Y' was created, if participants have low scores on age, income, and moral development, your equation predicts that they will engage in relatively few prosocial behaviors. However, if participants have high scores on these X variables, your equation predicts that they will engage in a relatively large number of prosocial behaviors. For example, Lars had relatively low scores on these X variables, and as a consequence, the equation predicts that he will perform only 46 prosocial acts over the next six months. In contrast, Eamonn displayed relatively high scores on age, income, and moral development; as a consequence, your equation predicts that he will engage in 338 prosocial acts.

So, in effect, you have created a new variable, Y'. Imagine that you now go out and gather data regarding the actual number of prosocial acts that these participants engage in over the following six months. This variable would be represented with the symbol Y, because it represents the participants' *actual* scores on the criterion (and not their predicted scores, Y').

Once you determine the actual number of prosocial acts performed by the participants, you can list them in a table alongside their predicted scores on prosocial behavior as in Table 14.3:

Table 14.3

Actual and Predicted Scores on the Prosocial Behavior Variable

Participant	Actual Scores On Prosocial Behavior (Y)	Predicted Scores On Prosocial Behavior (Y')
1. Lars	10	46
2. Hiro	40	84
3. Karim	70	128
.		
.		
.		
100. Eamonn	130	338

Notice that, in some respects, your predictions of the participants' scores on Y are not terribly accurate. For example, your equation predicted that Lars would engage in 46 prosocial activities, but in reality he engaged in only 10. Similarly, it predicted that Eamonn would engage in 338 prosocial behaviors while he actually engaged in only 130.

Despite this, you should not lose sight of the fact that your new variable Y' does appear to be correlated with the actual scores on Y. Notice that participants with low scores on Y' (such as Lars) also tend to have low scores on Y; notice that participants with high scores on Y' (such as Eamonn) also tend to have high scores on Y. If you compute a product-moment correlation (r) between Y and Y', you would probably observe a moderately large correlation coefficient. This trend is supportive of your model as it suggests that there really is a relationship between Y and the three X variables when taken as a group.

The procedures (and the predictive equation) described in this section are somewhat crude in nature, and do not describe the way that multiple regression is actually performed. However, they do illustrate some important basic concepts in multiple regression analysis. For example, in multiple regression analysis, you create an artificial variable, Y', to represent your best guess of participants' standings on the criterion variable. The relationship between this variable and participants' actual standing on the criterion (Y) is assessed to indicate the strength of the relationship between Y and the X variables when taken as a group.

Multiple regression as it is actually performed has many important advantages over the crude practice of simply adding together the X variables, as was illustrated in this section. With true multiple regression, the various X variables are multiplied by optimal weights before they are added together to create Y'. This generally results in a more accurate estimate of participants' standing on the criterion variable. In addition, you can use the results of a true multiple regression procedure to determine which of the X variables are relatively important and which are relatively unimportant predictors of Y. These issues are discussed in the following section.

An Equation with Weighted Predictors

In the preceding section, each predictor variable was given approximately equal weight when computing scores on Y′. You did not, for example, give twice as much weight to income as you gave to age when computing Y′ scores. Assigning equal weights to the various predictors might make sense in some situations, especially when all of the predictors are equally predictive of the criterion.

However, what if some X variables in a model are better predictors of Y than others? For example, what if your measure of moral development displayed a strong correlation with prosocial behavior ($r = .70$), income demonstrated a moderate correlation with prosocial behavior ($r = .40$), and age demonstrated only a weak correlation ($r = .20$)? In a situation such as this, it would make sense to assign different weights to the different predictors. For example, you might assign a weight of 1 to age, a weight of 2 to income, and a weight of 3 to moral development. The predictive equation that reflects this weighting scheme is

$$Y' = (1) X_1 + (2) X_2 + (3) X_3$$

In this equation, once again X_1 = age, X_2 = income, and X_3 = moral development. In calculating a given participant's score on Y′, you would multiply his or her score on each X variable by the appropriate weight and sum the resulting products. For example, Table 14.2 showed that Lars had a score of 19 on X_1, a score of 15 on X_2, and a score of 12 on X_3. His predicted prosocial behavior score, Y′, could be calculated in the following way:

```
Y' =  (1)  X₁   +  (2)  X₂  +  (3)  X₃
Y' =  (1)  19   +  (2)  15  +  (3)  12
Y' =       19   +       30  +       36
Y' =  85
```

This weighted equation predicts that Lars will engage in 85 prosocial acts over the next six months. You could use the same weights to compute the remaining participants' scores on Y′ in the same way. Although this example is again somewhat crude in nature, it is this concept of optimal weighting that is at the heart of multiple regression analysis.

The Multiple Regression Equation

Regression Coefficients and Intercepts

In linear multiple regression as performed by the SAS PROC REG (regression) procedure, optimal weights of the sort described in the preceding section are automatically calculated in the course of the analysis. The following symbols are used to represent the various components of an actual multiple regression equation:

$$Y' = b_1 X_1 + b_2 X_2 + b_3 X_3 \ ... + b_k X_k + a$$

where:

Y' = participant's predicted scores on the criterion variable;

b_k = the nonstandardized multiple regression coefficient for the kth predictor variable;

X_k = the kth predictor variable;

a = intercept constant.

Some components of this equation, such as Y', have already been discussed. However, some new components require additional explanation.

The term "b_k" represents the nonstandardized multiple regression coefficient for an X variable. A **multiple regression coefficient** for a given X variable represents the average change in Y that is associated with a one-unit change in that X variable, while holding constant the remaining X variables. This somewhat technical definition for a regression coefficient is explained in more detail in a later section. For the moment, however, it is useful to think of a regression coefficient as revealing the amount of *weight* that the X variable is given when computing Y'. For this reason, these are sometimes referred to as **b weights**.

The symbol "a" represents the **intercept constant** of the equation. The intercept is a fixed value that is either added to or subtracted from the weighted sum of X scores when computing Y'. The inclusion of this constant in the regression equation improves the accuracy of prediction.

To develop a true multiple regression equation using PROC REG, it is necessary to gather data on both the Y variable and the X variables. Assume that you do this in a sample of 100 participants. You analyze the data, and the results of your analyses indicate that the relationship between prosocial behavior and the three predictor variables can be described by the following equation:

$$Y' = b_1 X_1 + b_2 X_2 + b_3 X_3 + a$$

$$Y' = (.10) X_1 + (.25) X_2 + (1.10) X_3 + (-3.25)$$

The preceding equation indicates that your best guess of a given participant's score on prosocial behavior can be computed by multiplying his or her age by .10, multiplying his or her income by .25, multiplying his or her score on moral development by 1.10 then summing these products and subtracting the intercept of 3.25 from this sum. This process is illustrated by inserting Lars' scores on the X variables in the equation:

$$Y' = (.10)\ 19 + (.25)\ 15 + (1.10)\ 12 + (-3.25)$$

$$Y' = 1.9 \qquad + 3.75 \qquad + 13.2 \qquad + (-3.25)$$

$$Y' = 18.85 \quad + (-3.25)$$

$$Y' = 15.60$$

Your best guess is that Lars will perform 15.60 prosocial acts over the next six months. You can calculate the Y' scores for the remaining participants in Table 14.2 by inserting their X scores in this same equation.

The Principle of Least Squares

At this point, it is reasonable to ask, "How did PROC REG determine that the 'optimal' b weight for X_1 was .10? How did it determine that the 'optimal' weight for X_2 was .25? How did it determine that the 'optimal' intercept ("a" term) was –3.25?"

The answer is that these values are "optimal" in the sense that they minimize errors of prediction. An **error of prediction** refers to the difference between a participant's actual score on the criterion (Y), and his or her predicted score on the criterion (Y'). This difference can be illustrated as follows:

$$Y - Y'$$

Remember that you must gather actual scores on Y in order to perform multiple regression so it is, in fact, possible to compute the error of prediction (the difference between Y and Y') for each participant in the sample. For example, Table 14.4 reports the actual score for several participants on Y, their predicted scores on Y' based on the optimally weighted regression equation above, and their errors of prediction.

Table 14.4

Errors of Prediction Based on an Optimally Weighted Multiple Regression Equation

Participant	Actual Scores On Prosocial Behavior (Y)	Predicted Scores On Prosocial Behavior (Y')	Errors of Prediction (Y – Y')
1. Lars	10	15.60	-5.60
2. Hiro	40	37.95	2.05
3. Karim	70	66.30	3.70
.			
.			
.			
100. Eamonn	130	121.75	8.25

For Lars, the actual number of prosocial acts performed was 10, while your multiple regression equation predicted that he would perform 15.60 acts. The error of prediction for Lars is therefore 10 – 15.60 or –5.60. The errors of prediction for the remaining participants are calculated in the same manner.

Earlier, it was stated that the b weights and intercept calculated by PROC REG are optimal in the sense that they minimize errors of prediction. More specifically, these b weights and intercept are computed according to the principle of least squares. The **principle of least squares** says that Y′ values should be calculated so that the sum of the squared errors of prediction are minimal. The sum of the squared errors of prediction can be calculated using this formula:

$$\sum (Y - Y')^2$$

To compute the sum of the squared errors of prediction according to this formula, it is necessary only to:

- compute the error of prediction (Y – Y′) for a given participant;
- square this error;
- repeat this process for all remaining participants;
- sum the resulting squares. (The purpose of squaring the errors before summing them is to eliminate the minus sign that some of the difference scores will display.)

When a given dataset is analyzed using multiple regression, PROC REG applies a set of formulae that calculates the optimal b weights and the optimal intercept for that dataset. The formulas that do this calculate b weights and an intercept that best minimize these squared errors of prediction. That is why we say that multiple regression calculates "optimal" weights and intercepts. They are optimal in the sense that no other set of b weights or intercept could do a better job of minimizing squared errors of prediction for the current dataset.

With these points established, it is now possible to summarize what multiple regression actually allows you to do:

> Multiple regression allows you to examine the relationship between a single criterion variable and an optimally weighted linear combination of predictor variables.

In the preceding statement, "optimally weighted linear combination of predictor variables" refers to Y′. The expression "linear combination" refers to the fact that the various X variables are combined or added together (using the formula for a straight line), to arrive at Y′. The words "optimally weighted" refer to the fact that X variables are assigned weights that satisfy the principle of least squares.

Although we normally think of multiple regression as a procedure that examines the relationship between single criterion and multiple predictor variables, it also possible to view it in somewhat more simple terms: as a procedure that examines the relationship between just two variables, Y and Y′.

The Results of a Multiple Regression Analysis

The Multiple Correlation Coefficient

The **multiple correlation coefficient**, symbolized as R, represents the strength of the relationship between a criterion variable and an optimally weighted linear combination of predictor variables. Its values can range from 0 through 1.00; it is interpreted in the same way a Pearson product-moment correlation coefficient (r) is interpreted (except that R can only assume positive values). Values approaching zero indicate little relationship between the criterion and the predictors whereas values near 1.00 indicate strong relationships. An R value of 1.00 indicates perfect or complete prediction of criteria values.

Conceptually, R should be viewed as the product-moment correlation between Y and Y'. This can be symbolized in the following way:

$$R = r_{YY'}$$

In other words, if you were to obtain data from a sample of participants that included their scores on Y as well as a number of X variables, computed Y' scores for each participant, and then correlated predicted criterion scores (Y') with actual scores (Y), the resulting bivariate correlation coefficient would be equivalent to R.

With bivariate regression, it is possible to estimate the amount of variance in Y that is accounted for by X by simply squaring the correlation between the two variables. The resulting product is sometimes referred to as the **coefficient of determination**. For example, if $r = .50$ for a given pair of variables, then their coefficient of determination can be computed as:

$$\text{Coefficient of determination} = r^2$$
$$= (.50)^2$$
$$= .25$$

In this situation, therefore, it can be said the X variable accounts for 25% of the observed variance in the Y variable.

An analogous coefficient of determination can also be computed in multiple regression by simply squaring the observed multiple correlation coefficient, R. The resultant R^2 **value** (often referred to simply as R-squared) represents the percentage of variance in Y that is accounted for by the linear combination of predictor variables. The concept of "variance accounted for" is an extremely important one in multiple regression analysis and is therefore given detailed treatment in the following section.

Variance Accounted for by Predictor Variables: The Simplest Models

In multiple regression analyses, researchers often speak of "variance accounted for." By this, they mean the percent of observed variance in the criterion variable accounted for by the linear combination of predictor variables.

A Single Predictor Variable

This concept is easier to understand by beginning with a simple bivariate example. Assume that you compute the correlation between prosocial behavior and moral development and find that $r = .50$ for these two variables. As was previously discussed, you can determine the percentage of variance in prosocial behavior that is accounted for by moral development by squaring this correlation coefficient:

$$r^2 = (.50)^2$$

$$= .25$$

Thus, 25% of observed variance in prosocial behavior is accounted for by moral development. This can be illustrated graphically by using a Venn diagram in which the total variance in a variable is represented with a circle. The Venn diagram representing the correlation between prosocial behavior and moral development is in Figure 14.3:

Figure 14.3 Venn Diagram: Variance in Prosocial Behavior Accounted for by Moral Development

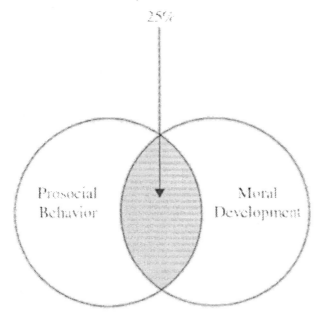

Notice that the circle representing moral development overlaps the circle representing prosocial behavior. This indicates that the two variables are correlated. More specifically, the figure shows that there is an overlap of 25% of the area of the prosocial behavior circle by the moral development circle, meaning that moral development accounts for 25% of observed variance in prosocial behavior.

Multiple Predictor Variables with Intercorrelations of Zero

It is now possible to expand the discussion to the situation in which there are multiple X variables. Assume that you obtained data on prosocial behavior, age, income, and moral development from a sample of 100 participants and observe correlations among the four variables as summarized in Table 14.5:

Table 14.5

Correlation Matrix: Zero Correlations among X Variables

	Variable	Y	X_1	X_2	X_3
Y	Prosocial behavior	1.00			
X_1	Age	.30	1.00		
X_2	Income	.40	.00	1.00	
X_3	Moral development	.50	.00	.00	1.00

When all possible correlations are computed for a set of variables, these correlations are usually presented in the form of a **correlation matrix** such as the one presented in Table 14.5. To find the correlation between two variables, you simply find the row for one variable, and the column for the second variable. The correlation between variables appears where the row intersects the column. For example, where the row for X_1 (age) intersects the column for Y (prosocial behavior), you see a correlation coefficient of $r = .30$.

You can use the preceding correlations to determine how much variance in Y is accounted for by the three X variables. For example, the correlation between age and prosocial behavior is $r = .30$; squaring this results in .09 (because .30 x .30 = .09). This means that age accounts for 9% of observed variance in prosocial behavior. Following the same procedure, you learn that income accounts for 16% of the variance in Y (because .40 x .40 = .16), while moral development continues to account for 25% of the variance in Y.

Notice another (somewhat unusual) fact concerning the correlations in this table: each of the X variables has a correlation of zero with all of the other X variables. For example, where the row for X_2 (income) intersects with the column for X_1 (age), you can see a correlation coefficient of .00. In the same way, Table 14.5 also shows correlations of .00 between X_2 and X_3, and between X_1 and X_3.

The correlations among the variables of Table 14.5 can be illustrated with the Venn diagram of Figure 14.4:

Figure 14.4 Venn Diagram: Variance in Prosocial Behavior Accounted for by Three Noncorrelated Predictors

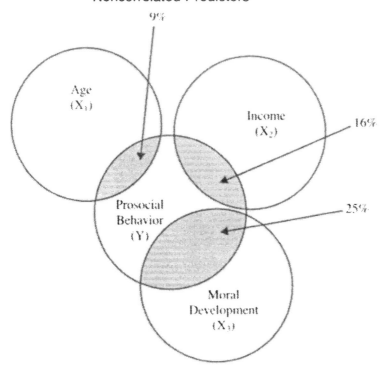

Notice two important points concerning the Venn diagram in Figure 14.4: each X variable accounts for some variance in Y but no X variable accounts for any variance in any other X variable. Because the preceding table showed that the X variables were uncorrelated, it is necessary to draw the Venn diagram so that there is no overlap between any of the X variables.

How much variance in Y is accounted for by the three predictors of Figure 14.4? This is easily determined by simply adding together the percent of variance accounted for by each of the three predictors individually:

$$\text{Total variance accounted for} = .09 + .16 + .25$$
$$= .50$$

The linear combination of X_1, X_2, and X_3 accounts for 50% of observed variance in prosocial behavior. In most areas of research in the social sciences, this would be considered a fairly large percentage of variance.

In the preceding example, the total variance in Y accounted for by the X variables was determined by simply summing the squared bivariate correlations between Y and the individual X variables. It is important to remember that you can use this procedure to determine the total variance accounted for in Y only when the X variables are completely uncorrelated with one another (which is rare). When there is any degree of correlation among the X variables, this approach gives misleading results. The reasons for this are discussed in the next section.

Variance Accounted for by Intercorrelated Predictor Variables

The preceding section shows that it is simple to determine the percent of variance in a criterion that is accounted for by a set of predictor variables when the predictor variables display zero correlation with one another. In that situation, the total variance accounted for will be equal to the sum of the squared bivariate correlations (i.e., the sum of r_{Y1}^2, $r_{Y2}^2 \dots r_{Yn}^2$). The situation becomes much more complex, however, when the predictor variables are correlated with one another. In this situation, it is not possible to make simple statements about how much variance will be accounted for by a set of predictors.

In part, this is because multiple regression equations with correlated predictors behave one way when the predictors include a suppressor variable and behave a very different way when the predictors do not contain a suppressor variable. A later section will explain just what a suppressor variable is, and will describe the complexities that are introduced by this somewhat rare phenomenon. First, consider the (relatively) simpler situation that exists when the predictors included in a multiple regression equation are intercorrelated but do not contain a suppressor variable.

When Intercorrelated Predictors Do Not Include a Suppressor Variable

The preceding example was fairly unrealistic in that each X variable was said to have a correlation of zero with all of the other X variables. In nonexperimental research in the social sciences, you will almost never observe a set of predictor variables that are mutually uncorrelated in this way. Remember that nonexperimental research involves measuring naturally occurring (nonmanipulated) variables, and naturally occurring variables will almost always display some degree of intercorrelation.

For example, consider the nature of the variables studied here. Given that people tend to earn higher salaries as they grow older, it is likely that participant age would be positively correlated with income. Similarly, people with higher incomes might well demonstrate higher levels of moral development since they do not experience the deprivation and related stresses of poverty. If these assertions are correct (and please remember that this is only speculation), you might expect to see a correlation of perhaps $r = .50$ between age and income as well as a correlation of $r = .50$ between income and moral development. These new correlations are displayed in Table 14.6. Notice that this table is similar to the preceding table in that all X variables display the same correlations with Y as were displayed earlier; age and moral development are still uncorrelated.

Table 14.6

Correlation Matrix: Nonzero Correlations among X Variables

	Variable	Y	X_1	X_2	X_3
Y	Prosocial behavior	1.00			
X_1	Age	.30	1.00		
X_2	Income	.40	.50	1.00	
X_3	Moral development	.50	.00	.50	1.00

The X variables of Table 14.6 display the same correlations with Y as previously displayed in Table 14.5. Does this mean that the linear combination of the three X variables will still account for the same total percentage of variance in Y? In most cases, the answer is no. Remember that, with Table 14.5, the X variables were not correlated with one another whereas in Table 14.6, there is now a substantial correlation between age and income, and between income and moral development. These correlations can decrease the total variance in Y that is accounted for by the X variables. The reasons for this are illustrated in the Venn Diagram of Figure 14.5.

Figure 14.5 Venn Diagram: Variance Accounted for by Three Correlated Predictors

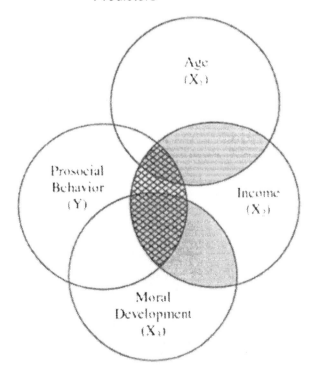

In Figure 14.5, the shaded area represents variance in income that is accounted for by age and moral development. The area with cross-hatching, on the other hand, represents variance in prosocial behavior that is accounted for by income.

Notice that each X variable individually still accounts for the same percentage of variance in prosocial behavior: age still accounts for 9%; income still accounts for 16%; and moral development still accounts for 25%. In this respect, Figure 14.5 is similar to Figure 14.4.

However, there is one important respect in which the two figures are different. Figure 14.5 shows that some of the X variables are now correlated with one another: the area of overlap between age and income (shaded with dots) indicates the fact that age and income now share about 25% of their variance in common (because the correlation between these variables was .50, and $.50^2 = .25$). In the same way, the area of overlap between income and moral development (also shaded) represents the fact that these variables share about 25% of their variance in common.

There is an important consequence of the fact that some of the X variables are now correlated with one another: some of the variance in Y that is accounted for by one X variable is now also accounted for by another X variable. For example, consider the X_2 variable, income. By itself, income accounted for 16% of the variance in prosocial behavior. But notice how the circle for age overlaps part of the variance in Y that is accounted for by income. (This area is shaded and cross-hatched.) This means that some of the variance in Y that is accounted for by income is also accounted for by age. In other words, because age and income are correlated, this has decreased the amount of variance in Y that is accounted for *uniquely* by income.

The same is true when one considers the correlation between income and moral development. The circle for moral development overlaps part of the variance in Y that is accounted for by income. (Again, this area is shaded and cross-hatched.) This shows that some of the variance in Y that was accounted for by income is now also accounted for by moral development.

In short, there is redundancy between age and income in the prediction of Y, and there is also some redundancy between income and moral development in the prediction of Y. The result of this correlation between the predictor variables is a net decrease in the total amount of variance in Y that is accounted for by the linear combination of X variables. Compare the Venn diagram for the situation in which the X variables were uncorrelated (Figure 14.4) to the Venn diagram in which there was some correlation among the X variables (Figure 14.5). When the X variables are not correlated among themselves, they account for 50% of the variance in Y. Now, notice the area in Y that overlaps with the X variables in Figure 14.5. The area of overlap is smaller in the second figure, showing that the X variables account for less of the total observed variance in Y when they are correlated with one another.

This is because X variables usually make a smaller unique contribution to the prediction of Y when they are correlated with one another. The greater the correlation among the X variables, the smaller the amount of unique variance in Y accounted for by each individual X variable, and hence the smaller the total variance in Y that is accounted for by the combination of X variables.

The meaning of unique variance can be understood with reference to the cross-hatched area in the Venn diagram of Figure 14.5. In the figure, the area that is both shaded and cross-hatched identifies the variance in Y that is accounted for by both income and age. Shading and cross-hatching are also used to identify the variance in Y accounted for by both income and moral development. The remaining variance in Y provided by income is the variance that is uniquely accounted for by income. In the figure, this is the area of overlap between income and prosocial behavior that is shaded in with only cross-hatching (not shading and cross-hatching). Obviously, this area is quite small, indicating that income accounts for very little variance in prosocial behavior that is not already accounted for by age and moral development.

There are several practical implications arising from this scenario. The first implication is that, in general, the amount of variance accounted for in a criterion variable will be larger to the extent that the following two conditions hold:

- The predictor variables demonstrate relatively strong correlations with the criterion variable.

- The predictor variables demonstrate relatively weak correlations with each other.

These conditions hold true "in general," because they do not apply to the special case of a suppressor variable. (More on this in the following section.)

The second implication is that there generally is a point of diminishing returns when it comes to adding new X variables to a multiple regression equation. Because so many predictor variables in social science research are correlated, only the first few predictors added to a predictive equation are likely to account for meaningful amounts of unique variance in a criterion. Variables that are subsequently added will tend to account for smaller and smaller percentages of unique variance. At some point, predictors added to the equation will account for only negligible amounts of unique variance. For this reason, most multiple regression equations in social science research contain a relatively small number of variables, usually 2 to 10.

When Intercorrelated Predictors Do Include a Suppressor Variable

The preceding section describes the results that you can normally expect to observe when regressing a criterion variable on multiple intercorrelated predictor variables. However, there is a special case in which the preceding generalizations do not hold; that is the case of the suppressor variable. Although reports of genuine suppressor variables are somewhat rare in the social sciences, it is important that you understand the concept, so that you will recognize a suppressor variable when you encounter one.

Briefly, a **suppressor variable** is a predictor variable that improves the predictive power of a multiple regression equation by controlling for unwanted variation that it shares with other predictors in the equation. Suppressor variables typically display the following characteristics:

- zero or near-zero correlations with the criterion;
- moderate-to-strong correlations with at least one other predictor variable.

Suppressor variables are interesting because, even though they can display a bivariate correlation with the criterion variable of zero, adding them to a multiple regression equation can result in a meaningful increase in R^2 for the model. This of course, violates the generalizations drawn in the preceding section.

To understand how suppressor variables work, consider this fictitious example: imagine that you want to identify variables that can be used to predict the success of firefighters. To do this, you conduct a study with a group of 100 firefighters. For each participant, you obtain a "Firefighter Success Rating" that indicates how successful this person has been as a firefighter. These ratings are on a scale of 1 to 100, with higher ratings indicating greater success.

To identify variables that might be useful in predicting these success ratings, you have each firefighter complete a number of "paper-and-pencil" tests. One of these is a "Firefighter Knowledge Test." High scores on this test indicate that the participant possesses the knowledge needed to operate a fire hydrant, enter a burning building safely, and perform other tasks related to firefighting. A second test is a "Verbal Ability Test." This test has nothing to do with firefighting; high scores simply indicate that the participant has a good vocabulary and other verbal skills.

The three variables in this study could be represented with the following symbols:

Y = Firefighter Success Ratings (the criterion variable);

X_p = Firefighter Knowledge Test (the predictor variable of interest);

X_s = Verbal Ability Test (the suppressor variable).

Imagine that you perform some analyses to understand the nature of the relationship among these three variables. First, you compute the Pearson correlation coefficient between the Firefighter Knowledge Test and the Verbal Ability Test, and find that $r = .40$. This is a moderately strong correlation (and it only makes sense that these two tests would be moderately correlated). This is because both firefighter knowledge and verbal ability are assessed by the same method: a paper-and-pencil questionnaire. To some extent, getting a high score on either of these tests requires that the participant be able to read instructions, read questions, read possible responses, and perform other verbal tasks. This means that the two tests are correlated because scores on both tests are influenced by participants' verbal ability.

Next, you perform a series of regressions in which Firefighter Success Ratings (Y) is the criterion to determine how much of its variance is accounted for by various regression equations. This is what you learn:

- When the regression equation contains only the Verbal Ability Test, it accounts for 0% of the variance in Y.

- When the regression equation contains only the Firefighter Knowledge Test, it accounts for 20% of the variance in Y.

- When the regression equation contains both the Firefighter Knowledge Test and the Verbal Ability test, it accounts for 25% of the variance in Y.

The finding that the Verbal Ability Test accounts for none of the variance in Firefighter Success Ratings makes sense, because it (presumably) does not require a good vocabulary or other verbal skills to be a good firefighter.

The second finding that the Firefighter Knowledge Test accounts for a respectable 20% of the variance in Firefighter Success Ratings also makes sense as it is reasonable to expect more knowledgeable firefighters to be rated as better firefighters.

However, you run into a difficulty when trying to make sense of the third finding: that the equation that contains both the Verbal Ability Test and the Firefighter Knowledge Test accounts for 25% of observed variance in Y. How is it possible that the combination of these two variables accounts for 25% of the variance in Y when one accounted for only 20% and the other accounted for 0%?

The answer is that, in this situation, the Verbal Ability Test is serving as a suppressor variable. It is suppressing irrelevant variance in scores on the Firefighter Knowledge Test, thus "purifying" the relationship between the Firefighter Knowledge Test and Y. Here is how it works. Scores on the Firefighter Knowledge Test are influenced by at least two factors: their actual knowledge about firefighting; and their verbal ability (e.g., ability to read instructions). Obviously, the first of these two factors is relevant for predicting Y whereas the second factor is not. Because scores on the Firefighter Knowledge Test are to some extent "contaminated" by the effects of participants' verbal ability, the actual correlation between the Firefighter Knowledge Test and Firefighter Success Ratings is somewhat lower than it would be if you could somehow purify Knowledge Test scores of this unwanted verbal factor. That is exactly what a suppressor variable does.

In most cases, a suppressor variable is given a negative regression weight in a multiple regression equation. (These weights will be discussed in more detail later in this chapter.) Partly because of this, including the suppressor variable in the equation adjusts each participant's predicted score on Y so that it comes closer to that participant's actual score on Y. In the present case, this means that, if a participant scores above the mean on the Verbal Ability Test, his or her predicted score on Y will be adjusted downward to penalize for scoring high on this irrelevant predictor. Alternatively, if a participant scores below the mean on the Verbal Ability Test, his or her predicted score on Y will be adjusted upward. Another way of thinking about this is to say that a person applying to be a firefighter who has a high score on the Firefighter's Knowledge Test but a low score on the Verbal Ability Test would be preferred over an applicant with a high score on the Knowledge Test and a high score on the Verbal Test. (This is because the second candidate's score on the Knowledge Test was probably inflated due to his or her good verbal skills.)

The net effect of these corrections is improved accuracy in predicting Y. This is why you earlier found that R^2 is .25 for the equation that contains the suppressor variable, but only .20 for the equation that does not contain it.

The possible existence of suppressor variables has implications for multiple regression analyses. For example, when attempting to identify variables that would make good predictors in a multiple regression equation, it is clear that you should not base the selection exclusively on the bivariate (Pearson) correlations between the variables. For example, even if two predictor variable are moderately or strongly correlated, it does not necessarily mean that they are always providing redundant information. If one of them is a suppressor variable, then the two variables are not entirely redundant. (That noted, predictor variables with coefficients greater than $r = |.89|$ create computational problems and should be avoided. See the description of **multicollinearity** in final section of this chapter, "Assumptions Underlying Multiple Regression.")

In the same way, a predictor variable should not be eliminated from consideration as a possible predictor just because it displays a low bivariate correlation with the criterion. This is because a suppressor variable might display a bivariate correlation with Y of zero, even though it could substantially increase R^2 if added to a multiple regression equation. When starting with a set of possible predictor variables, it is generally safer to begin the analysis with a multiple regression equation that contains all predictors, on the chance that one of them serves as a suppressor variable. (The topic of choosing an "optimal" subset of predictor variables from a larger set is a complex one that is beyond the scope of this book.)

To provide some sense of perspective, however, you should take comfort in the knowledge that true suppressor variables are somewhat rare in social science research. In most cases, you can expect your data to behave according to the generalizations made in the preceding section (dealing with regression equations that do not contain suppressors). That is, in most cases, you will find that R^2 is larger to the extent that the X variables are more strongly correlated with Y and less strongly correlated with one another. To learn more about suppressor variables, see Pedhazur (1982).

The Uniqueness Index

A **uniqueness index** represents the percentage of variance in a criterion that is accounted for by a given predictor variable, above and beyond the variance accounted for by the other predictor variables in the equation. A uniqueness index is one measure of an X variable's importance as a predictor: the greater the amount of unique variance accounted for by a predictor, the greater its usefulness.

The concept of a uniqueness index can be illustrated with reference to Figure 14.6 that illustrates the uniqueness index for the predictor variable, income. Figure 14.6 is identical to Figure 14.5 with respect to the correlations between income and prosocial behavior and the correlations among the three X variables. In Figure 14.6, however, only the area that represents the uniqueness index for income is shaded. It can be seen that this area is consistent with the previous definition which states that the uniqueness index for a given variable represents the percentage of variance in the criterion (prosocial behavior) that is accounted for by the predictor variable (income) over and above the variance accounted for by the other predictors in the equation (i.e., age and moral development).

Figure 14.6 Venn Diagram: Uniqueness Index for Income

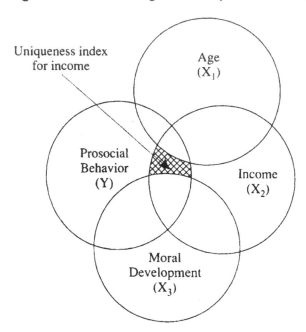

When performing multiple regression, it is often useful to compute the uniqueness index for each X variable in the equation. These indices, along with other information, can aid understanding of the nature of the relationship between the criterion and the predictor variables.

Computing the uniqueness index for a given X variable actually requires estimating two multiple regression equations. The first equation should include all of the X variables of interest. This can be referred to as the **full equation** because it contains the full set of predictors.

In contrast, the second multiple regression equation should include all of the X variables except for the X variable of interest. For this reason, this second equation can be called the **reduced equation**.

To calculate the uniqueness index for the X variable of interest, subtract the R^2 value for the reduced equation from the R^2 value for the full equation. The resulting difference is the uniqueness index for the X variable of interest.

This procedure can be illustrated by calculating the uniqueness index for income. Assume that two multiple regression equations are computed and that prosocial behavior is the criterion variable for both. With the full equation, the predictor variables include age, income, and moral development. With reduced equation, the predictors include only age and moral development. (Income was dropped because you wish to calculate the uniqueness index for income.) Assume that the following R^2 values are obtained for the two equations:

$$R^2_{Full} = .40$$
$$R^2_{Reduced} = .35$$

These R^2 values show that, when all three predictor variables were included in the equation, they accounted for 40% of the variance in prosocial behavior. However, when income was dropped from this equation, the reduced equation accounted for only 35% of the variance in prosocial behavior. The uniqueness index for income (symbolized as U) can now be calculated by subtracting the reduced-equation R^2 from the full-equation R^2:

$$U = R^2_{\text{Full}} - R^2_{\text{Reduced}}$$
$$U = .40 \quad - .35$$
$$U = .05$$

The uniqueness index for income is .05, meaning that income accounts for only 5% of the variance in prosocial behavior, beyond the variance accounted for by age and moral development.

To compute the uniqueness index for age, you would have to estimate yet another multiple regression equation: one in which age is dropped from the full equation (i.e., an equation containing only income and moral development). The R^2 for the new reduced equation is calculated and is subtracted from the R^2 for the full equation to arrive at the U for age. A similar procedure then is performed to compute the U for moral development. A later section of this chapter ("Computing Uniqueness Indices with PROC REG") describes how the results of SAS' REG procedure can be used to compute uniqueness indices for a set of predictors.

Testing the Significance of the Difference between Two R^2 Values

The preceding section shows that you can calculate a uniqueness index by

- estimating two R^2 values: one for a full equation with all X variables included and one for a reduced equation with all X variables except for the X variable of interest;
- subtracting the R^2 for the reduced model from the R^2 for the full model.

The difference between the two R^2 values is the uniqueness index for the dropped X variable.

Once a uniqueness index is calculated, however, the next logical question is, "Is this uniqueness index significantly different from zero?" In other words, does the predictor variable represented by this index account for a significant amount of variance in the criterion beyond the variance accounted for by the other predictors? To answer this question, it is necessary to test the significance of the difference between the R^2 value for the full equation versus the R^2 for the reduced equation. This section shows how this is done.

One of the most important procedures in multiple regression analysis involves testing the significance of the difference between R^2 values. This test is conducted when you have estimated two multiple regression equations, and want to know whether the R^2 value for one equation is significantly larger than the R^2 for the other equation.

There is one important condition when performing this test, however: The two multiple regression equations must be nested. Two equations are said to be **nested** when they include the same criterion variable and the X variables included in the smaller equation are a subset of the X variables included in the larger equation.

To illustrate this concept, we refer again to the study in which

> Y = Prosocial behavior
> X_1 = Age
> X_2 = Income
> X_3 = Moral development

The following two multiple regression equations are nested:

> Full equation: $Y = X_1 + X_2 + X_3$
> Reduced equation: $Y = X_1 + X_2$

It can be said that the reduced equation is nested within the full equation because the predictors for the reduced equation (X_1 and X_2) are a subset of the predictors in the full equation (X_1, X_2, and X_3).

In contrast, the following two multiple regression equations are not nested:

> Full equation: $Y = X_1 + X_2 + X_3$
> Reduced equation: $Y = X_1 + X_4$

Even though the "reduced" equation in the preceding example contains fewer X variables than the "full" equation, it is not nested within the full equation. This is because the reduced equation contains the variable X_4, which does not appear in the full equation. The X variables of the reduced equation are, therefore, not a subset of the X variables in the full equation.

When two equations are nested, it is possible to test the significance of the difference between the R^2 values for the equations. This is sometimes referred to as "testing the significance of variables added to the equation." To understand this expression, think about a situation in which the full equation contains X_1, X_2, and X_3, and the reduced equation contains only X_1 and X_2. The only difference between these two equations is the fact that the full equation contains X_3 while the reduced equation does not. By testing the significance of the difference between the R^2 values for the two equations, you are determining whether adding X_3 to the reduced equation results in a significant increase in R^2.

Here is the equation for the *F* ratio that tests the significance of the difference between two R^2 values:

$$F = \frac{(R^2_{Full} - R^2_{Reduced}) / (K_{Full} - K_{Reduced})}{(1 - R^2_{Full}) / (N - K_{Full} - 1)}$$

where

R^2_{Full} = the obtained value of R^2 for the full multiple regression equation (i.e., the equation containing the larger number of predictor variables)

$R^2_{Reduced}$ = the obtained value of R^2 for the reduced multiple regression equation (i.e., the equation containing the smaller number of predictor variables)

K_{Full} = the number of predictor variables in the full multiple regression equation

$K_{Reduced}$ = the number of predictor variables in the reduced multiple regression equation

N = the total number of participants in the sample

The preceding equation can be used to test the significance of any number of variables added to a regression equation. The examples in this section involve testing the significance of adding just one variable to an equation. This section focuses on the one-variable situation because of its relevance to the concept of the uniqueness index, as previously discussed. Remember that a uniqueness index represents the percentage of variance in a criterion that is accounted for by a single X variable over and above the variance accounted for by the other X variables. Earlier, it was stated that a uniqueness index is calculated by estimating a full equation with all X variables, along with a reduced equation in which one X variable has been deleted. The difference between the two resulting R^2 values is the uniqueness index for the deleted X variable.

The relevance of the preceding formula to the concept of the uniqueness index should now be evident. This formula allows you to test the statistical significance of a uniqueness index. It allows you to test the statistical significance of the variance in Y accounted for by a given X variable beyond the variance accounted for by other X variables. For this reason, the formula is helpful in determining which predictors are relatively important and which are comparatively unimportant.

To illustrate the formula, imagine that you conduct a study in which you use age, income, and moral development to predict prosocial behavior in a sample of 104 participants. You now want to calculate the uniqueness index for the predictor, moral development. To do this, you estimate two multiple regression equations: a "full model" that includes all three X variables as predictors and a "reduced model" that includes all predictors except for moral development. Here are the R^2 values obtained for the two models:

Table 14.7

Variance Accounted for by Full and Reduced Models, Prosocial Behavior Study

R^2 Obtained for this model	Predictor variables included in the equation
.30	Age, income, moral development (full model)
.20	Age, income (reduced model)

Note: $N = 104$.

The preceding table shows that the reduced equation (including only age and income) accounts for only 20% of the variance in prosocial behavior. Adding moral development to this equation, however, increases R^2 to .30. The uniqueness index for moral development is therefore .30 − .20 or .10.

To compute the F test that tests the significance of the difference between these two R^2 values, insert the appropriate figures in the following equation:

$$F = \frac{(R^2_{Full} - R^2_{Reduced}) / (K_{Full} - K_{Reduced})}{(1 - R^2_{Full}) / (N - K_{Full} - 1)}$$

$$F = \frac{(.30 - .20) / (3 - 2)}{(1 - .30) / (104 - 3 - 1)}$$

$$F = \frac{(.10) / (1)}{(.70) / (100)}$$

$$F = \frac{(.10)}{(.007)}$$

$$F = 14.29$$

The F ratio for this test is 14.29. To determine whether this F is significant, you turn to the table of critical values of F in Appendix C. To find the appropriate critical value of F, it is necessary to first establish the degrees of freedom for the numerator and the degrees of freedom for the denominator that are associated with the current analysis.

When comparing the significance of the difference between two R^2 values, the degrees of freedom for the numerator are equal to $K_{Full} - K_{Reduced}$ (which appears in the numerator of the preceding equation). In the present analysis, the number of predictors in the full equation is 3, and the number in the reduced equation is 2; therefore, the degrees of freedom (*df*) for the numerator is 3 – 2 or 1.

The degrees of freedom for the denominator of this *F* test is $N - K_{Full} - 1$ (which appears in the denominator of the preceding equation). The sample size in the present study was 104, so the degrees of freedom for the denominator is 104 – 3 –1 = 100.

The table of *F* values shows that, with 1 *df* for the numerator and 100 *df* for the denominator, the critical value of *F* is approximately 3.94 ($p < .05$). The preceding formula produced an observed *F* value of 14.29, which of course is much larger than this critical value. You can therefore reject the null hypothesis of no difference and conclude that there is a significant difference between the two R^2 values. In other words, adding moral development to an equation already containing age and income results in a significant increase in observed variance in prosocial behavior. Therefore, the uniqueness index for moral development is proven to be statistically significant.

Multiple Regression Coefficients

The b weights discussed earlier in this chapter were referred to as nonstandardized multiple regression coefficients. It was said that a **multiple regression coefficient** for a given X variable represents the average change in Y that is associated with a one-unit change in that X variable while holding constant the remaining X variables. "Holding constant" means that the multiple regression coefficient for a given predictor variable is an estimate of the average change in Y that would be associated with a one-unit change in that X variable if all participants had identical scores on the remaining X variables.

When conducting multiple regression analyses, researchers often seek to determine which of the X variables are relatively important predictors of the criterion, and which are relatively unimportant. By doing this, you might be tempted to review the multiple regression coefficients estimated in the analysis and use these as indicators of importance. According to this logic, a regression coefficient represents the amount of *weight* that is given to a given X variable in the prediction of Y. If a variable is given much weight, it must be an important predictor.

Nonstandardized versus Standardized Coefficients

However, considerable caution must be exercised when using multiple regression coefficients in this manner for at least two reasons. First, you must remember that two types of multiple regression coefficients are produced in the course of an analysis: nonstandardized coefficients and standardized coefficients. **Nonstandardized multiple regression coefficients** (sometimes called b weights and symbolized with a lowercase letter b) are the coefficients that are produced when the data analyzed are in raw score form. "Raw score" form means that the variables have not been standardized in any way: the different variables might have very different means and standard deviations. For example, the standard deviation for X_1 can be 1.35, while the standard deviation for X_2 can be 584.20.

Generally speaking, it is not appropriate to use the relative size of nonstandardized regression coefficients in assessing the relative importance of predictor variables. This is because the relative size of a nonstandardized coefficient for a given predictor variable is influenced by the size of that predictor's standard deviation. Other things being equal, the variables with larger standard deviations tend to have smaller nonstandardized regression coefficients while variables with smaller standard deviations tend to have larger regression coefficients. Therefore, the size of nonstandardized coefficients generally indicates nothing about the relatively important predictor variables.

If this is the case, then what are the nonstandardized coefficients useful for? They are most frequently used to calculate participants' predicted scores on Y. For example, an earlier section presented a multiple regression equation for the prediction of prosocial behavior in which X_1 = age, X_2 = income, and X_3 = moral development. That equation is reproduced here:

$$Y' = (.10) X_1 + (.25) X_2 + (1.10) X_3 + (-3.25)$$

In this equation, the nonstandardized multiple regression coefficient for X_1 is .10, the coefficient for X_2 is .25, and so forth. If you had a participant's raw scores on the three predictor variables, these values could be inserted in the preceding formula to compute that participant's estimated score on Y. The resulting Y' value would also be in raw score form. It would be an estimate of the number of prosocial acts you expect the participant to perform over a six-month period.

In summary, you should not refer to the nonstandardized regression coefficients to assess the relative importance of predictor variables. A better alternative (though still not perfect) is to refer to the standardized coefficients. **Standardized multiple regression coefficients** (sometimes called beta weights and symbolized by the Greek letter β) are the coefficients that would be produced if the data analyzed were in standard score form. "Standard score" form (or "z score form") means that the variables are standardized so that each has a mean of zero and a standard deviation of 1. This is important because all variables (Y variables and X variables alike) now have the same standard deviation (i.e., a standard deviation of 1); they are now measured on the same scale of magnitude or metric. You no longer have to worry that some variables display large regression coefficients simply because they have small standard deviations. To some extent, the size of standardized regression coefficients does reflect the relative importance of the various predictor variables; these coefficients should therefore be among the results that are consulted when interpreting the results of a multiple regression analysis.

For example, assume that the analysis of the prosocial behavior study produces the following multiple regression equation with standardized coefficients:

$$Y' = (.70) X_1 + (.20) X_2 + (.20) X_3$$

In the preceding equation, X_1 displays the largest standardized coefficient. This could be interpreted as evidence that it is a relatively important predictor variable compared to X_2 and X_3.

You can see that the preceding equation, like all regression equations with standardized coefficients, does not contain an intercept constant. This is because the intercept is always equal to zero in a standardized equation. This is a useful fact to know; if a researcher has presented a multiple regression equation in a research article but has not indicated whether it is a standardized or a nonstandardized equation, look for the intercept constant. If an intercept is included in the equation, it is almost certainly a nonstandardized equation. If there is no intercept, it is probably a standardized equation. Remember also that the lowercase letter b is typically used to represent nonstandardized regression coefficients, whereas the Greek letter β is used to represent standardized coefficients.

The Reliability of Multiple Regression Coefficients

The preceding section notes that standardized coefficients reflect the importance of predictors only to some extent. This qualification is necessary because of the unreliability that is often demonstrated by multiple regression weights. In this case, unreliability refers to the fact that, when multiple regression using the same variables is performed on data from more than one sample, very different estimates of the multiple regression coefficients are often obtained with the different samples. This is the case for standardized as well as nonstandardized coefficients.

For example, assume that you recruit a sample of 50 participants, measure the variables discussed in the preceding section (prosocial behavior, age, income, moral development), and compute a multiple regression equation in which prosocial behavior is the criterion and the remaining variables are predictors (assume that age, income, and moral development are X_1, X_2, and X_3, respectively). With the analysis completed, it is possible that your output would reveal the following standardized regression coefficients for the three predictors:

$$Y' = (.70) X_1 + (.20) X_2 + (.20) X_3$$

The relative size of the coefficients in the preceding equation suggests that X_1 (with a beta weight of .70) is the most important predictor of Y, while X_2 and X_3 (each with beta weights of .20) are much less important.

However, what if you were then to attempt to replicate your study with a different group of 50 participants? It is unfortunately possible (if not likely) that would obtain very different beta weights for the X variables. For example, you might obtain the following:

$$Y' = (.30) X_1 + (.50) X_2 + (.10) X_3$$

In the second equation, X_2 has emerged as the most important predictor of Y, followed by X_1 and X_3.

This is what is meant by the unreliability of standardized regression coefficients. When the same study is performed on different samples, researchers sometimes obtain coefficients of very different sizes. This means that the interpretation of these coefficients must always be made with caution.

This problem of unreliability is more likely in some situations than in others. Specifically, multiple regression coefficients become increasingly unreliable as the analysis is based on increasingly smaller samples and as the X variables become increasingly correlated with one another. Unfortunately, much of the research that is carried out in the social sciences involves the use of small samples and correlated X variables. For this reason, the standardized regression coefficients (i.e., beta weights) are only some of the pieces of information that should be reviewed when assessing the relative importance of predictor variables. The use of these coefficients should be supplemented with a review of the simple bivariate correlations between the X variables and Y and the uniqueness indices for the X variables. The following sections show how to do this.

Example: A Test of the Investment Model

The statistical procedures of this chapter are illustrated by analyzing fictitious data from a correlational study based on the investment model (Rusbult, 1980a; 1980b). You will remember that the investment model was used to illustrate the use of a number of other statistical procedures earlier in this book (e.g., in Chapters 8 through 13).

The investment model identifies a number of variables that are believed to predict a person's level of commitment to a romantic relationship (as well as to other types of relationships). **Commitment** refers to the individual's intention to maintain the relationship and remain with a current partner. One version of the investment model asserts that commitment is affected by the following four variables:

Rewards: the number of "good things" that the participant associates with the relationship; the positive aspects of the relationship

Costs: the number of "bad things" or hardships associated with the relationship

Investment Size: the amount of time and personal resources that the participant has put into the relationship

Alternative Value: the attractiveness of the participant's alternatives to the relationship (e.g., the attractiveness of alternative romantic partners)

The hypothesized relationship among commitment and these four predictor variables is presented as Figure 14.7.

Figure 14.7 One Version of the Investment Model

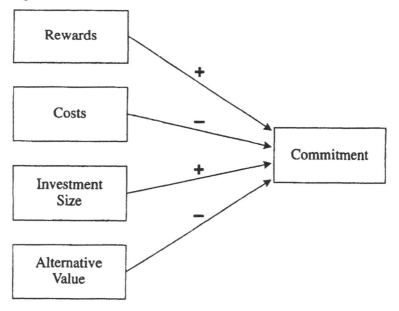

If these four variables actually have a causal effect on commitment, it is reasonable to assume that they should also be useful for *predicting* commitment in a simple correlational study. Therefore, the present study is correlational in nature and will use multiple regression procedures to assess the nature of the predictive relationship between these variables. If the model survives this test, you might want to follow it up by performing a new study that is more capable of testing the proposed causal relationships between the variables, perhaps a true experiment. This chapter, however, focuses only on multiple regression procedures.

Overview of the Analysis

To better understand the nature of the relationship among the five variables presented in Figure 14.7, the data are analyzed using two SAS procedures. First, PROC CORR will compute Pearson correlations between variables. This is useful for understanding the big picture: the simple bivariate relations among the variables.

Next, PROC REG is used to perform a multiple regression in which commitment is regressed on the four predictor variables simultaneously. This provides a number of important pieces of information. First, you will learn whether there is a significant relationship between commitment and the linear combination of predictors (i.e., whether there is a significant relationship between commitment and the four variables taken as a group). In addition, you will review the multiple regression coefficients for each of the four predictors to determine which are statistically significant and which standardized coefficients are relatively large.

Finally, the SELECTION=RSQUARE option is used with PROC REG in a separate analysis to determine the amount of variance in commitment that is accounted for by every possible combination of predictor variables. Results from this analysis are used to determine the uniqueness index for each of the four predictors. The formula for testing the significance of the difference between two R^2 values are then used to determine which of the uniqueness indices significantly differ from zero.

Gathering and Entering Data

The Questionnaire

Assume that you conduct a study in which the participants are 50 college students, each of whom is currently involved in a romantic relationship. Each participant completes a 24-item questionnaire designed to assess the six investment model constructs: commitment; satisfaction; rewards; costs; investment size; and alternative value. (One construct, satisfaction, is not included in the version of the investment model to be tested in this study and is therefore not referred to again in this chapter.)

Each investment model construct is assessed with four questions. For example, the four items that assess the "commitment" construct are reproduced here:

---

21.   How committed are you to your current relationship?

```
       Not at all    1 2 3 4 5 6 7 8 9    Extremely
        committed                          committed
```

22.   How long do you intend to remain in your current relationship?

```
        A very       1 2 3 4 5 6 7 8 9    A very
      short time                          long time
```

23.   How often do you think about breaking up with your current partner?

```
      Frequently     1 2 3 4 5 6 7 8 9    Never
```

24.   How likely is it that you will maintain your current relationship for a long time?

```
      Extremely      1 2 3 4 5 6 7 8 9    Extremely
       unlikely                           likely
```

Notice that, with each item, selecting a higher response number (such as 8 or 9) indicates a higher level of commitment whereas selecting a lower response number (such as 1 or 2) indicates a lower level of commitment. Participants' responses to these four items are summed to arrive at a single score that reflects their overall level of commitment. This cumulative score will serve as the commitment variable in your analyses. Scores on this variable can range from 4 to 36 with higher values indicating higher levels of commitment.

Each remaining investment model construct is assessed in the same way with responses to four survey items being summed to create a single overall measure of the construct. With each variable, scores can range from 4 to 36 and higher scores indicate higher levels of the construct being assessed.

Entering the Data

Analyzing Raw Data

In practice, the easiest way to create these summed scores would involve entering the raw data and then using data manipulation statements in the SAS program to create the summed scores. Raw data in this sense refers to participants' responses to the 24 items on the questionnaire. You could prepare a single SAS program that would input these raw data, create new variables (overall scale scores) from existing variables, and perform the regression analyses on the new variables.

For example, imagine that you entered the raw data so that there is one line of data for each participant, and this line contains the participant's responses to questions 1 through 24. Assume that items 1 through 4 assessed the rewards construct, items 5 through 8 assessed the costs construct, and so forth. The following program would input these data and create the necessary composite scores:

```
1       DATA D1;
2          INPUT    #1     @1    (Q1-Q24)    (1.)    ;
3
4             REWARD        = Q01 + Q02 + Q03 + Q04;
5             COST          = Q05 + Q06 + Q07 + Q08;
6             INVESTMENT    = Q09 + Q10 + Q11 + Q12;
7             ALTERNATIVES  = Q13 + Q14 + Q15 + Q16;
8             SATISFACTION  = Q17 + Q18 + Q19 + Q20;
9             COMMITMENT    = Q21 + Q22 + Q23 + Q24;
10
11            IF REWARD NE . AND COST NE . AND INVESTMENT NE . AND
12                ALTERNATIVES NE . AND COMMITMENT NE . ;
13
14      DATALINES;
15      343434346465364735748234
16      867565768654544354865767
17      434325524243536435366355
18      .
19      .
20      .
21      874374848763747667677467
22      534232343433232343524253
23      979876768968789796868688
24      ;
25      RUN;
26
27      Place PROC statements here
```

Lines 1 and 2 of the preceding program tell the system that there is one line of data for each participant, variables Q1 through Q24 begin at column 1, and each variable is one column wide.

Line 4 of the program tells the system to create a new variable called REWARD, and participants' scores on REWARD should be equal to the sum of their scores on variables Q1, Q2, Q3, and Q4 (questionnaire items 1 to 4). Lines 5 though 9 of the program create the study's remaining variables in the same way.

Lines 11 and 12 contain a subsetting IF statement that eliminates from the dataset any participant with missing data on any of the six variables to be analyzed. This ensures that each analysis is performed on exactly the same participants. You do not have to worry that some analyses may be based on 50 participants while others are performed on only 45 due to missing data. (Chapter 4 discusses the use of subsetting IF statements.)

Notice that SATISFACTION (the satisfaction variable) is not listed in this subsetting IF statement. This is because SATISFACTION is not included in any of the analyses (i.e., SATISFACTION will not be listed as a predictor variable or a criterion variable in any analysis). Including SATISFACTION in the subsetting IF statement might have unnecessarily eliminated some participants from your dataset. For example, if a participant named Luc Dion had missing data on SATISFACTION, and SATISFACTION were in fact included in this subsetting IF statement, then Luc Dion would have been dropped from the dataset and each analysis would have been based on 49 observations rather than 50. But dropping Luc Dion in this case would have been pointless, because SATISFACTION is not going to be used as a predictor or criterion variable in any analysis. To summarize, only variables that are to be used in your analyses should be listed in the subsetting IF statement that eliminates participants with missing data.

The DATALINES statement and some fictitious data are presented on lines 14 to 24. Line 27 shows where the PROC statements (discussed later) would be placed in the program.

How This Program Handles Missing Data When Creating Scale Scores

With the preceding program, if a participant left blank any one of the four questions that constitute a scale, that participant will receive score of "missing" (.) on that scale. For example, if Luc Dion completes items 1, 2, and 3 of the REWARDS scale but leaves item 4 blank, he is assigned a missing value on REWARDS. If the dataset were printed using PROC PRINT, his value on REWARDS would appear as ".".

Consider the problems that would arise if the system did not assign missing values in this way. What if the system simply summed each participant's responses to the items that s/he did complete, even when some had been left blank? Remember that Luc Dion completed only 3 items on the 4-item response scale. In his case, the highest score he could possibly receive on REWARDS would be 27. (This is because the 3 items use a 9-point response scale, and $3 \times 9 = 27$.) For those participants who completed all 4 items in the REWARDS scale, the highest score possible is 36 (because $4 \times 9 = 36$). Obviously, under these circumstances the scores from participants with incomplete responses would not be directly comparable to scores from those with complete data. It is therefore best that participants with incomplete responses be assigned missing values when scale scores are computed.

Analyzing Pre-Calculated Scale Scores

A second option is to compute each participant's score on each of the six constructs by hand and then prepare a SAS program to analyze these pre-calculated scores. In other words, you could review the questionnaire completed by participant 1, sum the responses to items 1 through 4 (perhaps by using a hand calculator), record this sum on a summary sheet, repeat this process for the remaining scales, and repeat the entire process for all remaining participants.

Later, the summed scores could be entered as data in a SAS program. This dataset would consist of 50 lines of data (one line for each participant). Instead of raw data, a given participant's summed score on each of the six questions would appear on his or her data line. Scores on the commitment variable might be keyed in columns 1 to 2, scores on the satisfaction variable might be keyed in columns 4 to 5, and so forth. The guide used in entering these data is presented here:

Line	Column	Variable Name	Explanation
1	1-2	COMMITMENT	Scores on commitment variable.
	3	blank	
	4-5	SATISFACTION	Scores on the satisfaction variable.
	6	blank	
	7-8	REWARD	Scores on rewards variable.
	9	blank	
	10-11	COST	Scores on costs variable.
	12	blank	
	13-14	INVESTMENT	Scores on investment size variable.
	15	blank	
	16-17	ALTERNATIVES	Scores on the alternative value variable.

Fictitious data from the 50 participants in your study have been entered according to these guidelines and are presented in Appendix B. The analyses discussed in this chapter are performed on this dataset.

Following is the SAS program that will input these data. Lines 9 and 10 again include a subsetting IF statement that would eliminate from the dataset any participant with missing data on any of the variables to be analyzed (as was explained in the preceding section). A few lines of the actual data are presented as lines 13 through 22. Line 26 indicates the position where the PROC statements (to be discussed later) would be placed in this program.

```
1        DATA D1;
2           INPUT   #1    @1   COMMITMENT      2.
3                          @4   SATISFACTION    2.
4                          @7   REWARD          2.
5                          @10  COST            2.
6                          @13  INVESTMENT      2.
7                          @16  ALTERNATIVES    2.    ;
8
9           IF REWARD NE . AND COST   NE . AND INVESTMENT NE .
10          AND ALTERNATIVES NE . AND COMMITMENT NE . ;
11
```

```
12          DATALINES;
13          34 30 25 13 25 12
14          32 27 27 14 32 13
15          34 23 24 21 30 14
16          .
17          .
18          .
19          .
20          36 32 28 13 15  5
21          32 29 30 21 32 22
22          30 32 33 16 34  9
23          ;
24          RUN;
25
26          Place PROC statements here
```

Computing Bivariate Correlations with PROC CORR

In most studies in which data are analyzed using multiple regression, it is appropriate to begin by computing all possible correlations among the study's variables. Reviewing these correlations will help you understand the big picture concerning the simple relationships between the criterion variable and the four predictor variables as well as among the four predictor variables themselves.

Writing the Program

Following is the general form for the SAS statements that will compute all possible correlations among the variables being investigated. (Because detailed information concerning use of the CORR procedure is in Chapter 6, most of that information is not repeated here.)

```
PROC CORR   DATA=dataset-name;
   VAR  criterion-variable-and-predictor-variables ;
RUN;
```

Technically, the variables can be presented in any sequence though there are some advantages to (a) listing the criterion variable first and (b) then listing the predictor variables in the same order that they are discussed in the text. This often makes it easier to interpret the results of the correlation matrix.

Below is the entire program (including a portion of the fictitious data) that would compute all possible correlations among the five variables of interest in the current study. Output 14.1 presents the results produced by this program.

```
1       DATA D1;
2          INPUT   #1    @1    COMMITMENT     2.
3                         @4    SATISFACTION   2.
4                         @7    REWARD         2.
5                         @10   COST           2.
6                         @13   INVESTMENT     2.
7                         @16   ALTERNATIVES   2.   ;
8
9          IF REWARD NE . AND COST   NE . AND INVESTMENT NE . AND
10              ALTERNATIVES NE . AND COMMITMENT NE . ;
11
12         DATALINES;
13         34 30 25 13 25 12
```

```
14          32  27  27  14  32  13
15          34  23  24  21  30  14
16          .
17          .
18          .
19          .
20          36  32  28  13  15   5
21          32  29  30  21  32  22
22          30  32  33  16  34   9
23          ;
24          RUN;
25
26          PROC  CORR    DATA=D1;
27             VAR  COMMITMENT  REWARD  COST  INVESTMENT  ALTERNATIVES;
28          RUN;
```

Output 14.1 Results of the CORR Procedure

```
                          The CORR Procedure

      5  Variables:    COMMITMENT  REWARD       COST       INVESTMENT  ALTERNATIVES


                            Simple Statistics

Variable         N         Mean      Std Dev       Sum       Minimum       Maximum

COMMITMENT       48      27.70833     10.19587      1330       4.00000      36.00000
REWARD           48      26.64583      5.05076      1279       9.00000      33.00000
COST             48      16.58333      5.49597   796.00000     5.00000      26.00000
INVESTMENT       48      25.33333      6.08218      1216      11.00000      34.00000
ALTERNATIVES     48      16.60417      7.50529   797.00000     4.00000      34.00000


                 Pearson Correlation Coefficients, N = 48
                      Prob > |r| under H0: Rho=0

               COMMITMENT      REWARD        COST      INVESTMENT   ALTERNATIVES

COMMITMENT      1.00000       0.57597     -0.24826      0.61403      -0.72000
                              <.0001       0.0889       <.0001        <.0001

REWARD          0.57597       1.00000     -0.45152      0.57117      -0.46683
                <.0001                     0.0013       <.0001        0.0008

COST           -0.24826      -0.45152      1.00000      0.02143       0.27084
                0.0889        0.0013                    0.8851        0.0626

INVESTMENT      0.61403       0.57117      0.02143      1.00000      -0.44776
                <.0001        <.0001       0.8851                     0.0014

ALTERNATIVES   -0.72000      -0.46683      0.27084     -0.44776       1.00000
                <.0001        0.0008       0.0626       0.0014
```

Interpreting the Results of PROC CORR

1. Make Sure That the Numbers Look Right

Before interpreting the meaning of the correlation coefficients, it is important to review descriptive statistics to help verify that no errors were made when entering data or writing the INPUT statement. Most of this information is presented in a table of simple statistics that appears at the top of the output page.

The simple statistics table at the top of Output 14.1 provides means, standard deviations, and other descriptive statistics for the five variables analyzed. Note that $N = 48$ for all six variables, meaning that there must have been missing data for two of the 50 participants in the original dataset. (The two with missing data were deleted by the subsetting IF statement discussed earlier; review the actual dataset in Appendix B to verify that two participants did, in fact, have missing data.)

The rest of the descriptive statistics should also be reviewed to verify that all of the figures are reasonable. In particular, check the "Minimum" and "Maximum" columns for evidence of problems. The lowest score that a participant could possibly receive on any variable was 4; if any variable displays a minimum score below 4, an error must have been made either when entering the data or in the program. Similarly, no score in the "Maximum" column should exceed 36.

---

2.   Determine the Size of the Sample Producing the Correlations

When all correlations are based on the same number of participants, the sample size for the correlations should appear on the line just below the table of simple statistics. In Output 14.1, this reads as

```
Pearson Correlation Coefficients, N = 48
        Prob > |r| under H0: Rho=0
```

The final entry in this line tells us that $N = 48$ for the analysis. If the various correlations had, instead, been based on different sample sizes, then the N associated with each coefficient would appear within the table of correlations coefficients, just below the p value for the corresponding correlation.

---

3.   Review the Correlation Coefficients

The correlations among the five variables appear in the 5 x 5 matrix in the bottom half of Output 14.1. In this matrix, where the row for one variable intersects with the column of a second variable, you find the cell that provides information about the correlation between these variables. The top figure in the cell is the Pearson correlation coefficient between the variables; the bottom figure is the p value associated with this coefficient.

For example, consider the vertical column under the "COMMITMENT" heading. Where the column headed "COMMITMENT" intersects with the row headed "REWARD," you find that the correlation between commitment and rewards is approximately .58. The p value associated with this correlation is less than .01, meaning that there is less than 1 chance in 100 of obtaining a coefficient of this size by chance alone.

Reviewing the correlations in the COMMITMENT column tells you something about the pattern of simple bivariate correlations between commitment and the four predictors. Notice that COMMITMENT demonstrates a positive correlation with REWARD and INVESTMENT. As you would expect, participants who reported higher levels of rewards and investment size also reported higher levels of commitment. It can also be seen that COMMITMENT demonstrates negative relationships with COST and ALTERNATIVES. This is also as you would expect since participants who report higher levels of costs and alternative values report lower levels of commitment. Each correlation is in the direction predicted by the investment model (see Figure 14.7).

However, notice that some predictors are more strongly related to COMMITMENT than others. Specifically, ALTERNATIVES displays the strongest correlation at –.72, followed by INVESTMENT (.61) and REWARD (.58)[1]. Each correlation is statistically significant at the .01 level.

The correlation between COMMITMENT and COST is much lower at approximately –.25. In fact, this correlation is not statistically significant; the p value associated with it is .09. Based on these results, you cannot reject the null hypothesis that commitment and costs are uncorrelated in the population.

The correlations in Output 14.1 also provide important information about the correlations among the four predictor variables. When using multiple regression, an ideal predictive situation is typically one in which each predictor variable displays a relatively strong correlation with the criterion while the predictor variables display relatively weak correlations among themselves. (The reasons for this are discussed in an earlier section.)

With this in mind, the correlations presented in Output 14.1 indicate that the association among variables in the current dataset is less than ideal. It is true that three of the predictors display relatively strong correlations with commitment, but it is also true that most of the predictors display relatively strong correlations with each other. Notice that $r = -.45$ for the correlation between REWARD and COST, that $r = .57$ for REWARD and INVESTMENT, and $r = -.47$ for REWARD and ALTERNATIVES. These correlations are moderately large as is the correlation between INVESTMENT and ALTERNATIVES. Moderate intercorrelations of this sort sometimes result in nonsignificant multiple regression coefficients for at least some predictor variables when the data are analyzed. The analyses reported in the following section should bear this out.

Estimating the Full Multiple Regression Equation with PROC REG

There are several SAS procedures that allow you to perform multiple regression analyses. The basic multiple regression procedure, however, is PROC REG. This procedure estimates multiple regression coefficients for the various predictors, calculates R^2 and tests it for significance, and prints additional information relevant to the analysis.

Writing the Program

This is the general form for using PROC REG to request a basic multiple regression analysis with standardized multiple regression coefficients:

```
PROC REG    DATA=dataset-name    options;
   MODEL   criterion  =  predictor-variables  /  STB   options;
RUN;
```

---

[1] Note, we cannot necessarily conclude that the difference between these coefficients is statistically significant without calculating Fishers' Z transformations (see Hopkins, Glass, & Hopkins, 1987).

Predictor variables in the preceding MODEL statement should be separated by at least one space. The name of the last predictor variable should be followed by a slash and a list of options, if any are desired. You should always specify "STB" in the options field of the MODEL statement since this requests that the standardized multiple regression coefficients be printed.

Here are some options for the REG statement that can be particularly useful in social science research; additional options can be found in "The REG Procedure" in the *SAS/STAT User's Guide*:

CORR
> requests that the correlation matrix for all variables in the MODEL statement be printed.

SIMPLE
> requests simple statistics for all variables in the analysis (mean, variance, standard deviation, and uncorrected sum of squares).

These are some options for the MODEL statement that might be particularly useful; more are listed in "The REG Procedure" in the *SAS/STAT User's Guide*. The notes appearing here that refer the reader to the user's guide are similarly referred to in "The REG Procedure."

COLLIN
> prints diagnostics regarding collinearity among the predictor variables. See the section on collinearity diagnostics in the PROC REG chapter of the *SAS/STAT User's Guide* for details.

INFLUENCE
> prints diagnostics regarding the influence of each observation on the parameter estimates and on the predicted Y values. See the section on influence diagnostics in the PROC REG chapter of the *SAS/STAT User's Guide* for details.

P
> prints actual Y scores, predicted Y scores, and residual scores (errors of prediction) for each observation. See the section on predicted and residual values in the PROC REG chapter of the *SAS/STAT User's Guide* for details.

R
> requests a detailed analysis of the residuals, including Cook's D statistic, which assesses the influence of each observation on parameter estimates. See the section on predicted and residual values in the PROC REG chapter of the *SAS/STAT User's Guide* for details.

SELECTION=model-selection-method
> requests a specific model-selection method. Model-selection methods are used to select an "optimal" group of predictor variables from a larger set. Keywords for available methods are FORWARD, BACKWARD, STEPWISE, MAXR, MINR, RSQUARE, ADJRSQ, CP, and NONE. This chapter shows how to use the SELECTION=RSQUARE option to obtain information needed to compute uniqueness indices.

STB

requests the printing of the standardized multiple regression coefficient (beta weight) for each predictor variable.

In the present analysis, it is necessary to estimate a multiple regression equation in which the criterion variable was commitment, and the predictor variables were rewards, costs, investment size, and alternative value. Here are the statements that requested this model. The STB option in the MODEL statement requests the standardized regression coefficients:

```
PROC REG    DATA=D1;
    MODEL COMMIT = REWARD COST INVESTMENT ALTERNATIVES / STB;
RUN;
```

The SAS output created by these statements is reproduced as Output 14.2:

Output 14.2 Results of the REG Procedure with the STB Option

```
                        The SAS System

                       The REG Procedure
                        Model: MODEL1
                   Dependent Variable: COMMITMENT

                       Analysis of Variance

                             Sum of         Mean
    Source           DF     Squares        Square     F Value    Pr > F

    Model             4    3154.83928     788.70982    19.59     <.0001
    Error            43    1731.07738      40.25761
    Corrected Total  47    4885.91667


             Root MSE             6.34489    R-Square     0.6457
             Dependent Mean      27.70833    Adj R-Sq     0.6127
             Coeff Var           22.89885


                       Parameter Estimates

                      Parameter    Standard                          Standardized
    Variable     DF    Estimate      Error     t Value   Pr > |t|      Estimate

    Intercept     1    20.03813     8.73671      2.29     0.0268           0
    REWARD        1     0.27937     0.27331      1.02     0.3124        0.13839
    COST          1    -0.10575     0.20800     -0.51     0.6138       -0.05700
    INVESTMENT    1     0.52347     0.21037      2.49     0.0168        0.31226
    ALTERNATIVES  1    -0.67943     0.14670     -4.63     <.0001       -0.50014
```

Interpreting the Results of PROC REG

1. Make Sure That Eeverything Looks Right

At the top of the page, verify that the name of the criterion variable is listed to the right of "Dependent Variable." In this case, the criterion is COMMITMENT. In the "Analysis of Variance" section, one of the headings is "DF" for degrees of freedom. Where this DF column intersects with the row headed "Corrected Total," you find the corrected total degrees of freedom. Verify that this number is equal to $N - 1$, where $N =$ the total number of participants who provided usable data for the analysis. In the present case, the total size of the sample was 50, but two of these participants did not provide complete data (as discussed in the section on PROC CORR), leaving usable data from 48 participants. Output 14.2 shows that corrected total degrees of freedom are 47, which is equal to $N - 1$ (where $N = 48$). These degrees of freedom therefore appear to be correct.

2. Review the Obtained Value of R^2

Toward the center of the page, you find the heading "R-square." The value is the observed R^2 for this multiple regression equation. Earlier, it was noted that this R^2 value indicates the percent of variance in the criterion variable that is accounted for by the linear combination of predictor variables. In the present case, $R^2 = 0.65$. This indicates that the linear combination of REWARD, COST, INVESTMENT, and ALTERNATIVES accounts for about 65% of observed variance in COMMITMENT.

There is a significance test associated with this R^2 which tests the null hypothesis that $R^2 = 0$. To test this null hypothesis, look in the "Analysis of Variance" section, under "F Value." In this case, you see an F value of 19.59. Under the heading "Pr > F" is the p value associated with this F. Remember that the p value gives us the probability that you would obtain an F value this large or larger if the null hypothesis were true. In this case, the p value is very small ($< .01$), so you reject the null hypothesis and conclude that the obtained value of R^2 is statistically significant.

The preceding analysis determines whether the linear combination of predictor variables accounts for a significant amount of variance in the criterion; this test should be reviewed each time that you conduct a multiple regression. In addition to determining whether the predictors account for a significant amount of variance, you should also determine whether your predictors account for a *meaningful* amount of variance (i.e., a relatively large amount of variance). How large must an R^2 value be to be considered meaningful? That depends, in part, on what has been found in prior research concerning the criterion variable being investigated. If, for example, predictor variables in earlier investigations have routinely accounted for 50% of variance in the criterion but the predictors in your study account for only 10%, your findings might not be viewed as being very important. On the other hand, if the predictors of earlier studies have routinely accounted for only 5% of variance but the variables of your study have accounted for 10%, this might be considered a meaningful amount of variance.

This issue of "statistical significance" versus "percentage of variance accounted for" is important because it is possible to obtain an R^2 value that is very small (say, .03), but is still statistically significant. This often occurs when analyzing data from very large samples. Therefore, always review both the statistical significance of the equation, as well as the total amount of variance accounted for, in assessing the substantive importance of your findings.

NOTE

3. Review the Adjusted Value of R^2

Below the heading "R-square" is the heading "Adj R-Sq"; it stands for "adjusted R^2." To the right of "Adj R-Sq" is a version of R^2 that is adjusted for degrees of freedom. In other words, this statistic adjusts for complexity of the regression model (i.e., favors more parsimonious solutions). This is provided because the actual value of R^2 obtained with a given sample often overestimates the population value of R^2. The adjusted R^2, however, is adjusted to more closely approximate the population value. For this reason, the "Adj R-Sq" value is normally smaller than the "R-square" value.

4. Review the Intercept and Nonstandardized Regression Coefficients

The bottom half of the output page provides information about the parameter estimates. These parameter estimates are the terms that constitute the multiple regression equation (i.e., the intercept and the nonstandardized multiple regression coefficients for the predictor variables).

To begin, notice that the first column of information is headed "Variable." Below this heading are the terms in the regression equation: the intercept and the names of the four predictor variables (REWARD, COST, INVESTMENT, and ALTERNATIVES). The third column from the left is headed "Parameter Estimate." This provides the intercept estimate along with the nonstandardized multiple regression coefficients for each predictor. In this case, the intercept is approximately 20.04, the nonstandardized regression coefficient for REWARD is .28, the nonstandardized coefficient for COST is $-.11$, and so forth. Based on these estimates, you can write the multiple regression equation in this way:

$$Y' = 0.28(REWARD) - 0.11(COST) + 0.52(INVESTMENT) - 0.68(ALTERNATIVES) + 20.04$$

Remember that the multiple regression coefficient for a given predictor indicates the amount of change in Y that is associated with a one-unit change in that predictor while holding the remaining predictors constant. Nonstandardized coefficients represent the change that would be observed when the variables are in nonstandardized, "raw score" form (i.e., the different variables have different means and standard deviations). The nonstandardized regression equation would be used to predict participants' scores on COMMITMENT so that the resulting scores would be on the same scale of magnitude as observed with the raw data. However, the coefficients in this equation cannot be used to assess the relative importance of predictor variables.

5. Review the Significance of the Regression Coefficients

Researchers usually want to determine whether regression coefficients for the various predictor variables are significantly different from zero. When given coefficients are statistically significant, this suggests that the corresponding predictor variable is a relatively important predictor of the criterion.

For each predictor variable, the output of PROC REG provides a t test that tests the null hypothesis that the regression coefficient is equal to zero. The obtained t value can be found in the column headed "t Value." The *p* value corresponding to this value of *t* is in the next column, headed "Pr > |t|."

For example, in the present case, the nonstandardized regression coefficient for REWARD is approximately 0.28. When testing the significance of this coefficient, the obtained value of *t* is 1.02; it has a corresponding *p* value of .31. Because this *p* value is greater than .05, you cannot reject the null hypothesis and must conclude that the regression coefficient for REWARDS is not significantly different from zero. A different finding is obtained for the predictor INVESTMENT, however, which had a nonstandardized regression coefficient of .52. The *t* value for this coefficient is 2.49 with a corresponding *p* value of .02. Because this value is less than .05, you reject the null hypothesis and tentatively conclude that the coefficient for INVESTMENT is significantly different from zero.

The first paragraph of this subsection indicates that the statistical significance of a regression coefficient suggests that variable's importance as a predictor. This statement includes a qualification to emphasize that caution must be used in interpreting statistical significance as evidence of predictor's importance. There are at least two reasons for this.

First, an earlier section in the chapter stated that multiple regression coefficients are often unreliable, especially under the conditions that are often encountered in social science research. Second, a multiple regression coefficient can prove to be statistically significant even when the standardized coefficient is relatively small in absolute magnitude and hence is of little predictive value. This is likely to be the case especially when sample sizes are very large. For these reasons, the statistical significance of regression coefficients should be viewed as only one indicator of a variable's importance and should always be combined with additional information such as the size of the standardized regression coefficients and uniqueness indices.

6.   Review the Standardized Regression Coefficients (Beta Weights)

A previous section in this chapter indicated that nonstandardized regression coefficients generally indicate little about the relative importance of predictor variables. This is because the different predictors normally have different standard deviations, and these differences affect the size of the nonstandardized coefficients. To avoid this difficulty, it is necessary to review the standardized multiple regression coefficients or beta weights. Beta weights are the regression coefficients that would be obtained if all the variables were standardized so that they had the same standard deviations. It is therefore more appropriate to review the beta weights when you want to compare the relative importance of predictor variables. (In many textbooks, beta weights are represented by the Greek letter β while nonstandardized regression coefficients are represented by the letter *B*.)

In the preceding program, you requested standardized regression coefficients (beta weights) by specifying STB in the options section of the MODEL statement. These beta weights appear toward the bottom of Output 14.2, below the "Standardized Estimate" heading. Note that the intercept for this equation is zero as is always the case with a standardized regression equation.

The results of Output 14.2 show that the beta weight for REWARD is approximately .14, the beta for COST is –.06, the beta for INVESTMENT is .31, and the beta for ALTERNATIVES is –.50. Based on these findings, you could rank the predictors from most important to least important as follows: ALTERNATIVES; INVESTMENT; REWARD; and COST.

This interpretation should be made with caution, however, because multiple regression coefficients, whether standardized or nonstandardized, tend to be somewhat unreliable under the conditions normally encountered in social science research. A more cautious approach to understanding the relative importance of predictor variables would involve combining information from a variety of sources, including bivariate correlations, standardized regression coefficients, and uniqueness indices.

You can see that Output 14.2 does not report significance tests for the standardized regression coefficients. This is because the significance of the nonstandardized coefficients has already been reported, and a separate test for the standardized coefficients is not necessary. More precisely, if the *t* test for the nonstandardized coefficient is significant, then the corresponding standardized coefficient for that variable is also significantly different from zero.

Computing Uniqueness Indices with PROC REG

When the SELECTION=RSQUARE option is included in the MODEL statement of PROC REG, it requests that all possible multiple regression equations be created, given the predictor variables that are specified in the MODEL statement. This means that it creates every possible regression equation that could be created by taking the predictors one at a time, every possible equation that could be created by taking the predictors two at a time, and so forth. The last equation it creates is the full equation: that which includes all predictor variables listed in the MODEL statement.

Output from this procedure reports only the R^2 obtained for each of these equations (and not information concerning the regression coefficients, sum of squares, and so forth). The SELECTION=RSQUARE option is therefore useful for quickly and efficiently determining the percentage of variance in a criterion that is accounted for by every possible combination of predictor variables.

Although this information can be used for a number of purposes, this chapter shows how to use it to determine the uniqueness index for each predictor and how to test the significance of each uniqueness index. Earlier, it was said that the uniqueness index for a given predictor indicates the percentage of observed variance in the criterion that is accounted for by this predictor, over and above the variance accounted for by the other predictors. It can also be defined as the squared semipartial correlation between a criterion variable and the predictor variable of interest, after statistically controlling for the variance that the predictor shares with the other predictors.

Writing the Program

Here is the general form for the program statements that request the SELECTION=RSQUARE option with PROC REG:

```
PROC REG    DATA=dataset-name;
   MODEL   criterion  =  predictor-variables  /  SELECTION=RSQUARE;
RUN;
```

The names of the predictor variables in the MODEL statement should be separated by at least one space. The SELECTION=RSQUARE option computes and prints the R^2 value for every possible regression equation that can be created from these variables. First, it will print R^2 for a series of 1-predictor models, in which each predictor variable is the sole X variable in its own equation. Next, a series of 2-predictor models is printed in which the various equations represent every possible combination of predictors taken two at a time. The procedure continues in this manner until a single equation containing all of the predictor variables is printed.

These are the program statements that request the REG procedure for the current analysis. Notice that the MODEL statement is identical to that used with the previous REG procedure except that the STB option is replaced with the SELECTION=RSQUARE option:

```
PROC REG    DATA=D1;
   MODEL COMMIT = REWARD COST INVESTMENT ALTERNATIVES /
      SELECTION=RSQUARE;
RUN;
```

The results produced by these statements appear as Output 14.3.

Output 14.3 Results of the REG Procedure with the SELECTION=RSQUARE Option

```
                        The SAS System

                      The REG Procedure
                       Model: MODEL1
                 Dependent Variable: COMMITMENT

                   R-Square Selection Method

      Number in
        Model      R-Square      Variables in Model

          1         0.5184       ALTERNATIVES
          1         0.3770       INVESTMENT
          1         0.3317       REWARD
          1         0.0616       COST
      --------------------------------------------------------
          2         0.6248       INVESTMENT ALTERNATIVES
          2         0.5920       REWARD ALTERNATIVES
          2         0.5215       COST ALTERNATIVES
          2         0.4523       REWARD INVESTMENT
          2         0.4454       COST INVESTMENT
          2         0.3319       REWARD COST
      --------------------------------------------------------
          3         0.6436       REWARD INVESTMENT ALTERNATIVES
          3         0.6371       COST INVESTMENT ALTERNATIVES
          3         0.5947       REWARD COST ALTERNATIVES
          3         0.4690       REWARD COST INVESTMENT
      --------------------------------------------------------
          4         0.6457       REWARD COST INVESTMENT ALTERNATIVES
```

Interpreting the Results of PROC REG with SELECTION=RSQUARE

1. Make Sure That Everything Looks Right

In Output 14.3, the fourth line from the top indicates that the criterion variable is COMMITMENT, here referred to as the Dependent Variable.

The table of results created by the SELECTION=RSQUARE option includes three columns of information. The first column indicates the number of predictor variables included in a given multiple regression equation. The second column provides the R^2 for that equation (i.e., the percent of variance in the criterion variable accounted for by this set of predictors). The third column specifies the names of the predictor variables included in each equation.

In the first section of this table (the section containing the 1-predictor models), you can see that the equation that includes only the predictor ALTERNATIVES has an R^2 of approximately 0.52 (meaning that ALTERNATIVES account for approximately 52% of the variance in COMMITMENT). In the second section of the table (the section containing 2-predictor models), you can see that the model containing INVESTMENT and ALTERNATIVES account for about 62% of the variance in COMMITMENT.

2. Determine the Uniqueness Index for the Predictor Variables of Interest

The formula for determining the uniqueness index for a predictor variable is as follows:

$$U = R^2_{\text{Full}} - R^2_{\text{Reduced}}$$

where:

U = The uniqueness index for the predictor variable of interest.

R^2_{Full} = The obtained value of R^2 for the full multiple regression equation; that is, the equation containing all predictor variables.

R^2_{Reduced} = The obtained value of R^2 for the reduced multiple regression equation; that is, the equation containing all predictor variables except for the predictor variable of interest.

For example, assume that you want to determine the uniqueness index for the predictor, ALTERNATIVES. First, you identify the observed value of R^2 for the full regression equation. This is the equation that contains all four predictors: REWARD; COST; INVESTMENT; and ALTERNATIVES. In Output 14.3, you first look under the heading "Variables in the Model" to find the single entry that contains the names of all four variables. This is the last entry on the page; under the heading "R-square," you see that the R^2 for this 4-variable model is approximately .65.

Next, you find the R^2 for the reduced model: the model that includes all variables except for the variable of interest, ALTERNATIVES. Under "Variables in the Model," you find the model containing only REWARD, COST, and INVESTMENT. To the left, the R^2 for this model (the reduced model) is approximately .47. It is now possible to calculate the uniqueness index for ALTERNATIVES by inserting these values in the formula:

$$U = R^2_{Full} - R^2_{Reduced}$$

$$U = .65 - .47$$

$$U = .18$$

The uniqueness index for ALTERNATIVES is 0.18. There are several ways that this finding could be stated in a report; a few of these are now presented:

- alternative value accounted for approximately 18% of the variance in commitment, beyond the variance accounted for by the other three predictor variables;

- alternative value accounted for approximately 18% of the incremental variance in commitment;

- the squared semi-partial correlation between alternative value and commitment was .18, while partialling variance that commitment shared with the other three predictors.

Once the uniqueness index is determined, it can be tested for statistical significance. This means testing the null hypothesis that the uniqueness index for the variable of interest is equal to zero. This can be done by using the formula for testing the significance of the difference between two R^2 values. This formula is again reproduced here:

$$F = \frac{(R^2_{Full} - R^2_{Reduced}) / (K_{Full} - K_{Reduced})}{(1 - R^2_{Full}) / (N - K_{Full} - 1)}$$

where:

R^2_{Full} = The obtained value of R^2 for the full multiple regression equation; that is, the equation containing the larger number of predictor variables.

$R^2_{Reduced}$ = The obtained value of R2 for the reduced multiple regression equation; that is, the equation containing the smaller number of predictor variables.

K_{Full} = The number of predictor variables in the full multiple regression equation.

$K_{Reduced}$ = The number of predictor variables in the reduced multiple regression equation.

N = The total number of participants in the sample.

Begin by testing the significance of the uniqueness index for ALTERNATIVES. Remember that the "full equation" in this analysis is the equation that contains all four predictor variables and produced an R^2 of approximately .65. The "reduced equation" contains only three variables (all variables except ALTERNATIVES) and produced an R^2 of approximately .47. These analyses were based on a sample of 48 responses. Here, these figures are paired with the appropriate symbols from the formula:

$$R^2_{Full} = .65$$

$$R^2_{Reduced} = .47$$

$$K_{Full} = 4$$

$$K_{Reduced} = 3$$

$$N = 48$$

These values are now inserted in the appropriate locations in the formula, and the F ratio is calculated:

$$F = \frac{(R^2_{Full} - R^2_{Reduced}) / (K_{Full} - K_{Reduced})}{(1 - R^2_{Full}) / (N - K_{Full} - 1)}$$

$$F = \frac{(.65 - .47) / (4 - 3)}{(1 - .65) / (48 - 4 - 1)}$$

$$F = \frac{(.18) / (1)}{(.35) / (43)}$$

$$F = \frac{(.18)}{(.008)}$$

$$F = 22.5$$

The obtained F for this test is 22.5. To determine whether this F is large enough to reject the null hypothesis, you must first find the critical value of F appropriate for the test. To do this, turn to the table of F values found in Appendix C in the back of this text. You need to locate the critical value of F that is appropriate when $p = .05$, and corresponds to the following degrees of freedom:

df for the numerator $= K_{Full} - K_{Reduced}$

df for the denominator $= N - K_{Full} - 1$

Notice that these degrees of freedom are already calculated in the preceding F formula: the df for the numerator ($K_{Full} - K_{Reduced}$) is in the numerator of the F formula and the df for the denominator ($N - K_{Full} - 1$) is in the denominator of the formula. There, it is determined that the df statistic for the numerator was $4 - 3 = 1$, and the df statistic for the denominator was $48 - 4 - 1 = 43$.

A table of F values shows that the critical value of F with 1 and 43 degrees of freedom is approximately 4.07 ($p < .05$). Your obtained F value of 22.5 is considerably larger than this critical value, so you can reject the null hypothesis that the uniqueness index is zero. Apparently, ALTERNATIVES account for a significant amount of variance in COMMITMENT in excess of the variance accounted for by the remaining X variables.

At this point, you would proceed to test the uniqueness index for each of the remaining X variables in the equation. In each case, calculating the uniqueness index for a given variable involves finding the percentage of variance in COMMITMENT that was accounted for by the equation excluding that predictor (from Output 14.3) and subtracting this value from the R^2 value for the 4-variable model (.65). Here, the uniqueness index for each of the remaining three X variables is calculated:

REWARD: $.65 - .64 = .01$

COST: $.65 - .64 = .01$

INVESTMENT: $.65 - .60 = .05$

The preceding shows that REWARD accounts for approximately 1% of the variance in COMMITMENT, beyond the variance accounted for by the other three variables. The same is true for COST. INVESTMENT, on the other hand, has a somewhat larger uniqueness index, accounting for approximately 5% of the incremental variance in COMMITMENT.

Testing the statistical significance of the remaining three uniqueness indices is relatively straightforward because most of the necessary calculations have already been performed when testing the significance of ALTERNATIVES. Here, again, is the formula:

$$F = \frac{(R^2_{Full} - R^2_{Reduced}) / (K_{Full} - K_{Reduced})}{(1 - R^2_{Full}) / (N - K_{Full} - 1)}$$

Notice that most of the components of this formula remain unchanged when you test the significance of the uniqueness index of a different predictor variable. For example, regardless of which predictor variable's uniqueness index is being tested,

- $(K_{Full} - K_{Reduced})$ will be equal to $4 - 3 = 1$.

- $(1 - R^2_{Full})$ will be equal to $1 - .65 = .35$.

- $(N - K_{Full} - 1)$ will be equal to $48 - 4 - 1 = 43$.

This means that the only component of the equation that will vary for different predictor variables is the quantity $(R^2_{Full} - R^2_{Reduced})$. Therefore, for the current regression equation, the F formula simplifies in the following way:

$$F = \frac{(R^2_{Full} - R^2_{Reduced}) / (K_{Full} - K_{Reduced})}{(1 - R^2_{Full}) / (N - K_{Full} - 1)}$$

$$F = \frac{(R^2_{Full} - R^2_{Reduced}) / (4 - 3)}{(1 - .65) / (48 - 4 - 1)}$$

$$F = \frac{(R^2_{Full} - R^2_{Reduced}) / (1)}{(.35) / (43)}$$

$$F = \frac{(R^2_{Full} - R^2_{Reduced})}{(.008)}$$

Remember that $(R^2_{Full} - R^2_{Reduced})$ is the formula for the uniqueness index for the variable of interest. Therefore, the preceding shows that, for the current regression equation, the F ratio that tests the significance of the uniqueness index for a given X variable can be calculated by dividing that uniqueness index by .008. The degrees of freedom for each of these tests will continue to be 1 and 43; this means that the critical value of F will continue to be 4.07.

You can now use this simplified formula to test the significance of the uniqueness indices for the remaining X variables.

For REWARD:

$$F = \frac{(R^2_{Full} - R^2_{Reduced})}{(.008)} = \frac{(.65 - .64)}{(.008)} = \frac{(.01)}{(.008)}$$

$$F = 1.25$$

For COST:

$$F = \frac{(R^2_{Full} - R^2_{Reduced})}{(.008)} = \frac{(.65 - .64)}{(.008)} = \frac{(.01)}{(.008)}$$

$$F = 1.25$$

For INVESTMENT:

$$F = \frac{(R^2_{Full} - R^2_{Reduced})}{(.008)} = \frac{(.65 - .60)}{(.008)} = \frac{(.05)}{(.008)}$$

$$F = 6.25$$

Remember that, with 1 and 43 *df*, the critical value of F ($p < .05$) is 4.07 for each of these F tests. Only the obtained F ratio for INVESTMENT (at 6.25) exceeds this critical value. Therefore, only INVESTMENT demonstrates a statistically significant uniqueness index. These findings make sense when you consider the size of the uniqueness indices. INVEST accounts for approximately 5% of the unique variance in COMMITMENT, but REWARD and COST account for only 1%.

In summary, only ALTERNATIVES and INVESTMENT account for significant amounts of variance in COMMITMENT beyond the variance accounted for by the other predictors. ALTERNATIVES exhibits a uniqueness index of .18 while INVESTMENT demonstrates a uniqueness index of .05.

An earlier section of this chapter states that it is important to determine whether your linear combination of predictor variables accounts for a statistically significant amount of variance in the criterion, as well as whether this combination of predictors accounts for a meaningful (relatively large) amount of variance in the criterion. The same concerns remain when reviewing the results of these tests of the significance of uniqueness indices. With a large sample, it is possible to obtain a uniqueness index that is statistically significant even though the predictor variable accounts for negligible amounts of unique variance in the criterion (e.g., 2% and 3%). When reviewing your results, it is important to assess whether a significant uniqueness index is sufficiently large to be of substantive importance.

But how large is "large enough?" That depends, in part, on what prior research with your criterion variable has found. When doing research with a criterion variable that traditionally has been difficult to predict, a uniqueness index of .05 (and possibly even smaller) can be viewed as being of substantive importance. When researching a criterion that is relatively easy to predict, the uniqueness index might have to be larger to be considered important.

Summarizing the Results in Tables

There are several ways to summarize the results of a multiple regression analysis in a paper. Summarized results of the present analyses follow, using table formats that are fairly representative of those appearing in journals of the American Psychological Association (APA, 2001).

The Results of PROC CORR

It is usually desirable to present simple descriptive statistics, such as means and standard deviations, along with the correlation matrix. If reliability estimates (such as Cronbach's alpha) are available for the predictors, they can be included on the diagonal of the correlation matrix, in parentheses. The results for the present fictitious study are summarized in Table 14.8:

Table 14.8

Means, Standard Deviations, Intercorrelations, and Cronbach's Alpha Estimates

			Intercorrelations				
Variable	*M*	*SD*	1	2	3	4	5
1. Commitment	27.71	10.20	(.84)				
2. Rewards	26.65	5.05	.58**	(.75)			
3. Costs	16.58	5.50	.-25	.-45*	(.72)		
4. Investment size	25.33	6.08	.61**	.57**	.02	(.83)	
5. Alternative value	16.60	7.50	.-72**	.-47**	.27	-.45*	(.91)

Note. $N = 48$. Reliability estimates (Cronbach's alpha) appear on the diagonal above correlation coefficients.

*$p < .01$., **$p < .001$

The Results of PROC REG

Depending on the nature of the research problem, it is often feasible to report standardized regression coefficients (beta weights) and the uniqueness indices in a single table as done in Table 14.9:

Table 14.9

Beta Weights and Uniqueness Indices Obtained in Multiple Regression Analyses Predicting Commitment

Predictor	B	SE B	β	t	Uniqueness Index
Rewards	.28	.27	0.14	1.02	.01
Costs	-.11	.21	-0.06	-0.51	.01
Investment size	.52	.21	0.31	2.49 *	.05 *
Alternative value	-.68	.15	-0.50	-4.64 **	.18 **

* $p < .05$ ** $p < .001$

Note. $R^2 = .65$ ($F[4,43] = 19.59$), $p < .01$ for predictor variables.

Getting the Big Picture

It is instructive to reflect on the big picture concerning your findings before proceeding to summarize them in text form. First, notice the bivariate correlations from Table 14.8. The correlations between commitment and rewards, between commitment and investment size, and between commitment and alternative value are each significant and in the predicted direction. Only the correlation between commitment and costs is nonsignificant. These findings provide partial support for the investment model.

A somewhat similar pattern of results can be seen in Table 14.9 which shows that the beta weights and uniqueness indices for investment size and alternative value are both significant and in the predicted direction. Unlike correlations, however, Table 14.9 shows that neither the beta weight nor the uniqueness index for rewards is statistically significant. This might come as a surprise because the correlation between commitment and rewards is moderately strong at .58. With such a strong correlation, how could the multiple regression coefficient and uniqueness index for rewards be nonsignificant?

A possible answer can be found in the correlations of Table 14.8. Notice that the correlation between Commitment and Rewards is somewhat weaker than the correlation between Commitment and either Investment size or Alternative value. In addition, it can be seen that the correlations between Rewards and both Investment size and Alternative value are fairly substantial at $r = .57$ and $r = -.47$, respectively. In short, REWARDS shares a great deal of variance in common with Investment size and Alternative value and is a poorer predictor of Commitment. In this situation, it is unlikely that a multiple regression equation that already contains Investment size and Alternative value would need a variable like REWARDS to improve the accuracy of prediction. In other words, any variance in commitment that is accounted for by REWARDS has probably already been accounted for by investment size and alternative value. As a result, the REWARDS variable is largely redundant and consequently displays a nonsignificant beta weight and uniqueness index.

Was this a real test of the investment model? It must be emphasized again that the results concerning the investment model presented here are entirely fictitious and should not be viewed as legitimate tests of that conceptual framework. Most published studies of the investment model are, in fact, very supportive of its predictions. For representative examples of this research, please refer to Rusbult (1980a, 1980b), Rusbult and Farrell (1983), and Rusbult, Johnson, and Morrow (1986).

Formal Description of Results for a Paper

Again, there are several ways to summarize the results of these analyses within the text of the paper. The amount of detail provided when describing the analyses should be dictated by the statistical sophistication of your audience; less detail is needed if the audience is likely to be familiar with the use of multiple regression. The following format is fairly typical:

Results were analyzed using both bivariate correlations and multiple regression. Means, standard deviations, Pearson correlations, and Cronbach's alpha estimates appear in Table 14.8. The bivariate correlations revealed three predictor variables were significantly related to commitment: rewards (r = .58); investment size (r = .61); and alternative value (r = -.72). All of these correlations were significant at $p < .01$, and all were in the predicted direction. The correlation between costs and commitment, on the other hand, was nonsignificant as r = -.25.

Using multiple regression, commitment scores were then regressed on the linear combination of rewards, costs, investment size, and alternative value. The equation containing these four variables accounted for approximately 65% of observed variance in commitment, $F(4, 43) = 19.60$, $p < .01$, adjusted R^2 = .61.

Beta weights (standardized multiple regression coefficients) and uniqueness indices were subsequently reviewed to assess the relative importance of the four variables in the prediction of commitment. The uniqueness index for a given predictor is the percentage of variance in the criterion accounted for by that predictor, beyond the variance accounted for by the other predictor variables. Beta weights and uniqueness indices are presented in Table 14.9.

The table shows that only investment size and alternative value displayed significant beta weights. Alternative value demonstrated somewhat larger beta weight at -.50 ($p < .01$), while the beta weight for investment size was .31 ($p < .05$). Both coefficients were in the predicted direction.

```
     Findings regarding uniqueness indices correspond to those
for beta weights in that only investment size and alternative
value displayed significant indices.  Alternative value
accounted for approximately 18% of the variance in commitment,
beyond the variance accounted for by the other three
predictors, F(1, 43) = 21.51, p < .01.  In contrast, investment
size accounted for only 5% of the unique variance in
commitment, F(1, 43) = 6.20, p < .05.
```

The preceding tables and text make reference to the degrees of freedom for the statistical tests that assessed the significance of the model R^2 as well as the significance of the individual regression coefficients. It might be helpful to summarize how these degrees of freedom are calculated.

The second paragraph indicates that "The equation containing these four variables accounted for 65% of observed variance in commitment, $F(4, 43) = 19.60, p < .01$, adjusted $R^2 = .61$." The information from this passage is in the results of PROC REG using the STB option, which appeared as Output 14.2. The F value that tests the significance of the model R^2 is in the analysis of variance section under "F Value." The p value associated with this test appears to the right of the F value. The degrees of freedom for the test are in the second column, headed "DF." Degrees of freedom for the numerator in the F ratio (4) can be found to the right of "Model"; the degrees of freedom for the denominator (43), is to the right of "Error."

The beta weights and their corresponding *t* values are in the lower half of Output 14.2. The beta weights themselves appear under the heading "Standardized Estimate," while the *t* values appear under "t Value." The degrees of freedom for these *t* tests are equal to

$$N - K_{\text{Full}} - 1$$

in which N = total sample size and K_{Full} = number of predictor variables in the full equation. In this case, $df = 48 - 4 - 1 = 43$.

Conclusion: Learning More about Multiple Regression

This chapter provides only an elementary introduction to multiple regression, one of the most flexible and powerful research tools in the social sciences. It discusses only the situation in which a criterion variable is being predicted from continuous predictor variables, all of which display a linear relationship with the criterion. At this time, you might be ready to move on to a more comprehensive treatment of the topic, one that deals with curvilinear relationships, interactions among predictors, dummy coding, effect coding, and regression equations that contain both continuous and categorical predictor variables. Cohen, Cohen, West, and Aiken (2003), and Pedhazur (1982) provide authoritative treatments of these and other advanced topics in multiple regression.

Assumptions Underlying Multiple Regression

- **Level of measurement.** For multiple regression, both the predictor variables and the criterion variable should be assessed at the interval or ratio level of measurement. Nominal-level predictor variables can be used if they have been appropriately dummy- or effect-coded.

- **Random sampling.** Each participant in the sample contributes one score on each predictor variable, and one score on the criterion variable. These sets of scores should represent a random sample drawn from the population of interest.

- **Normal distribution of the criterion variable.** For any combination of values on the predictor variables, the criterion variable should be normally distributed.

- **Homogeneity of variance.** For any combination of values of the predictor variables, the criterion variable should demonstrate a constant variance.

- **Independent observations.** A given observation should not be affected by (or related to) any other observation in the sample. For example, this assumption would be violated if the various observations represented repeated measurements taken from a single participant. It would also be violated if the study included multiple participants, some of whom contributed more that one observation to the dataset (i.e., some participants contributed more than one set of scores on the criterion variable and predictor variables).

- **Linearity.** The relationship between the criterion variable and each predictor variable should be linear. This means that the mean criterion scores at each value of a given predictor should fall on a straight line.

- **Errors of prediction.** Errors of prediction should be normally distributed and the distribution of errors should be centered at zero. Error of prediction associated with a given observation should not be correlated with those associated with the other observations. Errors of prediction should demonstrate consistent variance. Errors of prediction should not be correlated with the predictor variables.

- **Absence of measurement error.** The predictor variables should be measured without error. Pronounced violations of this assumption lead to underestimation of the regression coefficient for the corresponding predictor.

- **Absence of specification errors.** The term **specification error** generally refers to situations in which the model represented by the regression equation is not theoretically tenable. In multiple regression, specification errors most frequently result from omitting relevant predictor variables from the equation or including irrelevant predictor variables in the equation. Specification errors also result when researchers posit a linear relationship among variables that are actually curvilinear.

It is infrequent that all of these assumptions will be fully satisfied in applied research. Fortunately, regression analysis is generally robust against minor violations of most of these assumptions. However, it is less robust against violations of the assumptions involving independent observations, measurement error, or specification errors (Pedhazur, 1982).

In addition to considering the preceding assumptions, researchers are also advised to inspect their data for possible problems involving outliers or multicollinearity. An outlier is an unusual observation that does not fit the regression model well. Outliers are often the result of mistakes made when entering data, and can profoundly bias parameter estimates (such as regression coefficients). As previously noted, **multicollinearity** exists when two or more predictor variables demonstrate a high degree of correlation with one another (i.e., $r > |.90|$). Multicollinearity can cause regression coefficient estimates to fail to demonstrate statistical significance, be biased, or even demonstrate the incorrect sign. Cohen, Cohen, West, and Aiken (2003) discuss these problems, and show how to detect outliers, multicollinearity, and other problems sometimes encountered in regression analysis.

References

American Psychological Association (2001). *Publication manual of the American Psychological Association* (5th ed.). Washington, DC: Author.

Cohen, J. (1992). A power primer. *Psychological Bulletin, 112,* 155-159.

Cohen, J., Cohen, P., West, S. G., & Aiken, L. S. (2003). *Applied multiple regression/correlation analysis for the behavioral sciences* (3rd ed.). Hillsdale, NJ: Lawrence Erlbaum.

Hopkins, K. D., Glass, G. V., & Hopkins, B. R. (1987). *Basic statistics for the behavioral sciences.* Englewood Cliffs, NJ: Prentice-Hall.

Pedhazur, E. J. (1982). *Multiple regression in behavioral research* (2nd ed.). New York: Holt, Rinehart, and Winston.

Rusbult, C. E. (1980a). Commitment and satisfaction in romantic associations: A test of the investment model. *Journal of Experimental Social Psychology, 16,* 172-186.

Rusbult, C. E. (1980b). Satisfaction and commitment in friendships. *Representative Research in Social Psychology, 11,* 96-105.

Rusbult, C. E., & Farrell, D. (1983). A longitudinal test of the investment model: The impact on job satisfaction, job commitment, and turnover of variations in rewards, costs, alternatives, and investments. *Journal of Applied Psychology, 68,* 429-438.

Rusbult, C. E., Johnson, D. J., & Morrow, G. D. (1986). Predicting satisfaction and commitment in adult romantic involvements: An assessment of the generalizability of the investment model. *Social Psychology Quarterly, 49,* 81-89.

SAS Institute Inc. (2004). *SAS/STAT 9.1 User's Guide.* Cary, NC: Author.

Principal Component Analysis

> **Overview.** This chapter provides an introduction to principal component analysis: a variable-reduction procedure similar to factor analysis. The chapter provides guidelines regarding the necessary sample size and number of items per component. It shows how to determine the number of components to retain, interpret the rotated solution, create factor scores, and summarize the results. Fictitious data from two studies are analyzed to illustrate these procedures. The present chapter deals only with the creation of orthogonal (uncorrelated) components.

Introduction: The Basics of Principal Component Analysis

Principal component analysis is an appropriate procedure when you have obtained measures on a number of observed variables and want to develop a smaller number of variables (called principal components) that will account for most of the variance in the observed variables. The principal components can then be used as predictors or criterion variables in subsequent analyses.

A Variable Reduction Procedure

Principal component analysis is a variable reduction procedure. It is useful when you obtain data for a number of variables (possibly a large number of variables) and believe that there is *redundancy* among those variables. In this case, redundancy means that some of the variables are correlated with one another, possibly because they are measuring the same construct. Because of this redundancy, you believe that it should be possible to reduce the observed variables into a smaller number of principal components that account for most of the observed variance in the variables.

Because it is a variable reduction procedure, principal component analysis is similar in many respects to exploratory factor analysis. In fact, the steps followed when conducting a principal component analysis are virtually identical to those followed when conducting an exploratory factor analysis. However, there are significant conceptual differences between the two procedures, and it is important that you do not mistakenly claim that you are performing factor analysis when you are actually performing principal component analysis. The differences between these two procedures are described in greater detail in a later section, "Principal Component Analysis Is *Not* Factor Analysis."

An Illustration of Variable Redundancy

This is a fictitious example of research presented to illustrate the concept of variable redundancy. Imagine that you have developed a 7-item measure of job satisfaction. The instrument is reproduced here:

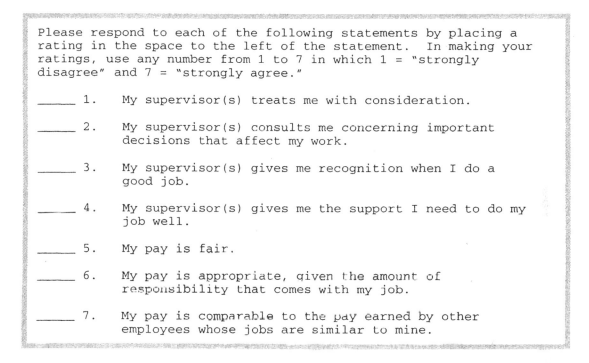

Please respond to each of the following statements by placing a rating in the space to the left of the statement. In making your ratings, use any number from 1 to 7 in which 1 = "strongly disagree" and 7 = "strongly agree."

_____ 1. My supervisor(s) treats me with consideration.

_____ 2. My supervisor(s) consults me concerning important decisions that affect my work.

_____ 3. My supervisor(s) gives me recognition when I do a good job.

_____ 4. My supervisor(s) gives me the support I need to do my job well.

_____ 5. My pay is fair.

_____ 6. My pay is appropriate, given the amount of responsibility that comes with my job.

_____ 7. My pay is comparable to the pay earned by other employees whose jobs are similar to mine.

Perhaps you began your investigation with the intention of administering this questionnaire to 200 employees using their responses to the seven items as seven separate variables in subsequent analyses.

There are several problems with conducting the study in this manner, however. One of the more important problems involves the concept of redundancy that was previously mentioned. Examine the content of the seven items in the questionnaire closely. Notice that items 1 to 4 all deal with the same topic: employees' satisfaction with their supervisors. In this way, items 1 to 4 are somewhat redundant. Similarly, notice that items 5 to 7 also all seem to deal with the same topic: employees' satisfaction with their pay.

Empirical findings can further support the notion that there is redundancy among items. Assume that you administer the questionnaire to 200 employees and compute all possible correlations between responses to the 7 items. The resulting fictitious correlations are presented in Table 15.1:

Table 15.1

Correlations among Seven Job Satisfaction Items

	Correlations						
Variable	1	2	3	4	5	6	7
1	1.00						
2	.75	1.00					
3	.83	.82	1.00				
4	.68	.92	.88	1.00			
5	.03	.01	.04	.01	1.00		
6	.05	.02	.05	.07	.89	1.00	
7	.02	.06	.00	.03	.91	.76	1.00

Note. N = 200.

When correlations among several variables are computed, they are typically summarized in the form of a **correlation matrix** such as the one presented in Table 15.1. This is an appropriate opportunity to review just how a correlation matrix is interpreted. The rows and columns of Table 15.1 correspond to the seven variables included in the analysis. Row 1 (and column 1) represents variable 1, row 2 (and column 2) represents variable 2, and so forth. Where a given row and column intersect, you will find the correlation between the two corresponding variables. For example, where the row for variable 2 intersects with the column for variable 1, you find a correlation of .75; this means that the correlation between variables 1 and 2 is .75.

The correlations in Table 15.1 show that the seven items seem to hang together in two distinct groups. First, notice that items 1 through 4 show relatively strong correlations with one another. This could be because items 1 through 4 are measuring the same construct. In the same way, items 5 through 7 correlate strongly with one another (a possible indication that they also measure a single construct). Even more interesting, notice that items 1 through 4 demonstrate very weak correlations with items 5 through 7. This is what you would expect to see if items 1 through 4 and items 5 through 7 were measuring two different constructs.

Given this apparent redundancy, it is likely that the seven items of the questionnaire are not really measuring seven different constructs. More likely, items 1 through 4 are measuring a single construct that could reasonably be labeled "satisfaction with supervision" whereas items 5 through 7 are measuring a different construct that could be labeled "satisfaction with pay."

If responses to the seven items actually displayed the redundancy suggested by the pattern of correlations in Table 15.1, it would be advantageous to reduce the number of variables in this dataset, so that (in a sense) items 1 through 4 are collapsed into a single new variable that reflects employees' satisfaction with supervision and items 5 through 7 are collapsed into a single new variable that reflects satisfaction with pay. You could then use these two new variables (rather than the seven original variables) as predictor variables in multiple regression or any other type of analysis.

In essence, this is what is accomplished by principal component analysis: it allows you to reduce a set of observed variables into a smaller set of variables called principal components. The resulting principal components can then be used in subsequent analyses.

What Is a Principal Component?

How Principal Components Are Computed

A **principal component** can be defined as a linear combination of optimally weighted observed variables. In order to understand the meaning of this definition, it is necessary to first describe how scores on a principal component are computed.

In the course of performing a principal component analysis, it is possible to calculate a score for each participant for a given principal component. For example, in the preceding study, each participant would have scores on two components: one score on the satisfaction with supervision component and one score on the satisfaction with pay component. Participants' actual scores on the seven questionnaire items would be optimally weighted and then summed to compute their scores for a given component.

Below is the general form for the formula to compute scores on the first component extracted (created) in a principal component analysis:

$$C_1 = b_{11}(X_1) + b_{12}(X_2) + ... b_{1p}(X_p)$$

where

C_1 = the participant's score on principal component 1 (the first component extracted);

b_{1p} = the regression coefficient (or weight) for observed variable p, as used in creating principal component 1;

X_p = the participant's score on observed variable p.

For example, assume that component 1 in the present study was "satisfaction with supervision." You could determine each participant's score on principal component 1 by using the following fictitious formula:

$$C_1 = .44 (X_1) + .40 (X_2) + .47 (X_3) + .32 (X_4) + .02 (X_5) + .01 (X_6) + .03 (X_7)$$

In the present case, the observed variables (the "X" variables) are responses to the seven job satisfaction questions: X_1 represents question 1; X_2 represents question 2; and so forth. Notice that different regression coefficients were assigned to the different questions in computing scores on component 1: questions 1 through 4 were assigned relatively large regression weights that range from .32 to .44 whereas questions 5 through 7 were assigned very small weights ranging from .01 to .03. This makes sense, because component 1 is the satisfaction with supervision component and satisfaction with supervision is assessed by questions 1 through 4. It is therefore appropriate that items 1 through 4 are given a good deal of weight in computing participant scores on this component, while items 5 through 7 have comparatively little weight.

Obviously, a different equation, with different regression weights, would be used to compute scores on component 2 (i.e., satisfaction with pay). Below is a fictitious illustration of this formula:

$$C_2 = \quad .01\,(X_1) + .04\,(X_2) + .02\,(X_3) + .02\,(X_4)$$
$$+ \quad .48\,(X_5) + .31\,(X_6) + .39\,(X_7)$$

The preceding shows that, in creating scores on the second component, much weight is given to items 5 through 7 and comparatively little is given to items 1 through 4. As a result, component 2 should account for much of the variability in the three satisfaction with pay items (i.e., it should be strongly correlated with those three items).

At this point, it is reasonable to wonder how the regression weights from the preceding equations are determined. The SAS PROC FACTOR procedure generates these weights by using a special type of equation called an **eigenequation**. The weights produced by these eigenequations are optimal weights in the sense that, for a given set of data, no other set of weights could produce components that are more effective in accounting for variance among observed variables. The weights are created so as to satisfy a principle of least squares similar (but not identical) to the principle of least squares used in multiple regression (see Chapter 14, "Multiple Regression"). Later, this chapter shows how PROC FACTOR can be used to extract (create) principal components.

It is now possible to understand the definition provided at the beginning of this section more fully. There, a principal component was defined as a linear combination of optimally weighted observed variables. The words "linear combination" refer to the fact that scores on a component are created by adding together scores on the observed variables being analyzed. "Optimally weighted" refers to the fact that the observed variables are weighted in such a way that the resulting components account for a maximal amount of observed variance in the dataset.

Number of Components Extracted

The preceding section might have created the impression that, if a principal component analysis were performed on data from the 7-item job satisfaction questionnaire, only two components would be created. However, such an impression would not be entirely correct.

In reality, the number of components extracted in a principal component analysis is equal to the number of observed variables being analyzed. This means that an analysis of your 7-item questionnaire would result in seven components, not two.

In most analyses, however, only the first few components account for meaningful amounts of variance so only these first few components are retained, interpreted, and used in subsequent analyses (such as in multiple regression). For example, in your analysis of the 7-item job satisfaction questionnaire, it is likely that only the first two components would account for a meaningful amount of variance. Therefore only these would be retained for interpretation. You would assume that the remaining five components accounted for only trivial amounts of variance. These latter components would therefore not be retained, interpreted, or further analyzed.

Characteristics of Principal Components

The first component extracted in a principal component analysis accounts for a maximal amount of total variance in the observed variables. Under typical conditions, this means that the first component is correlated with at least some observed variables. It is usually correlated with many.

The second component extracted will have two important characteristics. First, this component will account for a maximal amount of variance in the dataset that was not accounted for by the first component. Again under typical conditions, this means that the second component is correlated with some observed variables that did not display strong correlations with component 1.

The second characteristic of the second component is that it is *uncorrelated* with the first component. Literally, if you were to compute the correlation between components 1 and 2, that correlation would be zero. (For the exception, see the following section regarding oblique solutions.)

The remaining components extracted in the analysis display the same two characteristics: each component accounts for a maximal amount of variance in the observed variables that is not accounted for by the preceding components and is uncorrelated with all of the preceding components. A principal component analysis proceeds in this manner with each new component accounting for progressively smaller amounts of variance. This is why only the first few components are usually retained and interpreted. When the analysis is complete, the resulting components display varying degrees of correlation with the observed variables, but are completely uncorrelated with one another.

What is meant by "total variance" in the dataset? To understand the meaning of total variance as it is used in a principal component analysis, remember that the observed variables are standardized in the course of the analysis. This means that each variable is transformed so that it has a mean of zero and a standard deviation of one (and hence a variance of one). The "total variance" in the dataset is simply the sum of variances of these observed variables. Because they have been standardized to have a standard deviation of one, each observed variable contributes one unit of variance to "total variance" in the dataset. Because of this, total variance in a principal component analysis always equals the number of observed variables analyzed. For example, if seven variables are analyzed, the total variance equals seven. The components that are extracted in the analysis partition this variance. Perhaps the first component accounts for 3.2 units of total variance; perhaps the second component accounts for 2.1 units. The analysis continues in this way until all variance in the dataset is accounted for.

Orthogonal versus Oblique Solutions

This chapter discusses only principal component analyses that result in orthogonal solutions. An **orthogonal solution** is one in which the components are uncorrelated (orthogonal means "uncorrelated").

It is possible to perform a principal component analysis that results in correlated components. Such a solution is referred to as an **oblique solution**. In some situations, oblique solutions are preferred to orthogonal solutions because they produce cleaner, more easily interpreted results.

However, oblique solutions can also be somewhat more complicated to interpret. For this reason, the present chapter focuses only on the interpretation of orthogonal solutions.

Principal Component Analysis Is *Not* Factor Analysis

Principal component analysis is often confused with factor analysis. This is understandable because there are many important similarities between the two procedures. Both are methods that can be used to identify groups of observed variables that tend to hang together empirically. Both procedures can also be performed with the SAS FACTOR procedure and they generally provide similar results.

Nonetheless, there are some important conceptual differences between principal component analysis and factor analysis that should be understood at the outset. Perhaps the most important deals with the **assumption of an underlying causal structure**. Factor analysis assumes that the covariation in the observed variables is due to the presence of one or more latent variables (factors) that exert causal influence on these observed variables. An example of such a causal structure is presented in Figure 15.1:

Figure 15.1 Example of the Underlying Causal Structure That Is Assumed in Factor Analysis

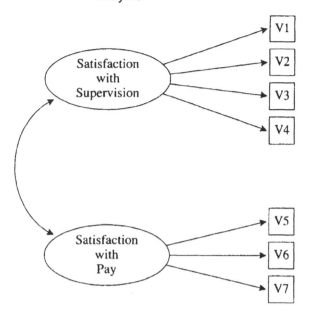

The ovals in Figure 15.1 represent the latent (unmeasured) factors of "satisfaction with supervision" and "satisfaction with pay." These factors are *latent* in the sense that they are assumed to actually exist in the employees' belief systems but cannot be measured directly. However, they do exert an influence on employees' responses to the seven items that constitute the job satisfaction questionnaire described earlier. (These seven items are represented as squares labeled V1 to V7 in the figure.) It can be seen that the "supervision" factor exerts influence on items V1 to V4 (the supervision questions) whereas the "pay" factor exerts influence on items V5 to V7 (the pay items).

Researchers use factor analysis when they believe that certain latent factors exist that exert causal influence on the observed variables they are studying. Exploratory factor analysis helps the researcher identify the number and nature of such latent factors.

In contrast, principal component analysis makes no assumption about an underlying causal structure. Principal component analysis is simply a variable reduction procedure that (typically) results in a relatively small number of components that account for most of the variance in a set of observed variables (i.e., groupings of observed variables vs. latent constructs).

In summary, both factor analysis and principal component analysis have important roles to play in social science research, but their conceptual foundations are quite distinct.

Example: Analysis of the Prosocial Orientation Inventory

Assume that you have developed an instrument called the Prosocial Orientation Inventory (POI) that assesses the extent to which a person has engaged in helping behaviors over the preceding six-month period. The instrument contains six items and is presented here.

Instructions: Below are a number of activities in which people sometimes engage. For each item, please indicate how frequently you have engaged in this activity during the past six months. Make your rating by circling the appropriate number to the left of each item using the following response format:

```
7 = Very Frequently
6 = Frequently
5 = Somewhat Frequently
4 = Occasionally
3 = Seldom
2 = Almost Never
1 = Never
```

1 2 3 4 5 6 7   1.   Went out of my way to do a favor for a coworker.

1 2 3 4 5 6 7   2.   Went out of my way to do a favor for a relative.

1 2 3 4 5 6 7   3.   Went out of my way to do a favor for a friend.

1 2 3 4 5 6 7   4.   Gave money to a religious charity.

1 2 3 4 5 6 7   5.   Gave money to a charity not associated with a religion.

1 2 3 4 5 6 7   6.   Gave money to a panhandler.

When this instrument was first developed, you intended to administer it to a sample of participants and use their responses to the six items as separate predictor variables in a multiple regression equation. As previously stated, however, you learned that this is a questionable practice and have decided, instead, to perform a principal component analysis on responses to the six items to see if a smaller number of components can successfully account for most of the variance in the dataset. If this is the case, you will use the resulting components as the predictor variables in your regression analyses.

At this point, it might be instructive to review the content of the six items that constitute the POI to make an informed guess as to what is likely to be observed from the principal component analysis. Imagine that, when you first constructed the instrument, you assumed that the six items were assessing six different types of prosocial behavior. Inspection of items 1 through 3, however, shows that these three items share something in common: they all deal with the activity of "going out of one's way to do a favor for someone else." It would not be surprising to learn that these three items hang together empirically in the principal component analysis to be performed. In the same way, a review of items 4 through 6 shows that all of these items involve the activity of giving money to those in need. Again, it is possible that these three items will also group together in the course of the analysis.

In summary, the nature of the items suggests that it might be possible to account for variance in the POI with just two components: a "helping others" component and a "financial giving" component. At this point, this is only speculation, of course; only a formal analysis can determine the number and nature of the components measured by the POI.

(Remember that the preceding instrument is fictitious and used for purposes of illustration only; it should not be regarded as an example of a good measure of prosocial orientation. Among other problems, this questionnaire obviously deals with very few forms of helping behavior.)

Preparing a Multiple-Item Instrument

The preceding section illustrates an important point about how *not* to prepare a multiple-item measure of a construct. Generally speaking, it is poor practice to throw together a questionnaire, administer it to a sample, and then perform a principal component analysis (or factor analysis) to determine what the questionnaire is measuring.

Better results are much more likely when you make *a priori* decisions about what you want the questionnaire to measure and then take steps to ensure that it does. For example, you would have been more likely to obtain desirable results if you:

- had begun with a thorough review of theory and research on prosocial behavior;
- used that review to determine how many types of prosocial behavior probably exist;
- wrote multiple questionnaire items to assess each type of prosocial behavior.

Using this approach, you could have made statements such as "there are three types of prosocial behavior: acquaintance helping; stranger helping; and financial giving." You could have then prepared a number of items to assess each of these three types, administered the questionnaire to a large sample, and performed a principal component analysis to see if the three components did, in fact, emerge.

Number of Items per Component

When a variable (such as a questionnaire item) is given a weight in constructing a principal component, we say that the variable **loads** on that component. For example, if the item "Went out of my way to do a favor for a coworker" is given a lot of weight in creating the "helping others" component, we say that this item loads on that component.

It is highly desirable to have at least three (and preferably more) variables loading on each retained component when the principal component analysis is complete (see Clark & Watson, 1995). Because some items might be dropped during the course of the analysis (for reasons to be discussed later), it is generally good practice to write at least five items for each construct that you want to measure. This increases the likelihood that at least three items per component will survive the analysis. Note that we have unfortunately violated this recommendation by apparently writing only three items for each of the two *a priori* components constituting the POI.

One additional note on scale length: the recommendation of three items per scale should be viewed as an absolute minimum and certainly not as an optimal number of items per scale. In practice, test and attitude scale developers normally desire that their scales contain many more than just three items to measure a given construct. It is not unusual to see individual scales that include 10, 20, or even more items to assess a single construct (e.g., O'Rourke & Cappeliez, 2002). Up to a point, the more items in the scale, the more reliable it will be. The recommendation of three items per scale should therefore be viewed as a rock-bottom lower limit, appropriate only if practical concerns (such as total questionnaire length) prevent you from including more items. For more information on scale construction, see Spector (1992).

Minimally Adequate Sample Size

Principal component analysis is a large-sample procedure. To obtain reliable results, the minimal number of participants providing usable data for the analysis should be the larger of 100 participants or five times the number of variables being analyzed (Streiner, 1994).

To illustrate, assume that you want to perform an analysis on responses to a 50-item questionnaire. (Remember that, when responses to a questionnaire are analyzed, the number of variables is equal to the number of items on that questionnaire.) Five times the number of items on the questionnaire equals 250. Therefore, your final sample should provide usable (complete) data from at least 250 participants. Of note, however, any participant who fails to answer just one item does not provide usable data for the principal component analysis and is therefore excluded from the final sample. A certain number of participants can always be expected to leave at least one question blank. To ensure that the final sample includes at least 250 usable responses, you would be wise to administer the questionnaire to perhaps 300 to 350 participants.

These rules regarding the number of participants per variable again constitute a lower limit, and some have argued that they should apply only under two optimal conditions for principal component analysis: when many variables are expected to load on each component; and when variable communalities are high. Under less optimal conditions, even larger samples might be required.

> **What is a communality?** A **communality** refers to the percent of variance in an observed variable that is accounted for by the retained components (or factors). A given variable displays a large communality if it loads heavily on at least one of the study's retained components. Although communalities are computed in both procedures, the *concept* of variable communality is more relevant in a factor analysis than in principal component analysis.

SAS Program and Output

You can perform a principal component analysis using either the PRINCOMP, CALIS, or FACTOR procedures. This chapter shows how to perform the analysis using PROC FACTOR since this is a somewhat more flexible SAS procedure. (It is also possible to perform an exploratory factor analysis with PROC FACTOR or CALIS.) Because the analysis is to be performed using the FACTOR procedure, the output will at times make reference to factors rather than to principal components (e.g., component 1 is referred to as FACTOR1 in the output). However, it is important to remember that you are performing a principal component analysis.

This section provides instructions on writing the SAS program and an overview of the SAS output. A subsequent section provides a more detailed treatment of the steps followed in the analysis as well as the decisions to be made at each step.

Writing the SAS Program

The DATA Step

To perform a principal component analysis, data can be entered in the form of raw data, a correlation matrix, a covariance matrix, and other types of data. (See Chapter 3, "Data Input," for a further description of these data input options.) Raw data are analyzed for this chapter's first example.

Assume that you administered the POI to 50 participants and entered their responses according to the following guide:

Line	Column	Variable Name	Explanation
1	1-6	V1-V6	Participants' responses to survey questions 1 through 6. Responses were made using a 7-point Likert-type scale.

Here are the statements that input these responses as raw data. The first three and the last three observations are reproduced here; for the entire dataset, see Appendix B.

```
 1      DATA D1;
 2         INPUT      #1      @1     (V1-V6)      (1.)   ;
 3
 4      DATALINES;
 5      556754
 6      567343
 7      777222
 8      .
 9      .
10      .
11      767151
12      455323
13      455544
14      ;
15      RUN;
```

The dataset in Appendix B includes only 50 cases so that it is relatively easy to enter the data and replicate the analyses presented here. However, it should be remembered that 50 observations normally constitute an unacceptably small sample for a principal component analysis. Earlier, it was said that a sample should provide usable data from the larger of either 100 cases or five times the number of observed variables. A small sample is being analyzed here for illustrative purposes only.

The PROC FACTOR Statement

The general form for the SAS program to perform a principal component analysis is presented here:

```
PROC FACTOR    DATA=dataset-name
               SIMPLE
               METHOD=PRIN
               PRIORS=ONE
               MINEIGEN=p
               SCREE
               ROTATE=VARIMAX
               ROUND
               FLAG=desired-size-of-"significant"-factor-loadings ;
    VAR   variables-to-be-analyzed ;
RUN;
```

Options Used with PROC FACTOR

The PROC FACTOR statement begins the FACTOR procedure; a number of options can be requested in this statement before it ends with a semicolon. Some options that can be especially useful in social science research are:

FLAG
> causes the output to flag (with an asterisk) any factor loading whose absolute value is greater than some specified size. For example, if you specify

> FLAG=.35

an asterisk appears next to any loading whose absolute value exceeds .35. This option can make it much easier to interpret a factor pattern. Negative values are not allowed in the FLAG option and the FLAG option can be used in conjunction with the ROUND option.

`METHOD=factor-extraction-method`
> specifies the method to be used in extracting the factors or components. The current program specifies `METHOD=PRIN` to request that the principal axis (principal factors) method be used for the initial extraction. This is the appropriate method for a principal component analysis.

`MINEIGEN=p`
> specifies the critical eigenvalue a component must display if that component is to be retained (here, p = the critical eigenvalue). For example, the current program specifies

> `MINEIGEN=1`

> This statement causes PROC FACTOR to retain and rotate any component with an eigenvalue of 1.00 or larger. Negative values are not allowed.

`NFACT=n`
> allows you to specify the number of components to be retained and rotated, where n = the number of components.

`OUT=name-of-new-dataset`
> creates a new dataset that includes all variables of the existing dataset, along with factor scores for the components retained in the present analysis. Component 1 is given the variable name FACTOR1, component 2 is given the name FACTOR2, and so forth. The name must be used in conjunction with the NFACT option, and the analysis must be based on raw data.

`PRIORS=prior-communality-estimates`
> specifies prior communality estimates. Users should always specify PRIORS=ONE to perform a principal component analysis.

`ROTATE=rotation-method`
> specifies the rotation method to be used. The preceding program requests a varimax rotation that provides orthogonal (uncorrelated) components. Oblique rotations can also be requested (correlated components).

`ROUND`
> causes all coefficients to be limited to two decimal places, multiplied by 100, and rounded to the nearest integer (thus eliminating the decimal point). This generally makes it easier to read coefficients because factor loadings and correlation coefficients in the matrices printed by PROC FACTOR are normally carried to several decimal places.

`SCREE`
> creates a plot that graphically displays the size of the eigenvalues associated with each component. This can be used to perform a scree test to visually determine how many components should be retained.

`SIMPLE`
> requests simple descriptive statistics: the number of usable cases on which the analysis was performed and the means and standard deviations of the observed variables.

The VAR Statement

The variables to be analyzed are listed on the VAR statement with each variable separated by at least one space. Remember that the VAR statement is a *separate* statement, not an option within the FACTOR statement; don't forget to end the FACTOR statement with a semicolon before beginning the VAR statement.

Example of an Actual Program

The following is an actual program, including the DATA step, which could be used to analyze fictitious study data. Only a few sample lines of data appear here; the entire dataset is in Appendix B.

```
1       DATA D1;
2          INPUT    #1    @1    (V1-V6)    (1.)   ;
3
4       DATALINES;
5       556754
6       567343
7       777222
8       .
9       .
10      .
11      767151
12      455323
13      455544
14      ;
15      RUN;
16
17      PROC FACTOR    DATA=D1
18                     SIMPLE
19                     METHOD=PRIN
20                     PRIORS=ONE
21                     MINEIGEN=1
22                     SCREE
23                     ROTATE=VARIMAX
24                     ROUND
25                     FLAG=.40    ;
26         VAR V1 V2 V3 V4 V5 V6;
27      RUN;
```

Results from the Output

If printer options are set so that LINESIZE=80 and PAGESIZE=60 (first line), the preceding program would produce four pages of output. This is a list of some of the most important output, and the page on which it appears:

- Page 1 includes simple statistics.
- Page 2 includes the eigenvalue table.
- Page 3 includes the scree plot of eigenvalues.
- Page 4 includes the unrotated factor pattern and final communality estimates.
- Page 5 includes the rotated factor pattern.

The output created by the preceding program is reproduced here as Output 15.1:

Output 15.1 Results of the Initial Principal Component Analysis of the Prosocial Orientation Inventory (POI) Data

```
                            The SAS System                              1

                          The FACTOR Procedure

              Means and Standard Deviations from 50 Observations

                  Variable         Mean        Std Dev

                     V1         5.1800000      1.3951812
                     V2         5.4000000      1.1065667
                     V3         5.5200000      1.2162170
                     V4         3.6400000      1.7929567
                     V5         4.2200000      1.6695349
                     V6         3.1000000      1.5551101
```

```
                                                                        2
                          The FACTOR Procedure
               Initial Factor Method: Principal Components

                   Prior Communality Estimates: ONE


      Eigenvalues of the Correlation Matrix: Total = 6  Average = 1

            Eigenvalue    Difference    Proportion    Cumulative

       1    2.26643553    0.29182092      0.3777        0.3777
       2    1.97461461    1.17731470      0.3291        0.7068
       3    0.79729990    0.35811605      0.1329        0.8397
       4    0.43918386    0.14791916      0.0732        0.9129
       5    0.29126470    0.06006329      0.0485        0.9615
       6    0.23120141                    0.0385        1.0000

      2 factors will be retained by the MINEIGEN criterion.
```

(continued on the next page)

Output 15.1 *(continued)*

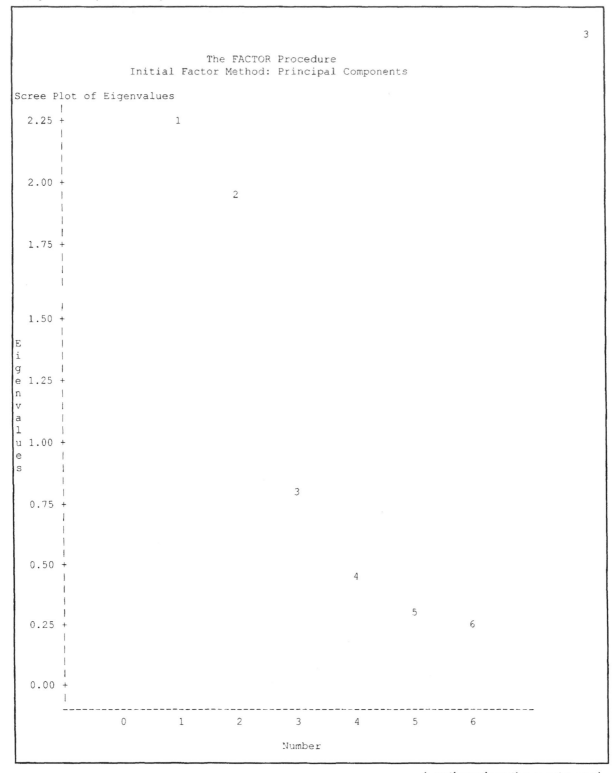

(continued on the next page)

Output 15.1 *(continued)*

```
                                                                        4
                              The FACTOR Procedure
                    Initial Factor Method: Principal Components

                                Factor Pattern

                             Factor1        Factor2

                    V1          58 *           70 *
                    V2          48 *           53 *
                    V3          60 *           62 *
                    V4          64 *          -64 *
                    V5          68 *          -45 *
                    V6          68 *          -46 *

                    Printed values are
                    multiplied by 100 and
                    rounded to the nearest
                    integer.  Values greater
                    than 0.4 are flagged by an
                    '*'.



                    Variance Explained by Each Factor

                         Factor1           Factor2

                        2.2664355          1.9746146


               Final Communality Estimates: Total = 4.241050

         V1             V2              V3             V4             V5             V6

   0.82341782     0.50852894      0.74399020     0.82257428     0.66596347     0.67657543
```

(continued on the next page)

Output 15.1 *(continued)*

```
                                                                    5
                        The FACTOR Procedure
                      Rotation Method: Varimax

          Orthogonal Transformation Matrix

                                 1                 2

             1            0.76914           0.63908
             2           -0.63908           0.76914


                     Rotated Factor Pattern

                       Factor1          Factor2

             V1             0               91  *
             V2             3               71  *
             V3             7               86  *
             V4            90  *            -9
             V5            81  *             9
             V6            82  *             8


             Printed values are
             multiplied by 100 and
             rounded to the nearest
             integer.  Values greater
             than 0.4 are flagged by an
             '*'.


             Variance Explained by Each Factor

                       Factor1          Factor2

                      2.1472475        2.0938026


             Final Communality Estimates: Total = 4.241050

       V1            V2            V3            V4            V5            V6
0.82341782    0.50852894    0.74399020    0.82257428    0.66596347    0.67657543
```

Page 1 from Output 15.1 provides simple statistics for the observed variables included in the analysis. Once the SAS log is checked to verify that no errors were made in the analysis, these simple statistics should be reviewed to determine how many usable observations were included in the analysis, and to verify that the means and standard deviations are in the expected range. The top line of Output 15.1, page 1, says "Means and Standard Deviations from 50 Observations," meaning that data from 50 participants are included in the analysis.

Steps in Conducting Principal Component Analysis

Principal component analysis is normally conducted in a sequence of steps, with somewhat subjective decisions being made at various points. Because this is an introductory treatment of the topic, it will not provide a comprehensive discussion of all options available at each step; instead, specific recommendations will be made, consistent with practices commonly followed in applied research. For a more detailed treatment of principal component analysis and factor analysis, see Kim and Mueller (1978a; 1978b), Rummel (1970), or Stevens (2002).

Step 1: Initial Extraction of the Components

In principal component analysis, the number of components extracted is equal to the number of variables analyzed. Because six variables are analyzed in the present study, six components are extracted. The first component can be expected to account for a fairly large amount of total variance. Each succeeding component accounts for progressively smaller amounts of variance. Although a large number of components can be extracted in this way, only the first few components are sufficiently important to be retained for interpretation.

Page 2 of Output 15.1 provides the eigenvalue table from the analysis. (This table appears just below the heading "Eigenvalues of the Correlation Matrix: Total = 6 Average = 1.") An **eigenvalue** represents the amount of variance that is captured by a given component. In the column headed "Eigenvalue," the eigenvalue for each component is presented. Each row in the matrix presents information about each of the six components. Row 1 provides information about the first component extracted, row 2 provides information about the second component extracted, and so forth.

Where the column headed EIGENVALUE intersects with rows "1" and "2," it can be seen that the eigenvalue for component 1 is approximately 2.27 while the eigenvalue for component 2 is 1.97. This pattern is consistent with our earlier statement that the first components tend to account for relatively large amounts of variance whereas the later components account for relatively smaller amounts.

Step 2: Determining How Many "Meaningful" Components to Retain

Earlier, it was stated that the number of components extracted is equal to the number of variables analyzed. This requires that you decide just how many of these components are truly meaningful and worthy of being retained for rotation and interpretation. In general, you expect that only the first few components will account for meaningful amounts of variance and that the later components will tend to account for only trivial variance. The next step of the analysis, therefore, is to determine how many meaningful components should be retained for interpretation. This section describes four criteria that can be used in making this decision: the eigenvalue-one criterion; the scree test; the proportion of variance accounted for; and the interpretability criterion.

A. The Eigenvalue-One Criterion. In principal component analysis, one of the most commonly used criteria for solving the number-of-components problem is the eigenvalue-one criterion, also known as the Kaiser-Guttman criterion (Kaiser, 1960). With this approach, you retain and interpret any component with an eigenvalue greater than 1.00.

The rationale for this criterion is straightforward. Each observed variable contributes one unit of variance to the total variance in the dataset. Any component that displays an eigenvalue greater than 1.00 is accounting for a greater amount of variance than had been contributed by one variable. Such a component therefore accounts for a meaningful amount of variance and is worthy of retention.

On the other hand, a component with an eigenvalue less than 1.00 accounts for less variance than contributed by one variable. The purpose of principal component analysis is to reduce a number of observed variables into a relatively smaller number of components. This cannot be effectively achieved if you retain components that account for less variance than had been contributed by individual variables. For this reason, components with eigenvalues less than 1.00 are viewed as trivial and are not retained.

The eigenvalue-one criterion has a number of positive features that contribute to its popularity. Perhaps the most important reason for its widespread use is its simplicity. You do not make subjective decisions but merely retain components with eigenvalues greater than one.

On the positive side, it has been shown that this criterion very often results in retaining the correct number of components, particularly when a small-to-moderate number of variables are analyzed, and the variable communalities are high. Stevens (2002) reviews studies that have investigated the accuracy of the eigenvalue-one criterion and recommends its use when fewer than 30 variables are being analyzed and communalities are greater than .70, or when the analysis is based on more than 250 observations and the mean communality is greater than .59.

There are several problems associated with the eigenvalue-one criterion, however. As suggested in the preceding paragraph, it can lead to retaining the wrong number of components under circumstances that are often encountered in research (e.g., when many variables are analyzed, when communalities are small). Also, the automatic application of this criterion can lead to retaining a certain number of components when the actual difference in the eigenvalues of successive components is trivial. For example, if component 2 displays an eigenvalue of 1.01 and component 3 displays an eigenvalue of 0.99, then component 2 is retained but component 3 is not. This might mislead you to believe that the third component was meaningless when, in fact, it accounted for almost exactly the same amount of variance as the second component. In short, the eigenvalue-one criterion can be helpful when used judiciously; yet, the automatic application of this approach can lead to serious errors of interpretation.

With SAS, the eigenvalue-one criterion can be implemented by including the MINEIGEN=1 option in the PROC FACTOR statement and not including the NFACT option. The use of MINEIGEN=1 causes PROC FACTOR to retain any component with an eigenvalue greater than 1.00.

The eigenvalue table from the current analysis appears on page 2 of Output 15.1. The eigenvalues for components 1, 2, and 3 were 2.27, 1.97, and 0.80, respectively. Only components 1 and 2 demonstrated eigenvalues greater than 1.00 so the eigenvalue-one criterion would lead you to retain and interpret only these two components.

Fortunately, the application of the criterion is fairly unambiguous in this case. The last component retained (2) displays an eigenvalue of 1.97, which is substantially greater than 1.00, and the next component (3) displays an eigenvalue of 0.80, which is clearly lower than 1.00. In this analysis, you are not faced with the difficult decision of whether to retain a component that demonstrates an eigenvalue approaching 1.00 (e.g., an eigenvalue of .99). In situations such as this, the eigenvalue-one criterion can be used with greater confidence.

B. The Scree Test. With the scree test (Cattell, 1966), you plot the eigenvalues associated with each component and look for a definitive "break" between the components with relatively large eigenvalues and those with small eigenvalues. The components that appear *before* the break are assumed to be meaningful and are retained for rotation whereas those appearing *after* the break are assumed to be unimportant and are not retained. Sometimes, a scree plot displays several large breaks. When this is the case, you should look for the last big break before the eigenvalues begin to level off. Only the components that appear before this last large break should be retained.

Specifying the SCREE option in the PROC FACTOR statement causes SAS to print an eigenvalue plot as part of the output. This appears as page 3 of Output 15.1.

You can see that the component numbers are listed on the horizontal axis, while eigenvalues are listed on the vertical axis. With this plot, notice that there is a relatively small break between components 1 and 2, and a relatively large break following component 2. The breaks between components 3, 4, 5, and 6 are all relatively small. It is often helpful to draw long lines with extended tails connecting successive pairs of eigenvalues so that these breaks are more apparent (e.g., measure degrees separating lines with a protractor).

Because the large break in this plot appears between components 2 and 3, the scree test would lead you to retain only components 1 and 2. The components appearing after the break (3 to 6) would be regarded as trivial.

The scree test can be expected to provide reasonably accurate results, provided that the sample is large (more than 200) and most of the variable communalities are large (Stevens, 2002). However, this criterion has its weaknesses as well, most notably the ambiguity that is often displayed by scree plots under typical research conditions. Very often, it is difficult to determine exactly where in the scree plot a break exists, or even if a break exists at all. In contrast to the eigenvalue-one criterion, the scree test is generally more subjective.

The break in the scree plot on page 3 of Output 15.1 was unusually obvious. In contrast, consider the plot in Figure 15.2.

Figure 15.2 A Scree Plot with No Obvious Break

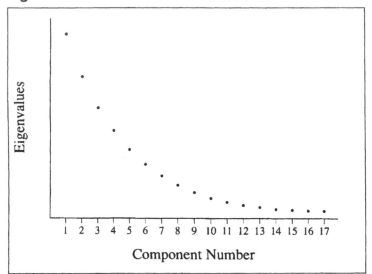

Figure 15.2 presents a fictitious scree plot from a principal component analysis of 17 variables. Notice that there is no obvious break in the plot that separates the meaningful components from the trivial components. Most researchers would agree that components 1 and 2 are probably meaningful whereas components 13 to 17 are probably trivial; however, it is difficult to decide exactly where you should draw the line. This example underscores the qualitative nature of judgments based solely on the scree test.

Scree plots such as the one presented in Figure 15.2 are common in social science research. When encountered, the use of the scree test must be supplemented with additional criteria such as the variance accounted for criterion and the interpretability criterion, to be described later.

> **Why do they call it a "scree" test?** The word "scree" refers to the loose rubble that lies at the base of a cliff or glacier. When performing a scree test, you normally hope that the scree plot will take the form of a cliff. At the top will be the eigenvalues for the few meaningful components, followed by a definitive break (the edge of the cliff). At the bottom of the cliff will lie the scree (i.e., eigenvalues for the trivial components).

C. Proportion of variance accounted for. A third criterion to address the number of factors problem involves retaining a component if it exceeds a specified proportion (or percentage) of variance in the dataset. For example, you might decide to retain any component that accounts for at least 5% or 10% of the total variance. This proportion can be calculated with a simple formula:

$$\text{Proportion} = \frac{\text{Eigenvalue for the component of interest}}{\text{Total eigenvalues of the correlation matrix}}$$

In principal component analysis, the "total eigenvalues of the correlation matrix" is equal to the total number of variables being analyzed (because each variable contributes one unit of variance to the analysis).

Fortunately, it is not necessary to manually compute these percentages since they are provided in the results of PROC FACTOR. The proportion of variance captured by each component is printed in the eigenvalue table (page 2) under the "Proportion" heading.

The eigenvalue table for the current analysis is on page 2 of Output 15.1. From the "Proportion" column, you can see that the first component alone accounts for 38% of the total variance, the second component alone accounts for 33%, the third component accounts for 13%, and the fourth component accounts for 7%. Assume that you have decided to retain any component that accounts for at least 10% of the total variance in the dataset. For the present results, using this criterion would cause you to retain components 1, 2, and 3. (Notice that use of this criterion would result in retaining more components than would be retained with the two preceding criteria.)

An alternative criterion is to retain enough components so that the *cumulative* percent of variance accounted for is equal to some minimal value. For example, remember that components 1, 2, 3, and 4 accounted for approximately 38%, 33%, 13% and 7% of the total variance, respectively. Adding these percentages together results in a sum of 91%. This means that the *cumulative* percent of variance accounted for by components 1, 2, 3 and 4 is 91%. When researchers use the "cumulative percent of variance accounted for" as the criterion for solving the number-of-components problem, they usually retain enough components so that the cumulative percent of variance accounted for at least 70% (and sometimes 80%).

With respect to the results of PROC FACTOR, the "cumulative percent of variance accounted for" is presented in the eigenvalue table (from page 2), below the "Cumulative" heading. For the present analysis, this information appears in the eigenvalue table on page 2 of Output 15.1. Notice the values that appear below the heading "Cumulative." Each value indicates the percent of variance accounted for by the present component as well as all preceding components. For example, the value for component 2 is approximately .71 (intersection of the column labeled "Cumulative" and the second row). This value of .71 indicates that approximately 71% of the total variance is accounted for by components 1 and 2. The corresponding entry for component 3 is approximately .84, indicating that 84% of the variance is accounted for by components 1, 2, and 3. If you were to use 70% as the "critical value" for determining the number of components to retain, you would retain components 1 and 2 in the present analysis.

The proportion of variance criterion has a number of positive features. For example, in most cases, you would not want to retain a group of components that, combined, account for only a fraction of variance in the dataset (say, 30%). Nonetheless, the critical values discussed earlier (10% for individual components and 70% to 80% for the combined components) are quite arbitrary. Because of these and related problems, this approach has sometimes been criticized for its subjectivity (Kim & Mueller, 1978b).

D. The interpretability criteria. Perhaps the most important criterion for solving the "number-of-components" problem is the **interpretability criterion**: interpreting the substantive meaning of the retained components and verifying that this interpretation makes sense in terms of what is known about the constructs under investigation. The following list provides four rules to follow when doing this. A later section ("Step 4: Interpreting the Rotated Solution") shows how to interpret the results of a principal component analysis. The following rules will be more meaningful after you complete that section.

1. **Are there at least three variables (items) with significant loadings on each retained component?** A solution is less satisfactory if a given component is measured by fewer than three variables.

2. **Do the variables that load on a given component share the same conceptual meaning?** For example, if three questions on a survey all load on component 1, do all three of these questions appear to be measuring the same construct?

3. **Do the variables that load on different components seem to be measuring different constructs?** For example, if three questions load on component 1 and three other questions load on component 2, do the first three questions seem to be measuring a construct that is conceptually distinct from the construct measured by the last three questions?

4. **Does the rotated factor pattern demonstrate "simple structure"? Simple structure** means that the pattern possesses two characteristics: (a) most of the variables have relatively high factor loadings on only one component and near-zero loadings on the other components; and (b) most components have relatively high factor loadings for some variables and near-zero loadings for the remaining variables. This concept of simple structure is explained in more detail in a later section, "Step 4: Interpreting the Rotated Solution."

Recommendations

Given the preceding options, what procedure should you follow in solving the number-of-components problem? We recommend combining all four in a structured sequence. First, use the MINEIGEN=1 options to implement the eigenvalue-one criterion. Review this solution for interpretability but use caution if the break between the components with eigenvalues above 1.00 and those below 1.00 is not clear-cut (i.e., if component 1 has an eigenvalue of 1.01, and component 2 has an eigenvalue of 0.99).

Next, perform a scree test and look for obvious breaks in eigenvalues. Because there often is more than one break in the scree plot, it might be necessary to examine two or more possible solutions.

Next, review the amount of common variance accounted for by each individual component. You probably should not rigidly use some specific but arbitrary cutoff point such as 5% or 10%. Still, if you are retaining components that account for as little as 2% or 4% of the variance, it might be wise to take a second look at the solution and verify that these latter components are truly of substantive importance. In the same way, it is best if the combined components account for at least 70% of cumulative variance. If less than 70% is captured, it might be prudent to consider alternate solutions that include a larger number of components.

Finally, apply the interpretability criteria to each solution. If more than one solution can be justified on the basis of the preceding criteria, which of these solutions is the most interpretable? By seeking a solution that is both interpretable and also satisfies one (or more) of the other three criteria, you maximize chances of retaining the optimal number of components.

Step 3: Rotation to a Final Solution

Factor Patterns and Factor Loadings

After extracting the initial components, PROC FACTOR creates an unrotated **factor pattern matrix**. The rows of this matrix represent the variables being analyzed, and the columns represent the retained components. (These components are referred to as FACTOR1, FACTOR2, and so forth in the output.)

The entries in the matrix are factor loadings. A **factor loading** is a general term for a coefficient that appears in a factor pattern matrix or a factor structure matrix. In an analysis that results in oblique (correlated) components, the definition for a factor loading is different depending on whether it is in a factor *pattern* matrix or in a factor *structure* matrix. However, the situation is simpler in an analysis that results in orthogonal components (as in the present chapter). In an orthogonal analysis, factor loadings are equivalent to bivariate correlations between the observed variables and the components.

For example, the factor pattern matrix from the current analysis is on page 4 of Output 15.1. Where the rows for observed variables intersect with the column for FACTOR1, you can see that the correlation between V1 and the first component is .58; the correlation between V2 and the first component is .48, and so forth.

Rotations

Ideally, you would like to review the correlations between the variables and the components, and use this information to *interpret* the components. That is, to determine what construct seems to be measured by component 1, what construct seems to be measured by component 2, and so forth. Unfortunately, when more than one component is retained in an analysis, the interpretation of an unrotated factor pattern is usually quite difficult. To facilitate interpretation, you normally perform an operation called a rotation. A **rotation** is a linear transformation that is performed on the factor solution for the purpose of making the solution easier to interpret.

PROC FACTOR allows you to request several different types of rotations. The preceding program that analyzed data from the POI study included the statement

```
ROTATE=VARIMAX
```

which requests a **varimax rotation**. A varimax rotation is an orthogonal rotation, meaning that it results in uncorrelated components. Compared to other types of rotations, a varimax rotation tends to maximize the variance of a column of the factor pattern matrix (as opposed to a row of the matrix). This rotation is probably the most commonly used orthogonal rotation in the social sciences. The results of the varimax rotation for the current analysis are on page 5 of Output 15.1.

Step 4: Interpreting the Rotated Solution

Interpreting a rotated solution means determining just what is measured by each of the retained components. Briefly, this involves identifying the variables that exhibit high loadings for a given component, and determining what these variables share in common. Usually, a brief name is assigned to each retained component to describe its content.

The first decision to be made at this stage is to decide how large a factor loading must be in order to be considered "large." Stevens (2002) discusses some of the issues relevant to this decision and provides guidelines for testing the statistical significance of factor loadings. Given that this is an introductory treatment of principal component analysis, consider a loading to be "large" if its absolute value exceeds .40.

The rotated factor pattern for the POI study appears on page 5 of Output 15.1. The following text provides a structured approach for interpreting this factor pattern.

A. Read across the row for the first variable. All "meaningful loadings" (i.e., loadings greater than .40) are flagged with an asterisk ("*"). This was accomplished by including the FLAG=.40 option in the preceding program. If a given variable has a meaningful loading on more than one component, scratch that variable out and ignore it in your interpretation. In many situations, researchers want to drop variables that load on more than one component because the variables are not pure measures of any one construct. (These are sometimes referred to as *complex items*.) In the present case, this means looking at the row headed "V1," and reading to the right to see if it loads on more than one component. In this case it does not, so you can retain this variable.

B. Repeat this process for the remaining variables, scratching out any variable that loads on more than one component. In this analysis, none of the variables have high loadings on more than one component, so none has to be deleted. In other words, there are no complex items.

C. Review all of the surviving variables with high loadings on component 1 to determine the nature of this component. From the rotated factor pattern, you can see that only items 4, 5, and 6 load on component 1 (note the asterisks). Now, turn to the questionnaire itself and review the content in order to decide what a given component should be named. What do questions 4, 5, and 6 have in common? What common construct do they appear to be measuring? For illustration, the questions being analyzed in the present case are reproduced here. Remember that question 4 is represented as V4 in the SAS program, question 5 is V5, and so forth. Read questions 4, 5, and 6 to see what they have in common.

```
1 2 3   5 6   7     1.      Went out of my way to do a favor for a
                            coworker.

1 2 3 4 5 6 7       2.      Went out of my way to do a favor for a
                            relative.

1 2 3 4 5 6 7       3.      Went out of my way to do a favor for a
                            friend.

1 2 3 4 5 6 7       4.      Gave money to a religious charity.

1 2 3 4 5 6 7       5.      Gave money to a charity not associated with
                            a religion.

1 2 3 4 5 6 7       6.      Gave money to a panhandler.
```

Questions 4, 5, and 6 all seem to deal with "giving money to those in need." It is therefore reasonable to label component 1 the "financial giving" component.

D. Repeat this process to name the remaining retained components. In the present case, there is only one remaining component to name: component 2. This component has high loadings for questions 1, 2, and 3. In reviewing these items, it becomes clear that each seems to deal with helping friends, relatives, or other acquaintances. It is therefore appropriate to name this the "helping others" component.

E. Determine whether this final solution satisfies the interpretability criteria. An earlier section indicated that the overall results of a principal component analysis are satisfactory only if they meet a number of interpretability criteria. In the following list, the adequacy of the rotated factor pattern presented on page 5 of Output 15.1 is assessed in terms of these criteria.

1. **Are there at least three variables (items) with significant loadings on each retained component?** In the present example, three variables load on component 1 and three load on component 2, so this criterion is met.

2. **Do the variables that load on a given component share similar conceptual meaning?** All three variables loading on component 1 measure giving to those in need, while all three loading on component 2 measure prosocial acts performed for others. Therefore, this criterion is met.

3. **Do the variables that load on different components seem to be measuring different constructs?** The items loading on component 1 measure respondents' financial contributions, while the items loading on component 2 measure helpfulness toward others. Because these seem to be conceptually distinct constructs, this criterion appears to be met as well.

4. **Does the rotated factor pattern demonstrate "simple structure"?** Earlier, it was said that a rotated factor pattern demonstrates simple structure when it has two characteristics. First, most of the variables should have high loadings on one component and near-zero loadings on the other components. It can be seen that the pattern obtained here meets that requirement: items 1 through 3 have high loadings on component 2 and near-zero loadings on component 1. Similarly,

items 4 through 6 have high loadings on component 1 and near-zero loadings on component 2. The second characteristic of simple structure is that each component should have high loadings for some variables and near-zero loadings for the others. Again, the pattern obtained here meets this requirement: component 1 has high loadings for items 4 through 6 and near-zero loadings for the other items whereas component 2 has high loadings for items 1 through 3 and near-zero loadings on the remaining items. In short, the rotated component pattern obtained in this analysis appears to demonstrate simple structure.

Step 5: Creating Factor Scores or Factor-Based Scores

Once the analysis is complete, it is often desirable to assign scores to participants to indicate where they stand on the retained components. For example, the two components retained in the present study were interpreted as financial giving and helping others. You might want to now assign one score to each participant to indicate that participant's standing on the financial giving component and a different score to indicate his/her standing on the helping others component. With this done, these component scores could be used either as predictor variables or as criterion variables in subsequent analyses.

Before discussing the options for assigning these scores, it is important to first draw a distinction between factor scores versus factor-based scores. In principal component analysis, a **factor score** (or **component score**) is a linear composite of the optimally weighted observed variables. If requested, PROC FACTOR computes each participant's factor scores for the two components by:

- determining the optimal regression weights;
- multiplying participant responses to the questionnaire items by these weights;
- summing the products.

The resulting sum is a given participant's score on the component of interest. Remember that a separate equation with different weights is developed for each retained component.

A **factor-based score**, on the other hand, is merely a linear composite of the variables that demonstrate meaningful loadings for the component in question. In the preceding analysis, for example, items 4, 5, and 6 demonstrated meaningful loadings for the financial giving component. Therefore, you could calculate the factor-based score on this component for a given participant by simply adding together his or her responses to items 4, 5, and 6. Notice that, with a factor-based score, the observed variables are not multiplied by optimal weights before they are summed.

Computing Factor Scores

Factor scores are requested by including the NFACT and OUT options in the PROC FACTOR statement. Here is the general form for a SAS program that uses the NFACT and OUT option to compute factor scores:

```
PROC FACTOR    DATA=dataset-name
               SIMPLE
               METHOD=PRIN
               PRIORS=ONE
               NFACT=number-of-components-to-retain
               ROTATE=VARIMAX
               ROUND
               FLAG=desired-size-of-"significant"-factor-loadings
               OUT=name-of-new-SAS-dataset ;
    VAR  variables-to-be-analyzed ;
RUN;
```

Here are the actual program statements (minus the DATA step) that could be used to perform a principal component analysis and compute factor scores for the POI study.

```
1     PROC FACTOR    DATA=D1
2                    SIMPLE
3                    METHOD-PRIN
4                    PRIORS=ONE
5                    NFACT=2
6                    ROTATE=VARIMAX
7                    ROUND
8                    FLAG=.40
9                    OUT=D2    ;
10       VAR V1 V2 V3 V4 V5 V6;
11    RUN;
```

Notice how this program differs from the original program presented earlier in the chapter (in "SAS Program and Output"). The MINEIGEN=1 option has been dropped and is replaced with the NFACT=2 option. The OUT=D2 option also is added.

Line 9 of the preceding programs asks that an output dataset be created and given the name D2. This name is arbitrary; any name consistent with SAS requirements is acceptable. The new dataset named D2 will contain all of the variables contained in the previous dataset (D1), as well as new variables named FACTOR1 and FACTOR2. FACTOR1 will contain factor scores for the first retained component and FACTOR2 will contain scores for the second. The number of new "FACTOR" variables created will be equal to the number of components retained by the NFACT statement.

The OUT option can be used to create component scores only if the analysis is performed on a raw dataset (as opposed to a correlation or covariance matrix). The use of the NFACT statement is also required.

Having created the new variables named FACTOR1 and FACTOR2, you might be interested in seeing how they relate to the study's original observed variables. This can be done by appending PROC CORR statements to the SAS program, following the last of the PROC FACTOR statements. The full program (minus the DATA step) is:

```
 1      PROC FACTOR    DATA=D1
 2                     SIMPLE
 3                     METHOD=PRIN
 4                     PRIORS=ONE
 5                     NFACT=2
 6                     ROTATE=VARIMAX
 7                     ROUND
 8                     FLAG=.40
 9                     OUT=D2    ;
10         VAR V1 V2 V3 V4 V5 V6;
11      RUN;
12
13      PROC CORR   DATA=D2;
14         VAR FACTOR1 FACTOR2;
15         WITH V1 V2 V3 V4 V5 V6 FACTOR1 FACTOR2;
16      RUN;
```

Notice that the PROC CORR statement on line 13 specifies DATA=D2. This dataset (D2) is the name of the output dataset created on line 9 in the PROC FACTOR statement. The PROC CORR statements request that the factor score variables (FACTOR 1 and FACTOR2) be correlated with participants' responses to questionnaire items 1 through 6 (V1 to V6).

With printer options of LINESIZE=80 and PAGESIZE=60 (first line), the preceding program would again produce four pages of output. Pages 1 to 2 provide simple statistics, the eigenvalue table, and the unrotated factor pattern, identical to those produced with the first program. Page 3 provides the rotated factor pattern and final communalities (same as before), along with the standardized scoring coefficients used in creating factor scores. Finally, page 4 provides the correlations requested by the CORR procedure. Pages 3 and 4 of the output created by the preceding program are presented here as Output 15.2.

Output 15.2 Output Pages 3 and 4 from the Analysis of POI Data in Which Factor Scores Were Created

```
                            The SAS System                              3

                         The FACTOR Procedure
                        Rotation Method: Varimax

                   Orthogonal Transformation Matrix

                                  1                 2

                 1             0.76914           0.63908
                 2            -0.63908           0.76914


                       Rotated Factor Pattern

                          Factor1         Factor2

                  V1          0              91  *
                  V2          3              71  *
                  V3          7              86  *
                  V4         90  *           -9
                  V5         81  *            9
                  V6         82  *            8


                   Printed values are
                   multiplied by 100 and
                   rounded to the nearest
                   integer.  Values greater
                   than 0.4 are flagged by an
                   '*'.


                   Variance Explained by Each Factor

                       Factor1            Factor2

                     2.1472475          2.0938026


             Final Communality Estimates: Total = 4.241050

     V1             V2              V3             V4             V5             V6

0.82341782     0.50852894      0.74399020     0.82257428     0.66596347     0.67657543


             Scoring Coefficients Estimated by Regression


    Squared Multiple Correlations of the Variables with Each Factor

                       Factor1            Factor2

                     1.0000000          1.0000000
```

(continued on the next page)

Output 15.2 *(continued)*

```
                    Standardized Scoring Coefficients

                          Factor1              Factor2

                V1        -0.03109              0.43551
                V2        -0.00726              0.34071
                V3         0.00388              0.41044
                V4         0.42515             -0.07087
                V5         0.37618              0.01947
                V6         0.38020              0.01361
```

```
                                                                          4
                          The CORR Procedure

  8 With Variables:    V1       V2       V3       V4       V5       V6
                       Factor1  Factor2
  2     Variables:     Factor1  Factor2


                          Simple Statistics

Variable        N        Mean     Std Dev        Sum      Minimum    Maximum

V1             50     5.18000     1.39518    259.00000    1.00000    7.00000
V2             50     5.40000     1.10657    270.00000    3.00000    7.00000
V3             50     5.52000     1.21622    276.00000    2.00000    7.00000
V4             50     3.64000     1.79296    182.00000    1.00000    7.00000
V5             50     4.22000     1.66953    211.00000    1.00000    7.00000
V6             50     3.10000     1.55511    155.00000    1.00000    7.00000
Factor1        50           0     1.00000          0     -1.87908    2.35913
Factor2        50           0     1.00000          0     -2.95892    1.58951


                 Pearson Correlation Coefficients, N = 50
                     Prob > |r| under H0: Rho=0

                          Factor1              Factor2

                V1       -0.00429              0.90741
                          0.9764               <.0001

                V2        0.03328              0.71234
                          0.8185               <.0001

                V3        0.06720              0.85993
                          0.6429               <.0001

                V4        0.90274             -0.08740
                          <.0001               0.5462

                V5        0.81055              0.09474
                          <.0001               0.5128

                V6        0.81934              0.08303
                          <.0001               0.5665

                Factor1   1.00000              0.00000
                                               1.0000

                Factor2   0.00000              1.00000
                          1.0000
```

The simple statistics for the CORR procedure appear on page 4 in Output 15.2. Notice that the simple statistics for the observed variables (V1 to V6) are identical to those at the beginning of the FACTOR output discussed earlier (at the top of Output 15.1, page 1). In contrast, note the simple statistics for FACTOR1 and FACTOR2 (the factor score variables for components 1 and 2, respectively). Both have means of 0 and standard deviations of 1. Obviously, these variables were constructed to be standardized variables.

The correlations between FACTOR1 and FACTOR2 and the original observed variables appear on the bottom half of page 4. You can see that the correlations between FACTOR1 and V1 through V6 on page 4 of Output 15.2 are identical to the factor loadings of V1 through V6 on FACTOR1 on page 5 of Output 15.1, under "Rotated Factor Pattern." This makes sense, as the elements of a factor pattern (in an orthogonal solution) are simply correlations between the observed variables and the components themselves. Similarly, you can see that the correlations between FACTOR2 and V1 through V6 from page 4 of Output 15.2 are also identical to the corresponding factor loadings from page 5 of Output 15.1.

Of special interest is the correlation between FACTOR1 and FACTOR2, as computed by PROC CORR. This appears on page 4 of Output 15.2, where the row for FACTOR2 intersects with the column for FACTOR1. Notice the observed correlation between these two components is zero. This is as expected; the rotation method used in this principal component analysis is the varimax method which produces orthogonal, or uncorrelated, components.

Computing Factor-Based Scores

A second (and less sophisticated) approach to scoring involves the creation of new variables that contain factor-based scores rather than true factor scores. A variable that contains factor-based scores is sometimes referred to as a **factor-based scale**.

Although factor-based scores can be created several ways, the following method has the advantage of being relatively straightforward and is commonly used:

1. To calculate factor-based scores for component 1, determine which questionnaire items had high loadings on that component.
2. For a given participant, add together that participant's responses to these items. The result is that participant's score on the factor-based scale for component 1.
3. Repeat these steps to calculate each participant's score on the remaining retained components.

Although this might sound like a cumbersome task, it is actually made quite simple through the use of data manipulation statements contained in a SAS program. For example, assume that you have performed the principal component analysis on your survey responses and have obtained the findings reported in this chapter. Specifically, you found that survey items 4, 5, and 6 loaded on component 1 (the financial giving component), while items 1, 2, and 3 loaded on component 2 (helping others component).

You would now like to create two new SAS variables. The first variable, called FINANCE, will include each participant's factor-based score for financial giving. The second variable, called HELPING, will include each participant's factor-based score for helping others. Once these variables are created, they can be used as criterion or predictor variables in subsequent analyses. To keep things simple, assume that you are simply interested in determining whether or not there is a significant correlation between FINANCE and HELPING.

At this time, it might be useful to review Chapter 4, "Working with Variables and Observations in SAS Datasets," particularly the section on creating new variables from existing variables. Such a review should make it easier to understand the data manipulation statements used here.

Assume that earlier statements in the SAS program have already input responses to the six questionnaire items. These variables are included in a dataset called D1. The following are the subsequent lines that would go on to create a new dataset called D2. The new dataset will include all of the variables in D1, as well as the newly created factor-based scales called FINANCE and HELPING.

```
14
15      DATA D2;
16         SET D1;
17
18      FINANCE   = (V4 + V5 + V6);
19      HELPING   = (V1 + V2 + V3);
20
21      PROC CORR   DATA=D2;
22         VAR FINANCE  HELPING;
23      RUN;
```

Lines 15 and 16 request that a new dataset called D2 be created, and that it be set up as a duplicate of existing dataset D1. In line 18, the new variable called FINANCE is created. For each participant, his or her responses to items 4, 5, and 6 are added together. The result is the participant's score on the factor-based scale for the first component. These scores are stored as a variable called FINANCE. The component-based scale for the helping others component is created on line 19, and these scores are stored as the variable called HELPING. Lines 21 to 23 request that the correlations between FINANCE and HELPING be determined. FINANCE and HELPING can now be used as predictor or criterion variables in subsequent analyses. To save space, the results of this program are not reproduced here.

However, note that this output would probably display a nonzero correlation between FINANCE and HELPING. This might come as a surprise, because earlier it was shown that the factor scores contained in FACTOR1 and FACTOR2 (counterparts to FINANCE and HELPING) were completely uncorrelated. The reason for this apparent contradiction is simple: FACTOR1 and FACTOR2 are true principal components; true principal components (created in an orthogonal solution) are always created with optimally weighted equations so that they will be mutually uncorrelated.

In contrast, FINANCE and HELPING are not true principal components that consist of true factor scores; they are merely variables *based* on the results of a principal component analysis. Optimal weights (that would ensure orthogonality) were not used in the creation of FINANCE and HELPING. This is why factor-based scales will often demonstrate nonzero correlations with one another while true principal components (from an orthogonal solution) will not.

Recoding Reversed Items prior to Analysis

It generally is best to recode any reversed items before conducting any of the analyses described here. In particular, it is essential that reversed items be recoded prior to the program statements that produce factor-based scales. For example, the three questionnaire items that assess financial giving appear again here:

```
1 2 3 4 5 6 7    4.    Gave money to a religious charity.

1 2 3 4 5 6 7    5.    Gave money to a charity not associated with
                       a religion.

1 2 3 4 5 6 7    6.    Gave money to a panhandler.
```

None of these items are reversed. With each item, a response of "7" indicates a high level of financial giving. In the following, however, item 4 is a reversed item: with item 4, a response of "7" indicates a *low* level of giving:

```
1 2 3 4 5 6 7    4.    Refused to give money to a religious
                       charity.

1 2 3 4 5 6 7    5.    Gave money to a charity not associated with
                       a religion.

1 2 3 4 5 6 7    6.    Gave money to a panhandler.
```

If you were to perform a principal component analysis on responses to these items, the factor loading for item 4 would most likely have a sign that is opposite of the sign of the loadings for items 5 and 6 (e.g., if items 5 and 6 had positive loadings, item 4 would have a negative loading). This would complicate the creation of a component-based scale: with items 5 and 6, higher scores indicate greater giving whereas with item 4, lower scores indicate greater giving. You would not want to sum these three items as they are presently coded. First, it is necessary to reverse item 4. Notice how this is done in the following program (assume that the data have already been entered in a SAS dataset named D1):

```
15      DATA D2;
16          SET D1;
17
18      V4 = 8 - V4;
19
20      FINANCE  = (V4 + V5 + V6);
21      HELPING  = (V1 + V2 + V3);
22
23      PROC CORR    DATA=D2;
24          VAR FINANCE HELPING;
25      RUN;
```

Line 18 creates a new, recoded version of variable V4. Values on this new version of V4 are equal to the quantity 8 minus the value of the old version of V4. For participants whose score on the old version of V4 is 1, their value on the new version of V4 is 7 (because $8 - 1 = 7$). For those whose score is 7, their value on the new version of V4 is 1 ($8 - 7 = 1$). See Chapter 4 for further description of this procedure.

The general form of the formula used when recoding reversed items is

```
Variable-name  =  constant  -  variable-name ;
```

In this formula, the "constant" is the following quantity:

the number of points on the response scale used with the questionnaire item plus 1.

Therefore, if you are using the 4-point response format, the constant is 5. If using a 9-point scale, the constant is 10.

If you have prior knowledge about which items are going to appear as reversed (with reversed component loadings) in your results, it is best to place these recoding statements early in your SAS program before the PROC FACTOR statements. This makes interpretation of the components more straightforward because it eliminates significant loadings with opposite signs from appearing on the same component. In any case, it is essential that the statements recoding reversed items appear before the statements that create any factor-based scales.

Step 6: Summarizing the Results in a Table

For reports that summarize the results of your analysis, it is generally desirable to prepare a table that presents the rotated factor pattern. When analyzed variables contain responses to questionnaire items, it can be helpful to actually reproduce the questionnaire items within this table. This is done in Table 15.2:

Table 15.2

Rotated Factor Pattern and Final Communality Estimates from
Principal Component Analysis of Prosocial Orientation Inventory

Component

1	2	h^2	Items
.00	.91	.82	1. Went out of my way to do a favor for a coworker.
.03	.71	.51	2. Went out of my way to do a favor for a relative.
.07	.86	.74	3. Went out of my way to do a favor for a friend.
.90	-.09	.82	4. Gave money to a religious charity.
.81	.09	.67	5. Gave money to a charity not associated with a religion.
.82	.08	.68	6. Gave money to a panhandler.

Note. $N = 50$. Communality estimates appear in column headed h^2.

The final communality estimates from the analysis are presented under the heading "h^2" in the table. These estimates appear in the SAS output following the "Variance Explained By Each Factor" (page 3 of Output 15.2).

Very often, the items that constitute the questionnaire are so lengthy, or the number of retained components is so large that it is not possible to present the factor pattern, the communalities, and the items themselves in the same table. In such situations, it might be preferable to present the factor pattern and communalities in one table and the items in a second (or in the text of the paper). Shared item numbers can then be used to associate each item with its corresponding factor loadings and communality.

Step 7: Preparing a Formal Description of the Results for a Paper

The preceding analysis could be summarized in the following way:

> Principal component analysis was applied to responses to the 6-item questionnaire using 1s as prior communality estimates. The principal axis method was used to extract the components, and this was followed by a varimax (orthogonal) rotation.
>
> Only the first two components exhibited eigenvalues greater than 1; results of a scree test also suggested that only the first two were meaningful. Therefore, only the first two components were retained for rotation. Combined, components 1 and 2 accounted for 71% of the total variance.
>
> Questionnaire items and corresponding factor loadings are presented in Table 15.2. In interpreting the rotated factor pattern, an item was said to load on a given component if the factor loading was .40 or greater for that component and less than .40 for the other. Using these criteria, three items were found to load on the first component, which was subsequently labeled financial giving. Three items also loaded on the second component labeled helping others.

An Example with Three Retained Components

The Questionnaire

The next example involves fictitious research that examines the investment model (Rusbult, 1980). As stated in earlier chapters, this model identifies variables believed to affect a person's commitment to a romantic relationship. In this context, **commitment** refers to a person's intention to maintain the relationship and stay with his/her current romantic partner.

One version of the investment model predicts that commitment is affected by three antecedent variables: satisfaction; investment size; and alternative value. **Satisfaction** refers to a participant's affective response to the relationship. Among other things, participants report high levels of satisfaction when their current relationship comes close to their perceived ideal relationship. **Investment size** refers to the amount of time, energy, and personal resources that an individual has put into the relationship. For example, participants report high investments when they have spent a lot of time with their current partner and have developed mutual friends that might be lost if the relationship were to end. Finally, **alternate value** refers to the attractiveness of alternatives to one's current partner. A participant would score high on alternate value if, for example, it would be appealing to date someone else or perhaps not to date at all.

Assume that you want to conduct research on the investment model and are in the process of preparing a 12-item questionnaire to assess levels of satisfaction, investment size, and alternate value in a group of participants involved in romantic relationships. Part of the instrument used to assess these constructs is reproduced here:

Indicate the extent to which you agree or disagree with each of the following statements by writing the appropriate number in the space to the left of the statement. Please use the following response format to make these ratings:

    7 = Strongly Agree
    6 = Agree
    5 = Slightly Agree
    4 = Neither Agree Nor Disagree
    3 = Slightly Disagree
    2 = Disagree
    1 = Strongly Disagree

_____   1.   I am satisfied with my current relationship.

_____   2.   My current relationship comes close to my ideal
               relationship.

_____   3.   I am more satisfied with my relationship than the
               average person.

_____   4.   I feel good about my current relationship.

_____   5.   I have invested a great deal of time in my current
               relationship.

_____   6.   I have invested a great deal of energy in my current
               relationship.

_____   7.   I have invested a lot of my personal resources (e.g.,
               money) in developing my current relationship.

_____   8.   My partner and I have established mutual friends that I
               might lose if we were to break up.

_____   9.   There are plenty of other attractive people around for
               me to date if I were to break up with my current
               partner.

_____ 10.   It would be appealing to break up with my current
               partner and date someone else.

_____ 11.   It would be appealing to break up with my partner to be
               alone for a while.

_____ 12.   It would be appealing to break up with my partner and
               "play the field" for a while.

In the preceding questionnaire, items 1 through 4 were designed to assess satisfaction, items 5 through 8 were designed to assess investment size, and items 9 through 12 were designed to assess alternate value. Assume that you administer this questionnaire to 300 participants and now want to perform a principal component analysis on their responses.

Writing the Program

Earlier, it was mentioned that it is possible to perform a principal component analysis on a correlation matrix as well as on raw data (or covariance matrix). This section shows how the former is done. The following program enters the correlation matrix that provides all possible correlations between responses to the 12 items on the questionnaire, and performs a principal component analysis on these fictitious data:

```
1     DATA D1(TYPE=CORR)   ;
2        INPUT    _TYPE_    $
3                 _NAME_    $
4                 V1-V12   ;
5     DATALINES;
6     N    .    300  300  300  300  300  300  300  300  300  300  300  300
7     STD  .    2.48 2.39 2.58 3.12 2.80 3.14 2.92 2.50 2.10 2.14 1.83 2.26
8     CORR V1   1.00  .    .    .    .    .    .    .    .    .    .    .
9     CORR V2    .69 1.00  .    .    .    .    .    .    .    .    .    .
10    CORR V3    .60  .79 1.00  .    .    .    .    .    .    .    .    .
11    CORR V4    .62  .47  .48 1.00  .    .    .    .    .    .    .    .
12    CORR V5    .03  .04  .16  .09 1.00  .    .    .    .    .    .    .
13    CORR V6    .05 -.04  .08  .05  .91 1.00  .    .    .    .    .    .
14    CORR V7    .14  .05  .06  .12  .82  .89 1.00  .    .    .    .    .
15    CORR V8    .23  .13  .16  .21  .70  .72  .82 1.00  .    .    .    .
16    CORR V9   -.17 -.07 -.04 -.05 -.33 -.26 -.38 -.45 1.00  .    .    .
17    CORR V10  -.10 -.08  .07  .15 -.16 -.20 -.27 -.34  .45 1.00  .    .
18    CORR V11  -.24 -.19 -.26 -.28 -.43 -.37 -.53 -.57  .60  .22 1.00  .
19    CORR V12  -.11 -.07  .07  .08 -.10 -.13 -.23 -.31  .44  .60  .26 1.00
20    ;
21
22    PROC FACTOR    DATA=D1
23                   METHOD=PRIN
24                   PRIORS=ONE
25                   MINEIGEN=1
26                   SCREE
27                   ROTATE=VARIMAX
28                   ROUND
29                   FLAG=.40   ;
30        VAR  V1-V12;
31    RUN;
```

The PROC FACTOR statement in this program follows the general form recommended for the initial analysis of a dataset. Notice that the MINEIGEN=1 statement requests that all components with eigenvalues greater than one be retained, and the SCREE option requests a scree plot of eigenvalues. These options are particularly helpful for the initial analysis of data as they can help determine the correct number of components to retain. If the scree test (or the other criteria) suggests retaining some number of components other than what would be retained using the MINEIGEN=1 option, that option can be dropped and replaced with the NFACT option.

Results of the Initial Analysis

The preceding program produced four pages of output, with the following information appearing on each page:

- Page 1 includes the eigenvalue table.

- Page 2 includes the scree plot of eigenvalues.

- Page 3 includes the unrotated factor pattern and final communality estimates.

- Page 4 includes the rotated factor pattern.

The eigenvalue table from this analysis appears on page 1 of Output 15.3. The eigenvalues themselves appear in the left-hand column under "Eigenvalue." From these values, you can see that components 1, 2 and 3 provide eigenvalues of 4.47, 2.73, and 1.70, respectively. Furthermore, you can see that only these first three components provide eigenvalues greater than one. This means that three components are retained by the MINEIGEN criterion. Notice that the first nonretained component (component 4) displays an eigenvalue of approximately 0.85 which, of course, is well below 1.00. This is encouraging, as you have more confidence in the eigenvalue-one criterion when the solution does not contain "near-miss" eigenvalues (e.g., .98 or .99).

Output 15.3 Results of the Initial Principal Component Analysis of the Investment Model Data

```
                              The SAS System                              1

                            The FACTOR Procedure
                  Initial Factor Method: Principal Components

                      Prior Communality Estimates: ONE


       Eigenvalues of the Correlation Matrix: Total = 12  Average = 1

              Eigenvalue    Difference    Proportion    Cumulative

          1    4.47058134    1.73995858      0.3725        0.3725
          2    2.73062277    1.02888853      0.2276        0.6001
          3    1.70173424    0.85548155      0.1418        0.7419
          4    0.84625269    0.22563029      0.0705        0.8124
          5    0.62062240    0.20959929      0.0517        0.8642
          6    0.41102311    0.06600575      0.0343        0.8984
          7    0.34501736    0.04211948      0.0288        0.9272

          8    0.30289788    0.07008042      0.0252        0.9524
          9    0.23281745    0.04595812      0.0194        0.9718
         10    0.18685934    0.08061799      0.0156        0.9874
         11    0.10624135    0.06091129      0.0089        0.9962
         12    0.04533006                    0.0038        1.0000

          3 factors will be retained by the MINEIGEN criterion.
```

(continued on the next page)

Output 15.3 *(continued)*

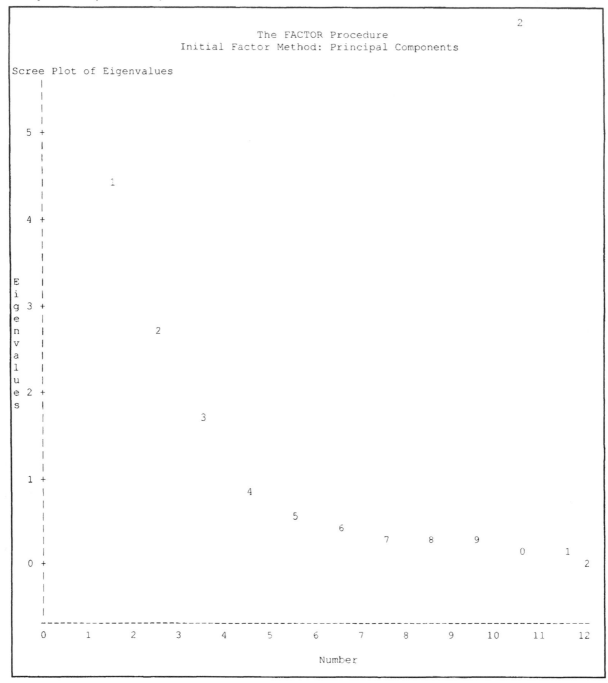

(continued on the next page)

Output 15.3 *(continued)*

```
                                                                          3
                          The FACTOR Procedure
              Initial Factor Method: Principal Components

                            Factor Pattern

                       Factor1        Factor2        Factor3

              V1          39             76 *         -14
              V2          31             82 *         -12
              V3          34             79 *           9
              V4          31             69 *          15
              V5          80 *          -26            41 *
              V6          79 *          -32            41 *
              V7          87 *          -27            26
              V8          88 *          -14             9
              V9         -61 *           14            47 *
              V10        -43 *           23            68 *
              V11        -72 *           -6            12
              V12        -40             19            72 *


         Printed values are multiplied by 100 and
         rounded to the nearest integer.  Values
         greater than 0.4 are flagged by an '*'.


               Variance Explained by Each Factor

                    Factor1        Factor2        Factor3

              4.4705813      2.7306228      1.7017342


           Final Communality Estimates: Total = 8.902938
```

V1	V2	V3	V4	V5	V6
0.75522085	0.78212254	0.74798227	0.59887827	0.87166830	0.89980383

V7	V8	V9	V10	V11	V12
0.89991828	0.79667991	0.61125027	0.69487666	0.53208443	0.71245274

```
                                                                          4
                          The FACTOR Procedure
                       Rotation Method: Varimax

                   Orthogonal Transformation Matrix

                            1              2              3

                 1      0.83136        0.34431       -0.43623
                 2     -0.29481        0.93864        0.17902
                 3      0.47110       -0.02022        0.88185
```

(continued on the next page)

Output 15.3 *(continued)*

```
                        Rotated Factor Pattern

                    Factor1        Factor2        Factor3

             V1         3            85 *          -16
             V2        -4            88 *          -10
             V3         9            86 *            8
             V4        13            75 *           12
             V5        93 *           2             -3
             V6        95 *          -4             -4
             V7        93 *           4            -19
             V8        81 *          17            -33
             V9       -32           -9             71 *
             V10      -11            6             82 *
             V11      -52 *         -30            41 *
             V12       -5            3             84 *

        Printed values are multiplied by 100 and
        rounded to the nearest integer.  Values
        greater than 0.4 are flagged by an '*'.


                Variance Explained by Each Factor

             Factor1        Factor2        Factor3

            3.7048597      2.9364774      2.2616012


        Final Communality Estimates: Total = 8.902938

      V1          V2          V3          V4          V5          V6

0.75522085  0.78212254  0.74798227  0.59887827  0.87166830  0.89980383

      V7          V8          V9         V10         V11         V12

0.89991828  0.79667991  0.61125027  0.69487666  0.53208443  0.71245274
```

The eigenvalue table in Output 15.3 also shows that the first three components combined account for approximately 74% of the total variance. (This variance value can be observed at the intersection of the column labeled "Cumulative" and row "3".) According to the "percentage of variance accounted for" criterion, this once again suggests that it might be appropriate to retain three components.

The scree plot from this solution appears on page 2 of Output 15.3. This scree plot shows that there are several large breaks in the data following components 1, 2, and 3, and then the line begins to flatten beginning with component 4. The last large break appears after component 3, suggesting that only components 1 through 3 account for meaningful variance. This suggests that only these first three components should be retained and interpreted. Notice how it is almost possible to draw a straight line through components 4 to 12. The components that lie along a semi-straight line such as this are typically assumed to be measuring only trivial variance (i.e., components 4 through 12 constitute the "scree" of your scree plot).

So far, the results from the eigenvalue-one criterion, the variance accounted for criterion, and the scree plot are in accord to suggest that a three-component solution might be most appropriate. It is now time to review the rotated factor pattern to see if such a solution is interpretable. This matrix is presented on page 4 of Output 15.3.

Following the guidelines provided earlier, you begin by looking for factorially complex items (i.e., items with meaningful loadings for more than one component). A review shows that item 11 (variable V11) is a complex item, loading on both components 1 and 3; item 11 should therefore be discarded. Except for this item, the solution is otherwise fairly straightforward.

To interpret component 1, you read down the column for FACTOR1 and see that items 5 through 8 exhibit significant loadings for this component. (Remember that item 11 has been discarded.) These items are shown here:

```
_____   5.  I have invested a great deal of time in my
              current relationship.

_____   6.  I have invested a great deal of energy in my
              current relationship.

_____   7.  I have invested a lot of my personal resources
              (e.g., money) in developing my current
              relationship.

_____   8.  My partner and I have established mutual friends
              that I might lose if we were to break up.
```

All of these items pertain to the investments that participants have made in their relationships so it makes sense to label this the "investment size" component.

The rotated factor pattern shows that items 1 through 4 displayed meaningful loadings for component 2. These items are shown here:

```
_____   1.  I am satisfied with my current relationship.

_____   2.  My current relationship comes close to my ideal
              relationship.

_____   3.  I am more satisfied with my relationship than
              the average person.

_____   4.  I feel good about my current relationship.
```

Given the content of the preceding items, it seems reasonable to label component 2 the "satisfaction" component.

Finally, component 3 displays large loadings for items 9, 10, and 12. (Again, remember that item 11 is discarded.) These items are shown here:

```
_____    9.   There are plenty of other attractive people
               around for me to date if I were to break up with
               my current partner.

_____   10.   It would be appealing to break up with my
               current partner and date someone else.

_____   12.   It would be appealing to break up with my
               partner and "play the field" for a while.
```

These items all seem to deal with the attractiveness of alternatives to one's current relationship, so it makes sense to label this the "alternate value" component.

You can now step back and determine whether this solution satisfies the interpretability criteria presented earlier.

1. Are there at least three variables with meaningful loadings on each retained component?

2. Do the variables that load on a given component share the same conceptual meaning?

3. Do the variables that load on different components seem to be measuring different constructs?

4. Does the rotated factor pattern demonstrate "simple structure"?

In general, the answer to each of these questions is "yes," indicating that the current solution is, in most respects, satisfactory. There is, however, a problem with item 11, which loads on both components 1 and 3. This problem prevented the current solution from demonstrating a perfectly "simple structure" (criterion 4 from above). To eliminate this problem, it might be desirable to repeat the analysis, this time analyzing all of the items *except* for item 11. This is done in the second analysis of the investment model data described below.

Results of the Second Analysis

To repeat the current analysis with item 11 deleted, it is necessary to modify only the VAR statement of the preceding program. This can be done by changing the VAR statement so that it appears as follows:

```
VAR V1-V10 V12;
```

All other aspects of the program remain as they were. The eigenvalue table, scree plot, the unrotated factor pattern, the rotated factor pattern, and final communality estimates obtained from this revised program appear in Output 15.4:

Output 15.4 Results of the Second Analysis of the Investment Model Data

```
                            The SAS System                              1

                          The FACTOR Procedure
                Initial Factor Method: Principal Components

                    Prior Communality Estimates: ONE


    Eigenvalues of the Correlation Matrix: Total = 11  Average = 1

              Eigenvalue   Difference   Proportion   Cumulative

         1    4.02408599   1.29704748     0.3658       0.3658
         2    2.72703851   1.03724743     0.2479       0.6137
         3    1.68979108   1.00603918     0.1536       0.7674
         4    0.68375190   0.12740106     0.0622       0.8295
         5    0.55635084   0.16009525     0.0506       0.8801
         6    0.39625559   0.08887964     0.0360       0.9161
         7    0.30737595   0.04059618     0.0279       0.9441
         8    0.26677977   0.07984443     0.0243       0.9683
         9    0.18693534   0.07388104     0.0170       0.9853
        10    0.11305430   0.06447359     0.0103       0.9956
        11    0.04858072                  0.0044       1.0000

        3 factors will be retained by the MINEIGEN criterion.
```

(continued on the next page)

Output 15.4 *(continued)*

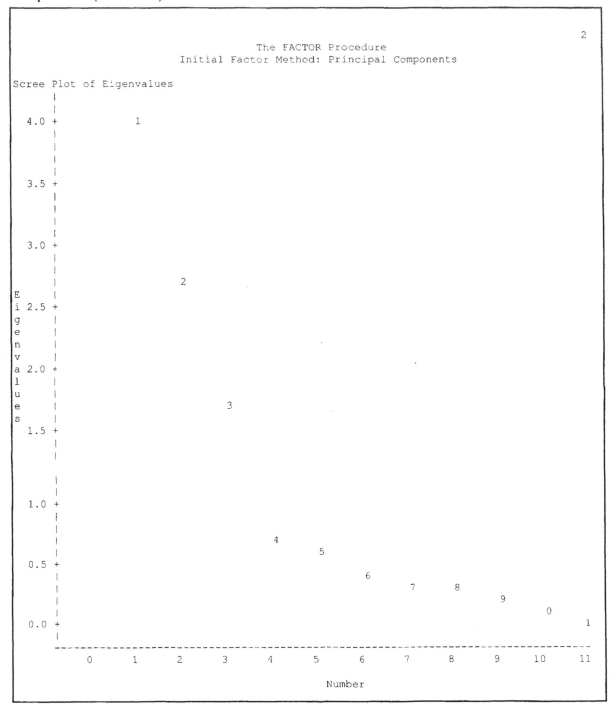

```
                                                                    3
                          The FACTOR Procedure
                 Initial Factor Method: Principal Components

                             Factor Pattern

                        Factor1        Factor2        Factor3

              V1           38            77  *         -17
              V2           30            83  *         -15
              V3           32            80  *           8
              V4           29            70  *          15
              V5           83  *        -23             38
              V6           83  *        -30             38
              V7           89  *        -24             24
              V8           88  *        -12              7
              V9          -56  *         13             47  *
              V10         -44  *         22             70  *
              V12         -40            18             74  *

         Printed values are multiplied by 100 and
         rounded to the nearest integer.  Values
         greater than 0.4 are flagged by an '*'.


                      Variance Explained by Each Factor

                  Factor1        Factor2        Factor3

               4.0240860      2.7270385      1.6897911


              Final Communality Estimates: Total = 8.440916
```

V1	V2	V3	V4	V5	V6
0.77238591	0.79828862	0.74823267	0.59192139	0.88254435	0.92134918

V7	V8	V9	V10	V12
0.90409576	0.79662327	0.55379952	0.73619279	0.73548214

(continued on the next page)

Output 15.4 *(continued)*

```
                                                              4
                        The FACTOR Procedure
                        Rotation Method: Varimax

                   Orthogonal Transformation Matrix

                            1              2              3

            1           0.84713        0.32918       -0.41716
            2          -0.27774        0.94354        0.18052
            3           0.45303       -0.03706        0.89073


                        Rotated Factor Pattern

                      Factor1        Factor2        Factor3

            V1           3             86 *          -17
            V2          -4             89 *          -11
            V3           8             86 *            8
            V4          12             75 *           14
            V5          94 *            4             -4
            V6          96 *           -2             -6
            V7          93 *            5            -20
            V8          81 *           18            -33
            V9         -30             -8             68 *
            V10        -12              4             85 *
            V12         -5              1             86 *

            Printed values are multiplied by 100 and
            rounded to the nearest integer.  Values
            greater than 0.4 are flagged by an '*'.


                   Variance Explained by Each Factor

                Factor1          Factor2          Factor3

               3.4449528        2.8661574        2.1298054


            Final Communality Estimates: Total = 8.440916

      V1            V2            V3            V4            V5            V6
 0.77238591    0.79828862    0.74823267    0.59192139    0.88254435    0.92134918

      V7            V8            V9            V10           V12

    0.90409576    0.79662327    0.55379952    0.73619279    0.73548214
```

The results obtained when item 11 is deleted from the analysis are very similar to those obtained when it was included. The eigenvalue table of Output 15.4 shows that the eigenvalue-one criterion would again result in retaining three components. The first three components account for almost 77% of the total variance which means that three components would also be retained if you used the variance-accounted-for criterion. Also, the scree plot from page 2 of Output 15.4 is cleaner than had been observed with the initial analysis; the break between components 3 and 4 is now more distinct, and the eigenvalues again level off after this break. This means that three components would also be retained if the scree test were used to answer the number-of-components question.

The biggest change can be seen in the rotated factor pattern on page 4 of Output 15.4. The solution is now cleaner in the sense that no item now loads on more than one component (i.e., no complex items). The current results now demonstrate a somewhat simpler structure than had been demonstrated by the initial analysis of the investment model data.

Conclusion

Principal component analysis is an effective procedure for reducing a number of observed variables into a smaller number that account for most of the variance in datasets. This technique is particularly useful when a data reduction procedure is required that makes no assumptions concerning an underlying causal structure responsible for covariation in the data. When such an underlying causal structure can be envisioned, it might be more appropriate to analyze the data using exploratory factor analysis.

Assumptions Underlying Principal Component Analysis

Because a principal component analysis is performed on a matrix of Pearson correlation coefficients, the data should satisfy the assumptions for this statistic. These assumptions were described in detail in Chapter 6, "Measures of Bivariate Association," and are briefly reviewed here:

- **Interval-level measurement.** All variables should be assessed on an interval or ratio level of measurement.

- **Random sampling.** Each participant will contribute one response for each observed variable. These sets of scores should represent a random sample drawn from the population of interest.

- **Linearity.** The relationship between all observed variables should be linear.

- **Bivariate normal distribution.** Each pair of observed variables should display a bivariate normal distribution (e.g., they should form an elliptical scattergram when plotted).

References

Cattell, R. B. (1966). The scree test for the number of factors. *Multivariate Behavioral Research, 1,* 245-276.

Clark, L. A., & Watson, D. (1995). Constructing validity: Basic issues in objective scale development. *Psychological Assessment, 7,* 309-319.

Kaiser, H. F. (1960). The application of electronic computers to factor analysis. *Educational and Psychological Measurement, 20,* 141-151.

Kim, J. O., & Mueller, C. W. (1978a). *Introduction to factor analysis: What it is and how to do it.* Beverly Hills, CA: Sage.

Kim, J. O., & Mueller, C. W. (1978b). *Factor analysis: Statistical methods and practical issues.* Beverly Hills, CA: Sage.

O'Rourke, N., & Cappeliez, P. (2002). Development and validation of a couples measure of biased responding: The Marital Aggrandizement Scale. *Journal of Personality Assessment, 78,* 301-320.

Rummel, R. J. (1970). *Applied factor analysis.* Evanston, IL: Northwestern University Press.

Rusbult, C.E. (1980). Commitment and satisfaction in romantic associations: A test of the investment model. *Journal of Experimental Social Psychology, 16,* 172-186.

Spector, P.E. (1992*). Summated rating scale construction: An introduction.* Newbury Park, CA: Sage.

Stevens, J. (2002). *Applied multivariate statistics for the social sciences* (4th ed.). Mahwah, NJ: Lawrence Erlbaum.

Streiner, D. L. (1994). Figuring out factors: The use and misuse of factor analysis. *Canadian Journal of Psychiatry, 39,* 135-140.

Appendix A

Choosing the Correct Statistic

Overview. This appendix provides a structured approach for choosing the correct statistical procedure to use when analyzing data. With this approach, the choice of a specific statistic is based on the number and scale of the criterion (dependent) variables in the study considered in conjunction with the number and scale of the predictor (independent) variables. Commonly used statistics are grouped into three tables based on the number of criterion and predictor variables in the analysis.

Introduction: Thinking about the Number and Scale of Your Variables

Researchers are often confused by the task of choosing the correct statistical procedure for analyzing a given dataset. There are so many procedures from which to choose that it is easy to become frustrated, not knowing where to begin. This appendix addresses this problem by providing a relatively simple system for classifying statistics. It provides a structured approach that should make it easier to find the appropriate statistical procedure for a wide variety of circumstances.

In a sense, most statistical procedures involve examining the relationship between two variables (or two sets of variables). In a given study, the outcome variable that you are interested in is called either a criterion variable (in nonexperimental research) or a **dependent variable** (in experimental research). In nonexperimental research, you study the relationship between the criterion variable and some **predictor variable** with values that are used to predict scores on the criterion (in experimental research, a manipulated **independent variable** is the counterpart to this predictor variable). In general, nonexperimental research involves examining the relationship between a criterion variable and a predictor variable whereas experimental research examines the relationship between a dependent variable and an independent variable or variables. To simplify matters, this appendix blurs the distinction between nonexperimental versus experimental research, and uses "criterion variable" to represent criterion variables as well as dependent variables, and uses "predictor variable" to represent predictor variables as well as independent variables.

Thinking about Your Criterion Variables

Two primary factors that determine the selection of an appropriate statistical procedure are the number and the scale of the criterion variables. The **number** merely refers to how many criterion variables appear in the dataset, while the **scale** refers to the level of measurement used in assessing these criterion variables (i.e., nominal, ordinal, ratio, or interval).

For example, assume that you want to conduct a study to learn about variables that predict success in college. In your study, you might choose to use just one criterion variable as an index of college success such as grade point average (GPA). Alternatively, you might choose to use several criterion variables so that you will have multiple indices of success such as college GPA, college class rank, and whether or not participants were inducted into some college honorary society (yes versus no). Here, you can see that the *number* of criterion variables varies. In the first case, there is only one criterion variable; in the second, there are three.

Notice, however, that the *scale* used to assess college success also varies. The criterion variable college GPA is assessed on an interval scale, college class rank is assessed on an ordinal scale, and induction into an honorary society is assessed on a nominal scale. The number of criterion variables used in the analysis, as well as the scale used to measure those variables, helps to determine which statistic you can use to analyze your data.

Thinking about Your Predictor Variables

However, you still do not have enough information to choose the appropriate statistic. Two additional factors that determine choice of the correct statistic are the number and scale of the *predictor* variables. Again, consider the college success study in which you are interested in learning about variables that predict college success. Assume that you have decided to use just one criterion variable as a measure of success: college GPA. You might choose to design a study that also includes just one predictor variable: high school GPA. Alternatively, you can design a study that includes multiple predictor variables such as high school GPA, scores on the SAT, high school rank, and whether the student received a scholarship (yes versus no).

In the previous paragraph, notice that the number of predictor variables that might be included in a study varies. The first study included just one predictor, while the second included multiple predictors. Note that the scale used to assess these predictors also varies. Predictors were assessed on an interval scale (high school GPA, SAT scores), an ordinal scale (high school rank), and a nominal scale (whether the student received a scholarship). The number of predictor variables included in your study as well as their scale also help determine the appropriate statistic.

Putting It Together

The preceding discussion has provided context for the following recommendation. When choosing the appropriate statistic for an analysis, always consider the following:

- the number and scale of the criterion variables, in conjunction with;
- the number and scale of the predictor variables.

For example, imagine that in your study, you used only one measure of college success (GPA) and one predictor variable (SAT scores). Since you have a single criterion variable assessed on an interval scale and a single predictor variable also on an interval scale, you know that the appropriate statistic is the Pearson correlation coefficient (assuming that a few additional assumptions are met). But what if you modify your study so that it still contained only one criterion variable but now contains two predictor variables, both assessed on an interval scale (e.g., SAT scores and high school GPA)? In the latter case, it would be more appropriate to analyze your data using multiple regression.

This is the approach recommended in this appendix. To select the right statistic, consider the number and nature of both your criterion and predictor variables. To facilitate this decision-making process, this appendix includes three tables: one that lists statistics for studies that involve a single criterion variable and a single predictor; one for studies that involve a single criterion variable and multiple predictors; and a final table for studies with multiple criterion variables.

A few words of caution are warranted before presenting the tables. First, these tables were not designed to present an *exhaustive* list of statistical procedures. They focus only on the tests that are the most commonly used in the social sciences. A good number of statistical procedures that did not fit neatly into this format (such as principal component analysis) do not appear. Second, these tables do not necessarily provide *all* of the information that you need to make the final selection of a statistical procedure. Many statistical procedures require that a number of assumptions be met concerning the data for the procedure to be appropriate, and these assumptions are often too numerous to include in a short appendix such as this. The purpose of this appendix is to help you locate the statistic that may be correct for your situation given the nature of the variables. It is then up to you to learn more about the assumptions for that statistic to determine whether your data satisfy those assumptions.

Guidelines for Choosing the Correct Statistic

Studies with a Single Criterion Variable and a Single Predictor Variable

These are among the simplest (and most common) studies conducted in the social sciences. Some of the statistics appropriate for this type of investigation are listed in Table A.1.

Table A.1

Studies with a Single Criterion Variable and a Single Predictor Variable

Criterion/ Dependent Variable	Predictor/ Independent Variable	Analysis
Nominal	Nominal	**Chi-square test of independence.***
Ordinal, Interval, Ratio	Nominal	**Kruskal-Wallis test**. Typically used with an ordinal-scale criterion; also used with interval/ratio scale criterion if markedly non-normal.
Interval, Ratio	Nominal	**t test.*** Appropriate only if predictor variable consists of no more than two values.
Interval, Ratio	Nominal	One-way analysis of variance (ANOVA).*
Ordinal, Interval, Ratio	Ordinal, Interval, Ratio	**Spearman correlation coefficient.*** Typically used if at least one variable is an ordinal scale. Also used with interval/ratio variables if markedly nonnormal.
Interval, Ratio	Interval, Ratio	**Pearson correlation coefficient.***

* Statistics covered in this text.

Table A.1 has three columns. The first column describes the nature of the criterion or dependent variable in the study, the second describes the predictor or independent variable, and the third describes the statistic that might be appropriate for that type of study. For example, the first entry indicates that, if your predictor variable is assessed on a nominal scale and your criterion is also assessed on a nominal scale, it might be appropriate to assess the relationship between these variables using the chi-square test of independence. To understand this, assume that in your study you use one index of college success and it is assessed on a nominal scale. You might have chosen "graduation" as this criterion (the graduation variable might have been coded with the value "Yes" if the student did graduate from college and "No" if s/he did not). Furthermore, assume further that you have used one nominal-scale predictor variable in your study: scholarship status. You have coded this variable so that the value "Athletic" represents students who received athletic scholarships, "Academic" represents students who received academic scholarships, and "None" represents students who received no scholarship. You analyze your dataset with a chi-square test of independence to determine whether there is a significant relationship between scholarship status and graduation. This can result in a significant value of chi-square, and inspection of the cells of your classification table might show that students in the academic scholarship group were more likely to graduate than students in the athletic scholarship or no scholarship groups.

Notice that in the Analysis column the entry "Chi-square test of independence" is flagged with an asterisk (*). Tests that are flagged with an asterisk in this way are described in this text. However, rest assured that even the procedures that are not flagged can still be analyzed with SAS. For help with these procedures, consult the *SAS/STAT User's Guide.*

The first row of Table A.1 deals with the chi-square test. The next row describes the appropriate conditions for a Kruskal-Wallis test. Notice that the entry in the Criterion/Dependent Variable column is "Ordinal, Interval, Ratio." This entry indicates that the Kruskal-Wallis test might be appropriate if you have a single criterion variable that is assessed on *either* an ordinal scale, an interval scale, or a ratio scale. (It is *not* meant to suggest that you should have three criterion variables, each assessed with different scales!)

Studies with a Single Criterion Variable and Multiple Predictor Variables

Table A.2 lists some procedures that are appropriate when the analysis includes a single criterion variable and multiple predictors. Notice that each entry in the Predictor/Independent Variable column is flagged with "(M)"; it stands for "multiple," as in "multiple predictor variables." It simply indicates that the analysis should include more than one predictor variable regardless of the scale on which the variables are measured.

For example, the last row of the table is for multiple regression. The entry in the predictor variable column is "Nominal, Interval, Ratio (M)"; this means that the predictor variables can be assessed on *either* a nominal scale, an interval scale, or a ratio scale (or some combination thereof) so long as you include more than one predictor variable. Earlier, we indicated that it would be appropriate to use multiple regression to analyze data that included college GPA as the criterion variable, and SAT scores and high school GPA as predictors. Note how this is consistent with the guidelines of Table A.2. There is a single criterion variable assessed on an interval scale (college GPA), and there are multiple predictor variables, both assessed on an interval scale (SAT scores and high school GPA).

Table A.2

Studies with a Single Criterion Variable and Multiple Predictor Variables

Criterion/ Dependent Variable	Predictor/ Independent Variables	Analysis
Nominal, Ordinal	Nominal, Interval, Ratio (M)	**Logistic regression.** If the criterion variable is on a nominal scale, it might consist of no more than two values; for more values, consider discriminant analysis. If used, nominal-scale predictor variables must be dummy-coded or effect-coded.
Nominal	Interval, Ratio (M)	**Discriminant function analysis.**
Interval, Ratio	Nominal (M)	**Factorial analysis of variance (ANOVA).***
Interval, Ratio	Nominal, Interval, Ratio (M)	**Analysis of covariance (ANCOVA).** Used when predictors include at least one nominal-scale variable and at least one interval/ratio-scale variable.
Interval, Ratio	Nominal, Interval, Ratio (M)	**Multiple regression.*** If used, nominal-scale predictor variables must be dummy-coded or effect-coded.

* Statistics covered in this text.
(M) = Multiple variables

Studies with Multiple Criterion Variables

The "(M)" symbol in the criterion variable column of Table A.3 indicates that all of the procedures in this table are appropriate for studies that include multiple criterion variables. Note, however, that only the last three analytic procedures (factorial MANOVA, MANCOVA, and canonical correlation) involve multiple *predictor*

variables. The first procedure (one-way MANOVA) requires only a single predictor variable on a nominal scale.

Table A.3

Studies with Multiple Criterion Variables

Criterion/ Dependent Variables	Predictor/ Independent Variable(s)	Analysis
Interval, Ratio (M)	Nominal	**One-way multivariate analysis of variance (MANOVA).** *
Interval, Ratio (M)	Nominal (M)	**Factorial multivariate analysis of variance (MANOVA).**
Interval, Ratio (M)	Nominal, Interval, Ratio (M)	**Multivariate analysis of covariance (MANCOVA).** Used when predictors include at least one nominal-scale variable and at least one interval/ratio-scale variable.
Interval, Ratio (M)	Interval, Ratio (M)	**Canonical Correlation.**

* Statistics covered in this text.
(M) = Multiple variables

Conclusion

As stated earlier, this appendix is intended to serve as a starting point for choosing appropriate statistics. You can use the preceding tables to identify procedures that might be appropriate for your research design. It is then up to you, however, to learn more about the assumptions associated with the statistic (e.g., whether it requires data drawn from a normal population, whether it requires independent observations). These tables, when used in conjunction with the "Assumptions" sections included in the earlier chapters of this text, should help you find the right statistical procedure for analyzing the types of data that are most frequently encountered in social science research.

Reference

SAS Institute Inc. 2004. *SAS/STAT 9.1 User's Guide*. Cary, NC: Author.

Appendix

B

Datasets

> **Overview.** This appendix provides datasets that were used in analyses reported in Chapter 7, "Assessing Scale Reliability with Coefficient Alpha"; Chapter 14, "Multiple Regression"; and Chapter 15, "Principal Component Analysis." The datasets appear here rather than in the chapters in which they were discussed because they are longer than most datasets used in the text.

Dataset from Chapter 7: Assessing Scale Reliability with Coefficient Alpha

```
DATALINES;
556754
567343
777222
665243
666665
353324
767153
666656
334333
567232
445332
555232
546264
436663
265454
757774
635171
667777
657375
545554
557231
666222
656111
464555
465771
142441
675334
665131
666443
244342
464452
654665
775221
657333
666664
545333
353434
666676
667461
544444
666443
676556
676444
676222
545111
777443
566443
```

```
767151
455323
455544
;
RUN;
```

Dataset from Chapter 14: Multiple Regression

```
DATALINES;
34  30  25  13  25  12
32  27  27  14  32  13
34  23  24  21  30  14
26  26  22  18  27  19
 4  17  25  10  11  34
31  26  30  22  31  13
22  29  31  18  27  14
32  29  27   9  31   8
33  36  31  14  28  13
36  32  26  14  19  12
19  24  22  23  23   4
36  35  32  14  34   4
30  30  30  20  30  23
35  21  22  20  24  19
36  27  33   7  31   4
35  31  30  20  29   8
 4  20  20  22  14  28
36  35  29  14  34   8
36  32  28  14  28  15
19  18  17  25  18  20
20  30  26  20  23  28
36  34  31  21  32  18
29  27  24  19  19  13
 9  36  30   5  18  24
28  24  27  15  22  20
10  26  24  24  27  27
34  31  29   9  29  17
 7  12  16  26  20  33
35  36  33  13  27  16
36  33  30   5  29  13
 6  19  27  21  24  26
35   .  30  18  24  21
35  34  32  18  28  16
24  24  23  17  17  20
36  24  31  14  30  20
28  24  19  24  24  17
33  33  28  16  19  11
 5  12   9  15  11  32
21  24  20  24  22  14
16   9  12  30   .  15
36  34  32   6  20   9
29  27  27  20  30  21
 .   7  12  23  13  26
34  34  32  20  31  13
35  27  23  21  31  18
25  26  29   7  23  18
36  25  25  16  29  11
```

```
36  32  28  13  15   5
32  29  30  21  32  22
30  32  33  16  34   9
;
RUN;
```

Dataset from Chapter 15: Principal Component Analysis

The dataset described in Chapter 15 is identical to the dataset from Chapter 7 that appears in the first section.

Appendix C

Critical Values of the
F Distribution

Upper 5% points

ν_2 \ ν_1	1	2	3	4	5	6	7	8	9	10	12	15	20	24	30	40	60	120	∞
1	161.4	199.5	215.7	224.6	230.2	234.0	236.8	238.9	240.5	241.9	243.9	245.9	248.0	249.1	250.1	251.1	252.2	253.3	254.3
2	18.51	19.00	19.16	19.25	19.30	19.33	19.35	19.37	19.38	19.40	19.41	19.43	19.45	19.45	19.46	19.47	19.48	19.49	19.50
3	10.13	9.55	9.28	9.12	9.01	8.94	8.89	8.85	8.81	8.79	8.74	8.70	8.66	8.64	8.62	8.59	8.57	8.55	8.53
4	7.71	6.94	6.59	6.39	6.26	6.16	6.09	6.04	6.00	5.96	5.91	5.86	5.80	5.77	5.75	5.72	5.69	5.66	5.63
5	6.61	5.79	5.41	5.19	5.05	4.95	4.88	4.82	4.77	4.74	4.68	4.62	4.56	4.53	4.50	4.46	4.43	4.40	4.36
6	5.99	5.14	4.76	4.53	4.39	4.28	4.21	4.15	4.10	4.06	4.00	3.94	3.87	3.84	3.81	3.77	3.74	3.70	3.67
7	5.59	4.74	4.35	4.12	3.97	3.87	3.79	3.73	3.68	3.64	3.57	3.51	3.44	3.41	3.38	3.34	3.30	3.27	3.23
8	5.32	4.46	4.07	3.84	3.69	3.58	3.50	3.44	3.39	3.35	3.28	3.22	3.15	3.12	3.08	3.04	3.01	2.97	2.93
9	5.12	4.26	3.86	3.63	3.48	3.37	3.29	3.23	3.18	3.14	3.07	3.01	2.94	2.90	2.86	2.83	2.79	2.75	2.71
10	4.96	4.10	3.71	3.48	3.33	3.22	3.14	3.07	3.02	2.98	2.91	2.85	2.77	2.74	2.70	2.66	2.62	2.58	2.54
11	4.84	3.98	3.59	3.36	3.20	3.09	3.01	2.95	2.90	2.85	2.79	2.72	2.65	2.61	2.57	2.53	2.49	2.45	2.40
12	4.75	3.89	3.49	3.26	3.11	3.00	2.91	2.85	2.80	2.75	2.69	2.62	2.54	2.51	2.47	2.43	2.38	2.34	2.30
13	4.67	3.81	3.41	3.18	3.03	2.92	2.83	2.77	2.71	2.67	2.60	2.53	2.46	2.42	2.38	2.34	2.30	2.25	2.21
14	4.60	3.74	3.34	3.11	2.96	2.85	2.76	2.70	2.65	2.60	2.53	2.46	2.39	2.35	2.31	2.27	2.22	2.18	2.13
15	4.54	3.68	3.29	3.06	2.90	2.79	2.71	2.64	2.59	2.54	2.48	2.40	2.33	2.29	2.25	2.20	2.16	2.11	2.07
16	4.49	3.63	3.24	3.01	2.85	2.74	2.66	2.59	2.54	2.49	2.42	2.35	2.28	2.24	2.19	2.15	2.11	2.06	2.01
17	4.45	3.59	3.20	2.96	2.81	2.70	2.61	2.55	2.49	2.45	2.38	2.31	2.23	2.19	2.15	2.10	2.06	2.01	1.96
18	4.41	3.55	3.16	2.93	2.77	2.66	2.58	2.51	2.46	2.41	2.34	2.27	2.19	2.15	2.11	2.06	2.02	1.97	1.92
19	4.38	3.52	3.13	2.90	2.74	2.63	2.54	2.48	2.42	2.38	2.31	2.23	2.16	2.11	2.07	2.03	1.98	1.93	1.88
20	4.35	3.49	3.10	2.87	2.71	2.60	2.51	2.45	2.39	2.35	2.28	2.20	2.12	2.08	2.04	1.99	1.95	1.90	1.84
21	4.32	3.47	3.07	2.84	2.68	2.57	2.49	2.42	2.37	2.32	2.25	2.18	2.10	2.05	2.01	1.96	1.92	1.87	1.81
22	4.30	3.44	3.05	2.82	2.66	2.55	2.46	2.40	2.34	2.30	2.23	2.15	2.07	2.03	1.98	1.94	1.89	1.84	1.78
23	4.28	3.42	3.03	2.80	2.64	2.53	2.44	2.37	2.32	2.27	2.20	2.13	2.05	2.01	1.96	1.91	1.86	1.81	1.76
24	4.26	3.40	3.01	2.78	2.62	2.51	2.42	2.36	2.30	2.25	2.18	2.11	2.03	1.98	1.94	1.89	1.84	1.79	1.73
25	4.24	3.39	2.99	2.76	2.60	2.49	2.40	2.34	2.28	2.24	2.16	2.09	2.01	1.96	1.92	1.87	1.82	1.77	1.71
26	4.23	3.37	2.98	2.74	2.59	2.47	2.39	2.32	2.27	2.22	2.15	2.07	1.99	1.95	1.90	1.85	1.80	1.75	1.69
27	4.21	3.35	2.96	2.73	2.57	2.46	2.37	2.31	2.25	2.20	2.13	2.06	1.97	1.93	1.88	1.84	1.79	1.73	1.67
28	4.20	3.34	2.95	2.71	2.56	2.45	2.36	2.29	2.24	2.19	2.12	2.04	1.96	1.91	1.87	1.82	1.77	1.71	1.65
29	4.18	3.33	2.93	2.70	2.55	2.43	2.35	2.28	2.22	2.18	2.10	2.03	1.94	1.90	1.85	1.81	1.75	1.70	1.64
30	4.17	3.32	2.92	2.69	2.53	2.42	2.33	2.27	2.21	2.16	2.09	2.01	1.93	1.89	1.84	1.79	1.74	1.68	1.62
40	4.08	3.23	2.84	2.61	2.45	2.34	2.25	2.18	2.12	2.08	2.00	1.92	1.84	1.79	1.74	1.69	1.64	1.58	1.51
60	4.00	3.15	2.76	2.53	2.37	2.25	2.17	2.10	2.04	1.99	1.92	1.84	1.75	1.70	1.65	1.59	1.53	1.47	1.39
120	3.92	3.07	2.68	2.45	2.29	2.17	2.09	2.02	1.96	1.91	1.83	1.75	1.66	1.61	1.55	1.50	1.43	1.35	1.25
∞	3.84	3.00	2.60	2.37	2.21	2.10	2.01	1.94	1.88	1.83	1.75	1.67	1.57	1.52	1.46	1.39	1.32	1.22	1.00

$F = \dfrac{s_1^2}{s_2^2} = \dfrac{S_1/\nu_1}{S_2/\nu_2}$, where $s_1^2 = S_1/\nu_1$ and $s_2^2 = S_2/\nu_2$ are independent mean squares estimating a common variance σ^2 and based on ν_1 and ν_2 degrees of freedom, respectively.

Table C.1 Critical Values of the *F* Distribution with Alpha Level of .05

Upper 1% points

ν_1 \ ν_2	1	2	3	4	5	6	7	8	9	10	12	15	20	24	30	40	60	120	∞
1	4052	4999·5	5403	5625	5764	5859	5928	5981	6022	6056	6106	6157	6209	6235	6261	6287	6313	6339	6366
2	98·50	99·00	99·17	99·25	99·30	99·33	99·36	99·37	99·39	99·40	99·42	99·43	99·45	99·46	99·47	99·47	99·48	99·49	99·50
3	34·12	30·82	29·46	28·71	28·24	27·91	27·67	27·49	27·35	27·23	27·05	26·87	26·69	26·60	26·50	26·41	26·32	26·22	26·13
4	21·20	18·00	16·69	15·98	15·52	15·21	14·98	14·80	14·66	14·55	14·37	14·20	14·02	13·93	13·84	13·75	13·65	13·56	13·46
5	16·26	13·27	12·06	11·39	10·97	10·67	10·46	10·29	10·16	10·05	9·89	9·72	9·55	9·47	9·38	9·29	9·20	9·11	9·02
6	13·75	10·92	9·78	9·15	8·75	8·47	8·26	8·10	7·98	7·87	7·72	7·56	7·40	7·31	7·23	7·14	7·06	6·97	6·88
7	12·25	9·55	8·45	7·85	7·46	7·19	6·99	6·84	6·72	6·62	6·47	6·31	6·16	6·07	5·99	5·91	5·82	5·74	5·65
8	11·26	8·65	7·59	7·01	6·63	6·37	6·18	6·03	5·91	5·81	5·67	5·52	5·36	5·28	5·20	5·12	5·03	4·95	4·86
9	10·56	8·02	6·99	6·42	6·06	5·80	5·61	5·47	5·35	5·26	5·11	4·96	4·81	4·73	4·65	4·57	4·48	4·40	4·31
10	10·04	7·56	6·55	5·99	5·64	5·39	5·20	5·06	4·94	4·85	4·71	4·56	4·41	4·33	4·25	4·17	4·08	4·00	3·91
11	9·65	7·21	6·22	5·67	5·32	5·07	4·89	4·74	4·63	4·54	4·40	4·25	4·10	4·02	3·94	3·86	3·78	3·69	3·60
12	9·33	6·93	5·95	5·41	5·06	4·82	4·64	4·50	4·39	4·30	4·16	4·01	3·86	3·78	3·70	3·62	3·54	3·45	3·36
13	9·07	6·70	5·74	5·21	4·86	4·62	4·44	4·30	4·19	4·10	3·96	3·82	3·66	3·59	3·51	3·43	3·34	3·25	3·17
14	8·86	6·51	5·56	5·04	4·69	4·46	4·28	4·14	4·03	3·94	3·80	3·66	3·51	3·43	3·35	3·27	3·18	3·09	3·00
15	8·68	6·36	5·42	4·89	4·56	4·32	4·14	4·00	3·89	3·80	3·67	3·52	3·37	3·29	3·21	3·13	3·05	2·96	2·87
16	8·53	6·23	5·29	4·77	4·44	4·20	4·03	3·89	3·78	3·69	3·55	3·41	3·26	3·18	3·10	3·02	2·93	2·84	2·75
17	8·40	6·11	5·18	4·67	4·34	4·10	3·93	3·79	3·68	3·59	3·46	3·31	3·16	3·08	3·00	2·92	2·83	2·75	2·65
18	8·29	6·01	5·09	4·58	4·25	4·01	3·84	3·71	3·60	3·51	3·37	3·23	3·08	3·00	2·92	2·84	2·75	2·66	2·57
19	8·18	5·93	5·01	4·50	4·17	3·94	3·77	3·63	3·52	3·43	3·30	3·15	3·00	2·92	2·84	2·76	2·67	2·58	2·49
20	8·10	5·85	4·94	4·43	4·10	3·87	3·70	3·56	3·46	3·37	3·23	3·09	2·94	2·86	2·78	2·69	2·61	2·52	2·42
21	8·02	5·78	4·87	4·37	4·04	3·81	3·64	3·51	3·40	3·31	3·17	3·03	2·88	2·80	2·72	2·64	2·55	2·46	2·36
22	7·95	5·72	4·82	4·31	3·99	3·76	3·59	3·45	3·35	3·26	3·12	2·98	2·83	2·75	2·67	2·58	2·50	2·40	2·31
23	7·88	5·66	4·76	4·26	3·94	3·71	3·54	3·41	3·30	3·21	3·07	2·93	2·78	2·70	2·62	2·54	2·45	2·35	2·26
24	7·82	5·61	4·72	4·22	3·90	3·67	3·50	3·36	3·26	3·17	3·03	2·89	2·74	2·66	2·58	2·49	2·40	2·31	2·21
25	7·77	5·57	4·68	4·18	3·85	3·63	3·46	3·32	3·22	3·13	2·99	2·85	2·70	2·62	2·54	2·45	2·36	2·27	2·17
26	7·72	5·53	4·64	4·14	3·82	3·59	3·42	3·29	3·18	3·09	2·96	2·81	2·66	2·58	2·50	2·42	2·33	2·23	2·13
27	7·68	5·49	4·60	4·11	3·78	3·56	3·39	3·26	3·15	3·06	2·93	2·78	2·63	2·55	2·47	2·38	2·29	2·20	2·10
28	7·64	5·45	4·57	4·07	3·75	3·53	3·36	3·23	3·12	3·03	2·90	2·75	2·60	2·52	2·44	2·35	2·26	2·17	2·06
29	7·60	5·42	4·54	4·04	3·73	3·50	3·33	3·20	3·09	3·00	2·87	2·73	2·57	2·49	2·41	2·33	2·23	2·14	2·03
30	7·56	5·39	4·51	4·02	3·70	3·47	3·30	3·17	3·07	2·98	2·84	2·70	2·55	2·47	2·39	2·30	2·21	2·11	2·01
40	7·31	5·18	4·31	3·83	3·51	3·29	3·12	2·99	2·89	2·80	2·66	2·52	2·37	2·29	2·20	2·11	2·02	1·92	1·80
60	7·08	4·98	4·13	3·65	3·34	3·12	2·95	2·82	2·72	2·63	2·50	2·35	2·20	2·12	2·03	1·94	1·84	1·73	1·60
120	6·85	4·79	3·95	3·48	3·17	2·96	2·79	2·66	2·56	2·47	2·34	2·19	2·03	1·95	1·86	1·76	1·66	1·53	1·38
∞	6·63	4·61	3·78	3·32	3·02	2·80	2·64	2·51	2·41	2·32	2·18	2·04	1·88	1·79	1·70	1·59	1·47	1·32	1·00

$F = \dfrac{s_1^2}{s_2^2} = \dfrac{S_1/S_2}{\nu_1/\nu_2}$, where $s_1^2 = S_1/\nu_1$ and $s_2^2 = S_2/\nu_2$, are independent mean squares estimating a common variance σ^2 and based on ν_1 and ν_2 degrees of freedom, respectively.

Table C.2 Critical Values of the *F* Distribution with Alpha Level of .01

Upper 0·1% points

ν_2 \ ν_1	1	2	3	4	5	6	7	8	9	10	12	15	20	24	30	40	60	120	∞
1	4053*	5000*	5404*	5625*	5764*	5859*	5920*	5981*	6023*	6056*	6107*	6158*	6209*	6235*	6261*	6287*	6313*	6340*	6366*
2	998·5	999·0	999·2	999·2	999·3	999·3	999·4	999·4	999·9	999·4	999·4	999·4	999·4	999·5	999·5	999·5	999·5	999·5	999·5
3	167·0	148·5	141·1	137·1	134·6	132·8	131·6	130·6	129·9	129·2	128·3	127·4	126·4	125·9	125·4	125·0	124·5	124·0	123·5
4	74·14	61·25	56·18	53·44	51·71	50·53	49·66	49·00	48·47	48·05	47·41	46·76	46·10	45·77	45·43	45·09	44·75	44·40	44·05
5	47·18	37·12	33·20	31·09	29·75	28·84	28·16	27·64	27·24	26·92	26·42	25·91	25·39	25·14	24·87	24·60	24·33	24·06	23·79
6	35·51	27·00	23·70	21·92	20·81	20·03	19·46	19·03	18·69	18·41	17·99	17·56	17·12	16·89	16·67	16·44	16·21	15·99	15·75
7	29·25	21·69	18·77	17·19	16·21	15·52	15·02	14·63	14·33	14·08	13·71	13·32	12·93	12·73	12·53	12·33	12·12	11·91	11·70
8	25·42	18·49	15·83	14·39	13·49	12·86	12·40	12·04	11·77	11·54	11·19	10·84	10·48	10·30	10·11	9·92	9·73	9·53	9·33
9	22·86	16·39	13·90	12·56	11·71	11·13	10·70	10·37	10·11	9·89	9·57	9·24	8·90	8·72	8·55	8·37	8·19	8·00	7·81
10	21·04	14·91	12·55	11·28	10·48	9·92	9·52	9·20	8·96	8·75	8·45	8·13	7·80	7·64	7·47	7·30	7·12	6·94	6·76
11	19·69	13·81	11·56	10·35	9·58	9·05	8·66	8·35	8·12	7·92	7·63	7·32	7·01	6·85	6·68	6·52	6·35	6·17	6·00
12	18·64	12·97	10·80	9·63	8·89	8·38	8·00	7·71	7·48	7·29	7·00	6·71	6·40	6·25	6·09	5·93	5·76	5·59	5·42
13	17·81	12·31	10·21	9·07	8·35	7·86	7·49	7·21	6·98	6·80	6·52	6·23	5·93	5·78	5·63	5·47	5·30	5·14	4·97
14	17·14	11·78	9·73	8·62	7·92	7·43	7·08	6·80	6·58	6·40	6·13	5·85	5·56	5·41	5·25	5·10	4·94	4·77	4·60
15	16·59	11·34	9·34	8·25	7·57	7·09	6·74	6·47	6·26	6·08	5·81	5·54	5·25	5·10	4·95	4·80	4·64	4·47	4·31
16	16·12	10·97	9·00	7·94	7·27	6·81	6·46	6·19	5·98	5·81	5·55	5·27	4·99	4·85	4·70	4·54	4·39	4·23	4·06
17	15·72	10·66	8·73	7·68	7·02	6·56	6·22	5·96	5·75	5·58	5·32	5·05	4·78	4·63	4·48	4·33	4·18	4·02	3·85
18	15·38	10·39	8·49	7·46	6·81	6·35	6·02	5·76	5·56	5·39	5·13	4·87	4·59	4·45	4·30	4·15	4·00	3·84	3·67
19	15·08	10·16	8·28	7·26	6·62	6·18	5·85	5·59	5·39	5·22	4·97	4·70	4·43	4·29	4·14	3·99	3·84	3·68	3·51
20	14·82	9·95	8·10	7·10	6·46	6·02	5·69	5·44	5·24	5·08	4·82	4·56	4·29	4·15	4·00	3·86	3·70	3·54	3·38
21	14·59	9·77	7·94	6·95	6·32	5·88	5·56	5·31	5·11	4·95	4·70	4·44	4·17	4·03	3·88	3·74	3·58	3·42	3·26
22	14·38	9·61	7·80	6·81	6·19	5·76	5·44	5·19	4·99	4·83	4·58	4·33	4·06	3·92	3·78	3·63	3·48	3·32	3·15
23	14·19	9·47	7·67	6·69	6·08	5·65	5·33	5·09	4·89	4·73	4·48	4·23	3·96	3·82	3·68	3·53	3·38	3·22	3·05
24	14·03	9·34	7·55	6·59	5·98	5·55	5·23	4·99	4·80	4·64	4·39	4·14	3·87	3·74	3·59	3·45	3·29	3·14	2·97
25	13·88	9·22	7·45	6·49	5·88	5·46	5·15	4·91	4·71	4·56	4·31	4·06	3·79	3·66	3·52	3·37	3·22	3·06	2·89
26	13·74	9·12	7·36	6·41	5·80	5·38	5·07	4·83	4·64	4·48	4·24	3·99	3·72	3·59	3·44	3·30	3·15	2·99	2·82
27	13·61	9·02	7·27	6·33	5·73	5·31	5·00	4·76	4·57	4·41	4·17	3·92	3·66	3·52	3·38	3·23	3·08	2·92	2·75
28	13·50	8·93	7·19	6·25	5·66	5·24	4·93	4·69	4·50	4·35	4·11	3·86	3·60	3·46	3·32	3·18	3·02	2·86	2·69
29	13·39	8·85	7·12	6·19	5·59	5·18	4·87	4·64	4·45	4·29	4·05	3·80	3·54	3·41	3·27	3·12	2·97	2·81	2·64
30	13·29	8·77	7·05	6·12	5·53	5·12	4·82	4·58	4·39	4·24	4·00	3·75	3·49	3·36	3·22	3·07	2·92	2·76	2·59
40	12·61	8·25	6·60	5·70	5·13	4·73	4·44	4·21	4·02	3·87	3·64	3·40	3·15	3·01	2·87	2·73	2·57	2·41	2·23
60	11·97	7·76	6·17	5·31	4·76	4·37	4·09	3·87	3·69	3·54	3·31	3·08	2·83	2·69	2·55	2·41	2·25	2·08	1·89
120	11·38	7·32	5·79	4·95	4·42	4·04	3·77	3·55	3·38	3·24	3·02	2·78	2·53	2·40	2·26	2·11	1·95	1·76	1·54
∞	10·83	6·91	5·42	4·62	4·10	3·74	3·47	3·27	3·10	2·96	2·74	2·51	2·27	2·13	1·99	1·84	1·66	1·45	1·00

* Multiply these entries by 100.

This 0·1% table is based on the following sources: Colcord & Deming (1935); Fisher & Yates (1953. Table V) used with the permission of the authors and of Messrs Oliver and Boyd; Norton (1952).

Table C.3 Critical Values of the *F* Distribution with Alpha Level of .001

This table is abridged from Table 18 in E.S. Pearson and H.O. Hartley (Eds), *Biometrika tables for statisticians* (3rd ed., Vol. 1), Cambridge University Press. New York, 1970, by permission of the Biometrika Trustees.

Index

Symbols

Numerics

WILEY SERIES IN PROBABILITY AND STATISTICS

Established by WALTER A. SHEWHART and SAMUEL S. WILKS

Editors: *David J. Balding, Noel A. C. Cressie, Nicholas I. Fisher,
Iain M. Johnstone, J. B. Kadane, Geert Molenberghs, Louise M. Ryan,
David W. Scott, Adrian F. M. Smith, Jozef L. Teugels*
Editors Emeriti: *Vic Barnett, J. Stuart Hunter, David G. Kendall*

A complete list of the titles in this series appears at the end of this volume.

WILEY SERIES IN PROBABILITY AND STATISTICS
ESTABLISHED BY WALTER A. SHEWHART AND SAMUEL S. WILKS

Editors: *David J. Balding, Noel A. C. Cressie, Nicholas I. Fisher,*
Iain M. Johnstone, J. B. Kadane, Geert Molenberghs. Louise M. Ryan,
David W. Scott, Adrian F. M. Smith, Jozef L. Teugels
Editors Emeriti: *Vic Barnett, J. Stuart Hunter, David G. Kendall*

The *Wiley Series in Probability and Statistics* is well established and authoritative. It covers many topics of current research interest in both pure and applied statistics and probability theory. Written by leading statisticians and institutions, the titles span both state-of-the-art developments in the field and classical methods.

Reflecting the wide range of current research in statistics, the series encompasses applied, methodological and theoretical statistics, ranging from applications and new techniques made possible by advances in computerized practice to rigorous treatment of theoretical approaches.

This series provides essential and invaluable reading for all statisticians, whether in academia, industry, government, or research.

ABRAHAM and LEDOLTER · Statistical Methods for Forecasting
AGRESTI · Analysis of Ordinal Categorical Data
AGRESTI · An Introduction to Categorical Data Analysis
AGRESTI · Categorical Data Analysis, *Second Edition*
ALTMAN, GILL, and McDONALD · Numerical Issues in Statistical Computing for the
    Social Scientist
AMARATUNGA and CABRERA · Exploration and Analysis of DNA Microarray and
    Protein Array Data
ANDĚL · Mathematics of Chance
ANDERSON · An Introduction to Multivariate Statistical Analysis, *Third Edition*
* ANDERSON · The Statistical Analysis of Time Series
ANDERSON, AUQUIER, HAUCK, OAKES, VANDAELE, and WEISBERG ·
    Statistical Methods for Comparative Studies
ANDERSON and LOYNES · The Teaching of Practical Statistics
ARMITAGE and DAVID (editors) · Advances in Biometry
ARNOLD, BALAKRISHNAN, and NAGARAJA · Records
* ARTHANARI and DODGE · Mathematical Programming in Statistics
* BAILEY · The Elements of Stochastic Processes with Applications to the Natural
    Sciences
BALAKRISHNAN and KOUTRAS · Runs and Scans with Applications
BARNETT · Comparative Statistical Inference, *Third Edition*
BARNETT and LEWIS · Outliers in Statistical Data, *Third Edition*
BARTOSZYNSKI and NIEWIADOMSKA-BUGAJ · Probability and Statistical Inference
BASILEVSKY · Statistical Factor Analysis and Related Methods: Theory and
    Applications
BASU and RIGDON · Statistical Methods for the Reliability of Repairable Systems
BATES and WATTS · Nonlinear Regression Analysis and Its Applications
BECHHOFER, SANTNER, and GOLDSMAN · Design and Analysis of Experiments for
    Statistical Selection, Screening, and Multiple Comparisons
BELSLEY · Conditioning Diagnostics: Collinearity and Weak Data in Regression

*Now available in a lower priced paperback edition in the Wiley Classics Library.
†Now available in a lower priced paperback edition in the Wiley–Interscience Paperback Series.

† BELSLEY, KUH, and WELSCH · Regression Diagnostics: Identifying Influential
    Data and Sources of Collinearity
  BENDAT and PIERSOL · Random Data: Analysis and Measurement Procedures,
    Third Edition
  BERRY, CHALONER, and GEWEKE · Bayesian Analysis in Statistics and
    Econometrics: Essays in Honor of Arnold Zellner
  BERNARDO and SMITH · Bayesian Theory
  BHAT and MILLER · Elements of Applied Stochastic Processes, *Third Edition*
  BHATTACHARYA and WAYMIRE · Stochastic Processes with Applications
† BIEMER, GROVES, LYBERG, MATHIOWETZ, and SUDMAN · Measurement Errors
    in Surveys
  BILLINGSLEY · Convergence of Probability Measures, *Second Edition*
  BILLINGSLEY · Probability and Measure, *Third Edition*
  BIRKES and DODGE · Alternative Methods of Regression
  BLISCHKE AND MURTHY (editors) · Case Studies in Reliability and Maintenance
  BLISCHKE AND MURTHY · Reliability: Modeling, Prediction, and Optimization
  BLOOMFIELD · Fourier Analysis of Time Series: An Introduction, *Second Edition*
  BOLLEN · Structural Equations with Latent Variables
  BOROVKOV · Ergodicity and Stability of Stochastic Processes
  BOULEAU · Numerical Methods for Stochastic Processes
  BOX · Bayesian Inference in Statistical Analysis
  BOX · R. A. Fisher, the Life of a Scientist
  BOX and DRAPER · Empirical Model-Building and Response Surfaces
* BOX and DRAPER · Evolutionary Operation: A Statistical Method for Process
    Improvement
  BOX, HUNTER, and HUNTER · Statistics for Experimenters: An Introduction to
    Design, Data Analysis, and Model Building
  BOX and LUCEÑO · Statistical Control by Monitoring and Feedback Adjustment
  BRANDIMARTE · Numerical Methods in Finance: A MATLAB-Based Introduction
  BROWN and HOLLANDER · Statistics: A Biomedical Introduction
  BRUNNER, DOMHOF, and LANGER · Nonparametric Analysis of Longitudinal Data in
    Factorial Experiments
  BUCKLEW · Large Deviation Techniques in Decision, Simulation, and Estimation
  CAIROLI and DALANG · Sequential Stochastic Optimization
  CASTILLO, HADI, BALAKRISHNAN, and SARABIA · Extreme Value and Related
    Models with Applications in Engineering and Science
  CHAN · Time Series: Applications to Finance
  CHARALAMBIDES · Combinatorial Methods in Discrete Distributions
  CHATTERJEE and HADI · Sensitivity Analysis in Linear Regression
  CHATTERJEE and PRICE · Regression Analysis by Example, *Third Edition*
  CHERNICK · Bootstrap Methods: A Practitioner's Guide
  CHERNICK and FRIIS · Introductory Biostatistics for the Health Sciences
  CHILÈS and DELFINER · Geostatistics: Modeling Spatial Uncertainty
  CHOW and LIU · Design and Analysis of Clinical Trials: Concepts and Methodologies,
    Second Edition
  CLARKE and DISNEY · Probability and Random Processes: A First Course with
    Applications, *Second Edition*
* COCHRAN and COX · Experimental Designs, *Second Edition*
  CONGDON · Applied Bayesian Modelling
  CONGDON · Bayesian Statistical Modelling
  CONOVER · Practical Nonparametric Statistics, *Third Edition*
  COOK · Regression Graphics
  COOK and WEISBERG · Applied Regression Including Computing and Graphics

*Now available in a lower priced paperback edition in the Wiley Classics Library.
†Now available in a lower priced paperback edition in the Wiley–Interscience Paperback Series.

*Now available in a lower priced paperback edition in the Wiley Classics Library.

†Now available in a lower priced paperback edition in the Wiley–Interscience Paperback Series.

*Now available in a lower priced paperback edition in the Wiley Classics Library.

†Now available in a lower priced paperback edition in the Wiley–Interscience Paperback Series.

*Now available in a lower priced paperback edition in the Wiley Classics Library.
†Now available in a lower priced paperback edition in the Wiley-Interscience Paperback Series.

*Now available in a lower priced paperback edition in the Wiley Classics Library.

†Now available in a lower priced paperback edition in the Wiley–Interscience Paperback Series.

PORT · Theoretical Probability for Applications

POURAHMADI · Foundations of Time Series Analysis and Prediction Theory

PRESS · Bayesian Statistics: Principles, Models, and Applications

PRESS · Subjective and Objective Bayesian Statistics, *Second Edition*

PRESS and TANUR · The Subjectivity of Scientists and the Bayesian Approach

PUKELSHEIM · Optimal Experimental Design

PURI, VILAPLANA, and WERTZ · New Perspectives in Theoretical and Applied
   Statistics

† PUTERMAN · Markov Decision Processes: Discrete Stochastic Dynamic Programming

QIU · Image Processing and Jump Regression Analysis

* RAO · Linear Statistical Inference and Its Applications, *Second Edition*

RAUSAND and HØYLAND · System Reliability Theory: Models, Statistical Methods,
   and Applications, *Second Edition*

RENCHER · Linear Models in Statistics

RENCHER · Methods of Multivariate Analysis, *Second Edition*

RENCHER · Multivariate Statistical Inference with Applications

* RIPLEY · Spatial Statistics

RIPLEY · Stochastic Simulation

ROBINSON · Practical Strategies for Experimenting

ROHATGI and SALEH · An Introduction to Probability and Statistics, *Second Edition*

ROLSKI, SCHMIDLI, SCHMIDT, and TEUGELS · Stochastic Processes for Insurance
   and Finance

ROSENBERGER and LACHIN · Randomization in Clinical Trials: Theory and Practice

ROSS · Introduction to Probability and Statistics for Engineers and Scientists

† ROUSSEEUW and LEROY · Robust Regression and Outlier Detection

* RUBIN · Multiple Imputation for Nonresponse in Surveys

RUBINSTEIN · Simulation and the Monte Carlo Method

RUBINSTEIN and MELAMED · Modern Simulation and Modeling

RYAN · Modern Regression Methods

RYAN · Statistical Methods for Quality Improvement, *Second Edition*

SALTELLI, CHAN, and SCOTT (editors) · Sensitivity Analysis

* SCHEFFE · The Analysis of Variance

SCHIMEK · Smoothing and Regression: Approaches, Computation, and Application

SCHOTT · Matrix Analysis for Statistics, *Second Edition*

SCHOUTENS · Levy Processes in Finance: Pricing Financial Derivatives

SCHUSS · Theory and Applications of Stochastic Differential Equations

SCOTT · Multivariate Density Estimation: Theory, Practice, and Visualization

* SEARLE · Linear Models

SEARLE · Linear Models for Unbalanced Data

SEARLE · Matrix Algebra Useful for Statistics

SEARLE, CASELLA, and McCULLOCH · Variance Components

SEARLE and WILLETT · Matrix Algebra for Applied Economics

SEBER and LEE · Linear Regression Analysis, *Second Edition*

† SEBER · Multivariate Observations

† SEBER and WILD · Nonlinear Regression

SENNOTT · Stochastic Dynamic Programming and the Control of Queueing Systems

* SERFLING · Approximation Theorems of Mathematical Statistics

SHAFER and VOVK · Probability and Finance: It's Only a Game!

SILVAPULLE and SEN · Constrained Statistical Inference: Inequality, Order, and Shape
   Restrictions

SMALL and McLEISH · Hilbert Space Methods in Probability and Statistical Inference

SRIVASTAVA · Methods of Multivariate Statistics

STAPLETON · Linear Statistical Models

*Now available in a lower priced paperback edition in the Wiley Classics Library.

†Now available in a lower priced paperback edition in the Wiley–Interscience Paperback Series.

STAUDTE and SHEATHER · Robust Estimation and Testing

STOYAN, KENDALL, and MECKE · Stochastic Geometry and Its Applications, *Second Edition*

STOYAN and STOYAN · Fractals, Random Shapes and Point Fields: Methods of Geometrical Statistics

STYAN · The Collected Papers of T. W. Anderson: 1943–1985

SUTTON, ABRAMS, JONES, SHELDON, and SONG · Methods for Meta-Analysis in Medical Research

TANAKA · Time Series Analysis: Nonstationary and Noninvertible Distribution Theory

THOMPSON · Empirical Model Building

THOMPSON · Sampling, *Second Edition*

THOMPSON · Simulation: A Modeler's Approach

THOMPSON and SEBER · Adaptive Sampling

THOMPSON, WILLIAMS, and FINDLAY · Models for Investors in Real World Markets

TIAO, BISGAARD, HILL, PEÑA, and STIGLER (editors) · Box on Quality and Discovery: with Design, Control, and Robustness

TIERNEY · LISP-STAT: An Object-Oriented Environment for Statistical Computing and Dynamic Graphics

TSAY · Analysis of Financial Time Series

UPTON and FINGLETON · Spatial Data Analysis by Example, Volume II: Categorical and Directional Data

VAN BELLE · Statistical Rules of Thumb

VAN BELLE, FISHER, HEAGERTY, and LUMLEY · Biostatistics: A Methodology for the Health Sciences, *Second Edition*

VESTRUP · The Theory of Measures and Integration

VIDAKOVIC · Statistical Modeling by Wavelets

VINOD and REAGLE · Preparing for the Worst: Incorporating Downside Risk in Stock Market Investments

WALLER and GOTWAY · Applied Spatial Statistics for Public Health Data

WEERAHANDI · Generalized Inference in Repeated Measures: Exact Methods in MANOVA and Mixed Models

WEISBERG · Applied Linear Regression, *Third Edition*

WELSH · Aspects of Statistical Inference

WESTFALL and YOUNG · Resampling-Based Multiple Testing: Examples and Methods for *p*-Value Adjustment

WHITTAKER · Graphical Models in Applied Multivariate Statistics

WINKER · Optimization Heuristics in Economics: Applications of Threshold Accepting

WONNACOTT and WONNACOTT · Econometrics, *Second Edition*

WOODING · Planning Pharmaceutical Clinical Trials: Basic Statistical Principles

WOODWORTH · Biostatistics: A Bayesian Introduction

WOOLSON and CLARKE · Statistical Methods for the Analysis of Biomedical Data, *Second Edition*

WU and HAMADA · Experiments: Planning, Analysis, and Parameter Design Optimization

YANG · The Construction Theory of Denumerable Markov Processes

* ZELLNER · An Introduction to Bayesian Inference in Econometrics

ZHOU, OBUCHOWSKI, and McCLISH · Statistical Methods in Diagnostic Medicine

*Now available in a lower priced paperback edition in the Wiley Classics Library.

†Now available in a lower priced paperback edition in the Wiley–Interscience Paperback Series.

Books Available from SAS Press

Advanced Log-Linear Models Using SAS®
by **Daniel Zelterman**............................... Order No. A57496

Analysis of Clinical Trials Using SAS®: A Practical Guide
by **Alex Dmitrienko, Walter Offen, Christy Chuang-Stein,**
and **Geert Molenbergs** .... Order No. A59390

Annotate: Simply the Basics
by **Art Carpenter**.................................... Order No. A57320

*Applied Multivariate Statistics with SAS® Software,
Second Edition*
by **Ravindra Khattree**
and **Dayanand N. Naik** .... Order No. A56903

*Applied Statistics and the SAS® Programming Language,
Fourth Edition*
by **Ronald P. Cody**
and **Jeffrey K. Smith**............................... Order No. A55984

An Array of Challenges — Test Your SAS® Skills
by **Robert Virgile**.................................... Order No. A55625

*Carpenter's Complete Guide to the SAS® Macro Language,
Second Edition*
by **Art Carpenter**.................................... Order No. A59224

The Cartoon Guide to Statistics
by **Larry Gonick**
and **Woollcott Smith**............................... Order No. A55153

*Categorical Data Analysis Using the SAS® System,
Second Edition*
by **Maura E. Stokes, Charles S. Davis,**
and **Gary G. Koch** .... Order No. A57998

Cody's Data Cleaning Techniques Using SAS® Software
by **Ron Cody** ....Order No. A57198

*Common Statistical Methods for Clinical Research with
SAS® Examples, Second Edition*
by **Glenn A. Walker**................................. Order No. A58086

*Debugging SAS® Programs: A Handbook of Tools
and Techniques*
by **Michele M. Burlew** .... Order No. A57743

*Efficiency: Improving the Performance of Your
SAS® Applications*
by **Robert Virgile**.................................... Order No. A55960

*The Essential PROC SQL Handbook for
SAS® Users*
by **Katherine Prairie** .... Order No. A58546

*Fixed Effects Regression Methods for Longitudinal Data
Using SAS®*
by **Paul D. Allisoin** .... Order No. A58348

Genetic Analysis of Complex Traits Using SAS®
Edited by **Arnold M. Saxton** .... Order No. A59454

*A Handbook of Statistical Analyses Using SAS®,
Second Edition*
by **B.S. Everitt**
and **G. Der** .... Order No. A58679

Health Care Data and the SAS® System
by **Marge Scerbo, Craig Dickstein,**
and **Alan Wilson**...................................... Order No. A57638

The How-To Book for SAS/GRAPH® Software
by **Thomas Miron** .... Order No. A55203

*In the Know ... SAS® Tips and Techniques From
Around the Globe*
by **Phil Mason** .... Order No. A55513

*Instant ODS: Style Templates for the Output
Delivery System*
by **Bernadette Johnson** .... Order No. A58824

*Integrating Results through Meta-Analytic Review Using
SAS® Software*
by **Morgan C. Wang**
and **Brad J. Bushman** .... Order No. A55810

Learning SAS® in the Computer Lab, Second Edition
by **Rebecca J. Elliott** .... Order No. A57739

The Little SAS® Book: A Primer
by **Lora D. Delwiche**
and **Susan J. Slaughter** .... Order No. A55200

The Little SAS® Book: A Primer, Second Edition
by **Lora D. Delwiche**
and **Susan J. Slaughter** .... Order No. A56649
(updated to include Version 7 features)

The Little SAS® Book: A Primer, Third Edition
by **Lora D. Delwiche**
and **Susan J. Slaughter** .... Order No. A59216
(updated to include SAS 9.1 features)

*Logistic Regression Using the SAS® System:
Theory and Application*
by **Paul D. Allison** .... Order No. A55770

Longitudinal Data and SAS®: A Programmer's Guide
by **Ron Cody** .... Order No. A58176

Maps Made Easy Using SAS®
by **Mike Zdeb** .... Order No. A57495

Models for Discrete Data
by **Daniel Zelterman** .... Order No. A57521

Multiple Comparisons and Multiple Tests Using SAS®
Text and Workbook Set
(books in this set also sold separately)
by **Peter H. Westfall, Randall D. Tobias,
Dror Rom, Russell D. Wolfinger,**
and **Yosef Hochberg** .... Order No. A58274

Multiple-Plot Displays: Simplified with Macros
by **Perry Watts** .... Order No. A58314

Multivariate Data Reduction and Discrimination with
SAS® Software
by **Ravindra Khattree**
and **Dayanand N. Naik** .... Order No. A56902

Output Delivery System: The Basics
by **Lauren E. Haworth** .... Order No. A58087

Painless Windows: A Handbook for SAS® Users,
Third Edition
by **Jodie Gilmore** .... Order No. A58783
(updated to include Version 8 and SAS 9.1 features)

PROC TABULATE by Example
by **Lauren E. Haworth** .... Order No. A56514

Professional SAS® Programming Shortcuts
by **Rick Aster** .... Order No. A59353

Quick Results with SAS/GRAPH® Software
by **Arthur L. Carpenter**
and **Charles E. Shipp**............................... Order No. A55127

Quick Results with the Output Delivery System
by **Sunil K. Gupta** ....Order No. A58458

Quick Start to Data Analysis with SAS®
by **Frank C. Dilorio**
and **Kenneth A. Hardy** .... Order No. A55550

Reading External Data Files Using SAS®: Examples Handbook
by **Michele M. Burlew** .... Order No. A58369

Regression and ANOVA: An Integrated Approach Using
SAS® Software
by **Keith E. Muller**
and **Bethel A. Fetterman**......................... Order No. A57559

SAS® Applications Programming: A Gentle Introduction
by **Frank C. Dilorio** .... Order No. A56193

SAS® for Forecasting Time Series, Second Edition
by **John C. Brocklebank,**
and **David A. Dickey** .... Order No. A57275

SAS® for Linear Models, Fourth Edition
by **Ramon C. Littell, Walter W. Stroup,**
and **Rudolf J. Freund**............................... Order No. A56655

SAS® for Monte Carlo Studies: A Guide for Quantitative
Researchers
by **Xitao Fan, Ákos Felsővályi, Stephen A. Sivo,**
and **Sean C. Keenan**................................. Order No. A57323

SAS® Functions by Example
by **Ron Cody** .... Order No. A59343

SAS® Macro Programming Made Easy
by **Michele M. Burlew** .... Order No. A56516

SAS® Programming by Example
by **Ron Cody**
and **Ray Pass** .... Order No. A55126

SAS® Programming for Researchers and Social Scientists,
Second Edition
by **Paul E. Spector**................................... Order No. A58784

SAS® Survival Analysis Techniques for Medical Research,
Second Edition
by **Alan B. Cantor** ....Order No. A58416

SAS® System for Elementary Statistical Analysis,
Second Edition
by **Sandra D. Schlotzhauer**
and **Ramon C. Littell**................................ Order No. A55172

SAS® System for Mixed Models
by **Ramon C. Littell, George A. Milliken, Walter W. Stroup,**
and **Russell D. Wolfinger** .... Order No. A55235

SAS® System for Regression, Third Edition
by **Rudolf J. Freund**
and **Ramon C. Littell**................................ Order No. A57313

SAS® System for Statistical Graphics, First Edition
by **Michael Friendly** .... Order No. A56143

The SAS® Workbook and Solutions Set
(books in this set also sold separately)
by **Ron Cody** .... Order No. A55594

Selecting Statistical Techniques for Social Science Data:
A Guide for SAS® Users
by **Frank M. Andrews, Laura Klem, Patrick M. O'Malley,
Willard L. Rodgers, Kathleen B. Welch,**
and **Terrence N. Davidson** .... Order No. A55854

Statistical Quality Control Using the SAS® System
by **Dennis W. King** .... Order No. A55232

A Step-by-Step Approach to Using the SAS® System
for Factor Analysis and Structural Equation Modeling
by **Larry Hatcher**....................................... Order No. A55129

A Step-by-Step Approach to Using the SAS® System
for Univariate and Multivariate Statistics, Second Edition
by **Norm O'Rourke, Larry Hatcher,**
and **Edward J. Stepanski**......................... Order No. A58929

Step-by-Step Basic Statistics Using SAS®: Student Guide
and Exercises
(books in this set also sold separately)
by **Larry Hatcher**....................................... Order No. A57541

support.sas.com/pubs